σ- and π-Hole Interactions

σ- and π-Hole Interactions

Editor

Antonio Frontera

MDPI • Basel • Beijing • Wuhan • Barcelona • Belgrade • Manchester • Tokyo • Cluj • Tianjin

Editor
Antonio Frontera
Department de Química,
Universitat de les Illes Balears,
Palma de Mallorca
Spain

Editorial Office
MDPI
St. Alban-Anlage 66
4052 Basel, Switzerland

This is a reprint of articles from the Special Issue published online in the open access journal *Crystals* (ISSN 2073-4352) (available at: https://www.mdpi.com/journal/crystals/special_issues/noncovalent_interaction).

For citation purposes, cite each article independently as indicated on the article page online and as indicated below:

LastName, A.A.; LastName, B.B.; LastName, C.C. Article Title. *Journal Name* **Year**, *Volume Number*, Page Range.

ISBN 978-3-0365-0464-3 (Hbk)
ISBN 978-3-0365-0465-0 (PDF)

© 2021 by the authors. Articles in this book are Open Access and distributed under the Creative Commons Attribution (CC BY) license, which allows users to download, copy and build upon published articles, as long as the author and publisher are properly credited, which ensures maximum dissemination and a wider impact of our publications.

The book as a whole is distributed by MDPI under the terms and conditions of the Creative Commons license CC BY-NC-ND.

Contents

About the Editor vii

Antonio Frontera
σ- and π-Hole Interactions
Reprinted from: *Crystals* **2020**, *10*, 721, doi:10.3390/cryst10090721 1

Ibon Alkorta, José Elguero and Antonio Frontera
Not Only Hydrogen Bonds: Other Noncovalent Interactions
Reprinted from: *Crystals* **2020**, *10*, 180, doi:10.3390/cryst10030180 5

Edward R. T. Tiekink
A Survey of Supramolecular Aggregation Based on Main Group Element···Selenium Secondary Bonding Interactions—A Survey of the Crystallographic Literature
Reprinted from: *Crystals* **2020**, *10*, 503, doi:10.3390/cryst10060503 35

Yi Wang, Xinrui Miao and Wenli Deng
Halogen Bonds Fabricate 2D Molecular Self-Assembled Nanostructures by Scanning Tunneling Microscopy
Reprinted from: *Crystals* **2020**, *10*, 1057, doi:10.3390/cryst10111057 55

Ibon Alkorta, Cristina Trujillo, Goar Sánchez-Sanz and José Elguero
Regium Bonds between Silver(I) Pyrazolates Dinuclear Complexes and Lewis Bases (N_2, OH_2, NCH, SH_2, NH_3, PH_3, CO and CNH)
Reprinted from: *Crystals* **2020**, *10*, 137, doi:10.3390/cryst10020137 79

Alexey V. Kletskov, Diego M. Gil, Antonio Frontera, Vladimir P. Zaytsev, Natalia L. Merkulova, Ksenia R. Beltsova, Anna A. Sinelshchikova, Mikhail S. Grigoriev, Mariya V. Grudova and Fedor I. Zubkov
Intramolecular sp^2-sp^3 Disequalization of Chemically Identical Sulfonamide Nitrogen Atoms: Single Crystal X-ray Diffraction Characterization, Hirshfeld Surface Analysis and DFT Calculations of *N*-Substituted Hexahydro-1,3,5-Triazines
Reprinted from: *Crystals* **2020**, *10*, 369, doi:10.3390/cryst10050369 95

Pradeep R. Varadwaj, Arpita Varadwaj and Helder M. Marques
Does Chlorine in CH_3Cl Behave as a Genuine Halogen Bond Donor?
Reprinted from: *Crystals* **2020**, *10*, 146, doi:10.3390/cryst10030146 109

Yu Zhang, Jian-Ge Wang and Weizhou Wang
Unexpected Sandwiched-Layer Structure of the Cocrystal Formed by Hexamethylbenzene with 1,3-Diiodotetrafluorobenzene: A Combined Theoretical and Crystallographic Study
Reprinted from: *Crystals* **2020**, *10*, 379, doi:10.3390/cryst10050379 127

Cynthia S. Novoa-Ramírez, Areli Silva-Becerril, Fiorella L. Olivera-Venturo, Juan Carlos García-Ramos, Marcos Flores-Alamo and Lena Ruiz-Azuara
N/N Bridge Type and Substituent Effects on Chemical and Crystallographic Properties of Schiff-Base (*Salen*/*Salphen*) Ni^{ii} Complexes
Reprinted from: *Crystals* **2020**, *10*, 616, doi:10.3390/cryst10070616 137

Jeannette Carolina Belmont-Sánchez, Noelia Ruiz-González, Antonio Frontera, Antonio Matilla-Hernández, Alfonso Castiñeiras and Juan Niclós-Gutiérrez
Anion–Cation Recognition Pattern, Thermal Stability and DFT-Calculations in the Crystal Structure of $H_2dap[Cd(HEDTA)(H_2O)]$ Salt ($H_2dap = H_2(N3,N7)$-2,6-Diaminopurinium Cation)
Reprinted from: *Crystals* **2020**, *10*, 304, doi:10.3390/cryst10040304 . . . . **161**

Seth Yannacone, Marek Freindorf, Yunwen Tao, Wenli Zou, and Elfi Kraka
Local Vibrational Mode Analysis of π–Hole Interactions between Aryl Donors and Small Molecule Acceptors
Reprinted from: *Crystals* **2020**, *10*, 556, doi:10.3390/cryst10070556 . . . . **177**

About the Editor

Antonio Frontera is Full Professor of Organic Chemistry at the Universitat de les Illes Balears (UIB) and co-leader of the Supramolecular Chemistry research group, where he leads the Theoretical Chemistry Laboratory. He received a B.Sc. degree from the Universidat de les Illes Balears as well as a Ph.D. degree (1994) from the same institution. After 2 years of postdoctoral stay in the Chemistry Department at Yale University (New Haven, USA) under the auspices of Prof. William L. Jorgensen, he started his independent career in 2000 at the UIB. He has obtained successive promotions up to the current position. Since 2010, he has regularly done short visiting stays at Bonn University (Germany). His general research interests are focused on the study of noncovalent interactions, supramolecular chemistry, noncovalent catalysts, and crystal engineering. He is a member of the Royal Society of Chemistry in Spain (RSEQ) and president of the local delegation in Baleares (RSEQ-IB).

Editorial

σ- and π-Hole Interactions

Antonio Frontera

Departament de Química, Universitat de les Illes Baleares, Crta de Valldemossa km 7.5, 07122 Palma de Mallorca, Baleares, Spain; toni.frontera@uib.es

Received: 17 August 2020; Accepted: 18 August 2020; Published: 19 August 2020

Keywords: σ-hole; π-hole; crystal engineering; crystal growth; supramolecular chemistry

---

Supramolecular chemistry is a very active research field that was initiated in the last century [1–4]. It was defined as *chemistry beyond the molecule*, and the word *supermolecule* was invented by Lehn [3]. The *chemistry beyond the molecule* refers to organized entities of higher complexity resulting from the association of molecules that are held together by noncovalent interactions [5]. The organized supramolecular entities are built by the formation of various noncovalent forces, which are frequently working synergistically in the same supramolecular assembly. Therefore, precise control of the noncovalent interactions is needed to succeed in this field, as exemplified by many regulation processes in nature.

A deep understanding of noncovalent interactions is necessary to advance in many fields, especially in crystal growth and crystal engineering [6]. Theoreticians have demonstrated that the distribution of the electron density around covalently bonded atoms is not isotropic, revealing that the use of point charges to define the properties of an atom (electron-rich or electron-poor) is not valid [7]. That is, a single atom presents regions of higher and lower electron density, where the electrostatic potential can be negative and positive, respectively, in some cases. The positive area is usually defined as a σ- or π-hole, depending on its location. These holes of electron density are responsible for the formation of attractive interactions with any electron-rich site (anion, Lewis base, π-system, etc.). The halogen bond can be considered as the prototypical example of σ-hole interaction [8]. After the emergence of the halogen bond, the interest in σ- and π-hole interactions embracing elements of groups 12–16 [9–13] and 18 [14–17] of the Periodic Table has grown exponentially. Halogen and chalcogen bonding interactions have already been defined by the IUPAC [18,19]. They are well-recognized interactions that are used regularly by the scientific community in crystal engineering, supramolecular chemistry, and catalysis [20]. However, more experimental and theoretical work is probably needed to extend such a statement to the elements of groups 12–15, acting as Lewis acids.

This issue gathers nine excellent contributions. In reference [21], Alkorta et al. combined theoretical calculations and a search in the Cambridge Structural Database (CSD) to investigate the interaction of dinuclear Ag(I) pyrazolates with Lewis bases, as examples of regium bonding [22]. They studied the effect of the substituents and ligands on the aromaticity and found an interesting relationship between the intramolecular Ag–Ag distance and stability.

In reference [23], Varadwaj et al. studied theoretically the CH_3Cl molecule and its complexes with ten Lewis bases to demonstrate that CH_3Cl is a genuine halogen bond donor. They have evidenced that the electronic charge density distribution around the Cl is anisotropic. The negative belt is able to participate in halogen, chalcogen, or hydrogen bonding interactions. Moreover, they show that the positive σ-hole on the Cl atom in CH_3Cl is not induced by the electric field of the interacting species, as previously suggested in the literature. Instead, it is an inherent property of chlorine in this molecule.

In reference [24], Belmont-Sanchez et al. reported the synthesis and X-ray characterization of several out-of-sphere cadmium complexes with 2,6-diaminopurine. The crystal packing of these compounds is mostly dominated by H-bonds, which were analyzed by using DFT calculations.

Interestingly, the results were in clear contrast with those previously reported for similar complexes with adenine instead of 2,6-diaminopurine [25]. The factors contributing to such differences are discussed and rationalized on the basis of the additional exocyclic 2-amino group in 2,6-diaminopurine compared to adenine.

In reference [26], Kletsov et al. synthesized and X-ray characterized four N-substituted 1,3,5-triazinanes and focused on the crucial role of C–H···π and C–H···O H-bonding interactions determining their solid-state architecture. Quite remarkable is the fact that the XRD analysis demonstrated an unprecedented feature of the crystalline structure. That is, the symmetrically substituted 1,3,5-triazacyclohexanes have two chemically identical sulfonamide N-atoms in different sp^2 and sp^3 hybridizations.

In reference [27], Zhang et al. reported the synthesis and X-ray characterization of a cocrystal formed by hexamethylbenzene (HMB) combined with 1,3-diiodotetrafluorobenzene (1,3-DITFB) founding an unexpected sandwiched-layer structure. The formation of the alternating layer was further studied using DFT calculations showing that dispersion forces are very important in the formation of the HMB layer. In contrast, the formation of the 1,3-DITFB layer is induced by weak but cooperative C–I···F halogen bonds.

In reference [28], Yannacone et al. studied the nature of π-hole interaction in several fluorinated aromatic systems focusing on the effect of the substituents and the presence/absence of heteroatoms in the arene on the strength of the π-hole interaction. Moreover, the authors have also analyzed cooperativity effects with other interactions like hydrogen bonding.

In reference [29], Novoa-Ramírez et al. have used thirteen ligands (N,N'-bis(5-R-salicylidene)ethylenediamine (where R = MeO, Me, OH, H, Cl, Br, NO$_2$) and (N,N'-bis(5-R-salicylidene)-1,2-phenylenediamine (where R = MeO, Me, OH, H, Cl, Br) to synthesize and X-ray characterize thirteen nickel complexes. By using Hirshfeld surface analysis, they showed that their packaging was favored by C···H/H···C interactions, C–H···O hydrogen, and π-stacking interactions.

This Special Issue also includes two reviews, one written by Tiekink [30], who elegantly describes the results of a survey of X-ray structures of main group element compounds (M = Sn, Pb As, Sb, Bi and Te) exhibiting intermolecular M···Se noncovalent interactions. The second review written by Alkorta, Elguero, and I [31], provides a consistent description of noncovalent interactions, covering most groups of the Periodic Table. The interactions are described and discussed using their trivial names. That is, apart from hydrogen bonds, the following noncovalent interactions are described: alkali, alkaline earth, regium, spodium, triel, tetrel, pnictogen, chalcogen, halogen, and aerogen, thus covering a wide range of interactions. In this review, the possibility of extending the Cahn-Ingold-Prelog priority rules to noncovalent interactions is suggested.

In summary, this Special Issue gathers an interesting group of manuscripts devoted to the study of several types of σ- and π-hole noncovalent interactions and their importance in the solid-state of different compounds, including biologically relevant ones like diaminopurines, good halogen bond donors like 1,3-diiodotetrafluorobenzene, and several theoretical investigations devoted to π-hole interactions in arenes and regium bonds in Ag(I) derivatives. Moreover, two excellent and comprehensive reviews are published in this collection with the latest advances in noncovalent interactions that I believe make this Special Issue even more special.

To finish, I wish to thank all authors who have submitted their excellent papers to this Special Issue and also the reviewers who carefully read them, providing constructive and helpful suggestions and corrections on all manuscripts. I am especially thankful to the editorial staff at Crystals for their incredibly fast and professional work, dealing with all manuscripts and the selection of suitable referees.

Conflicts of Interest: The author declares no conflict of interest.

References

1. Ariga, K.; Kunitake, T. *Supramolecular Chemistry: Fundamentals and Applications*; Springer: Berlin/Heidelberg, Germany, 2006.
2. Pedersen, C.J. The Discovery of Crown Ethers (Noble Lecture). *Angew. Chem. Int. Ed. Engl.* **1988**, *27*, 1021–1027. [CrossRef]
3. Lehn, J.-M. Supramolecular Chemistry—Scope and Perspectives Molecules, Supermolecules, and Molecular Devices (Nobel Lecture). *Angew. Chem. Int. Ed. Engl.* **1988**, *27*, 89–112. [CrossRef]
4. Cram, D.J. The Design of Molecular Hosts, Guests, and Their Complexes (Nobel Lecture). *Angew. Chem. Int. Ed. Engl.* **1988**, *27*, 1009–1020. [CrossRef]
5. Lehn, J.-M. Supramolecular chemistry: Receptors, catalysts, and carriers. *Science* **1985**, *227*, 849–856. [CrossRef]
6. Desiraju, G.R. Crystal Engineering: From Molecule to Crystal. *J. Am. Chem. Soc.* **2013**, *135*, 9952–9967. [CrossRef]
7. Politzer, P.; Murray, J.S.; Clark, T. Halogen bonding and other σ-hole interactions: A perspective. *Phys. Chem. Chem. Phys.* **2013**, *15*, 11178–11189. [CrossRef]
8. Cavallo, G.; Metrangolo, P.; Milani, R.; Pilati, T.; Priimagi, A.; Resnati, G.; Terraneo, T. The Halogen Bond. *Chem. Rev.* **2016**, *116*, 2478–2601. [CrossRef]
9. Bauzá, A.; Alkorta, I.; Elguero, J.; Mooibroek, T.J.; Frontera, A. Spodium Bonds: Noncovalent Interactions Involving Group 12 Elements. *Angew. Chem. Int. Ed.* **2020**. [CrossRef]
10. Grabowski, S.J. Boron and other triel lewis acid centers: From hypovalency to hypervalency. *ChemPhysChem.* **2014**, *15*, 2985–2993. [CrossRef]
11. Bauzá, A.; Mooibroek, T.J.; Frontera, A. Tetrel-bonding interaction: Rediscovered supramolecular force? *Angew. Chem. Int. Ed.* **2013**, *52*, 12317–12321. [CrossRef]
12. Mahmudov, K.T.; Gurbanov, A.V.; Aliyeva, V.A.; Resnati, G.; Pombeiro, A.J.L. Pnictogen bonding in coordination chemistry. *Coord. Chem. Rev.* **2020**, *418*, 213381. [CrossRef]
13. Scilabra, P.; Terraneo, G.; Resnati, G. The Chalcogen Bond in Crystalline Solids: A World Parallel to Halogen Bond. *Acc. Chem. Res.* **2019**, *52*, 1313–1324. [CrossRef] [PubMed]
14. Bauzá, A.; Frontera, A. Aerogen bonding interaction: A new supramolecular force? *Angew. Chem. Int. Ed.* **2015**, *54*, 7340–7343. [CrossRef] [PubMed]
15. Bauzá, A.; Frontera, A. σ/π-Hole noble gas bonding interactions: Insights from theory and experiment. *Coord. Chem. Rev.* **2020**, *404*, 213112. [CrossRef]
16. Gomila, R.M.; Frontera, A. Covalent and Non-covalent Noble Gas Bonding Interactions in XeFn Derivatives (n = 2–6): A Combined Theoretical and ICSD Analysis. *Front. Chem.* **2020**, *8*, 395. [CrossRef] [PubMed]
17. Frontera, A. Noble Gas Bonding Interactions Involving Xenon Oxides and Fluorides. *Molecules* **2020**, *25*, 3419. [CrossRef]
18. Desiraju, G.R.; Ho, P.S.; Kloo, L.; Legon, A.C.; Marquardt, R.; Metrangolo, P.; Politzer, P.; Resnati, G.; Rissanen, K. Definition of the halogen bond (IUPAC Recommendations 2013). *Pure Appl. Chem.* **2013**, *85*, 1711–1713. [CrossRef]
19. Aakeroy, C.B.; Bryce, D.L.; Desiraju, G.R.; Frontera, A.; Legon, A.C.; Nicotra, F.; Rissanen, K.; Scheiner, S.; Terraneo, G.; Metrangolo, P.; et al. Definition of the chalcogen bond (IUPAC Recommendations 2019). *Pure Appl. Chem.* **2019**, *91*, 1889–1892. [CrossRef]
20. Taylor, M.S. Anion recognition based on halogen, chalcogen, pnictogen and tetrel bonding. *Coord. Chem. Rev.* **2020**, *413*, 213270. [CrossRef]
21. Alkorta, I.; Trujillo, C.; Sánchez-Sanz, G.; Elguero, J. Regium Bonds between Silver(I) Pyrazolates Dinuclear Complexes and Lewis Bases (N_2, OH_2, NCH, SH_2, NH_3, PH_3, CO and CNH). *Crystals* **2020**, *10*, 137. [CrossRef]
22. Frontera, A.; Bauzá, A. Regium–π bonds: An Unexplored Link between Noble Metal Nanoparticles and Aromatic Surfaces. *Chem. Eur. J.* **2018**, *24*, 7228–7234. [CrossRef] [PubMed]
23. Varadwaj, P.R.; Varadwaj, A.; Marques, H.M. Does Chlorine in CH_3Cl Behave as a Genuine Halogen Bond Donor? *Crystals* **2020**, *10*, 146. [CrossRef]

24. Belmont-Sánchez, J.C.; Ruiz-González, N.; Frontera, A.; Matilla-Hernández, A.; Castiñeiras, A.; Niclós-Gutiérrez, J. Anion–Cation Recognition Pattern, Thermal Stability and DFT-Calculations in the Crystal Structure of H$_2$dap[Cd(HEDTA)(H$_2$O)] Salt (H$_2$dap = H$_2$(N^3,N^7)-2,6-Diaminopurinium Cation). *Crystals* **2020**, *10*, 304. [CrossRef]
25. Serrano-Padial, E.; Choquesillo-Lazarte, D.; Bugella-Altamirano, E.; Castineiras, A.; Carballo, R. Niclos-Gutierrez, New copper(II) compound having protonated forms of ethylenediaminetetraacetate(4−) ion (EDTA) and adenine (AdeH): Synthesis, crystal structure, molecular recognition and physical properties of (AdeH$_2$)[Cu(HEDTA)(H$_2$O)]·2H$_2$O. *J. Polyhedron.* **2002**, *21*, 1451. [CrossRef]
26. Kletskov, A.V.; Gil, D.M.; Frontera, A.; Zaytsev, V.P.; Merkulova, N.L.; Beltsova, K.R.; Sinelshchikova, A.A.; Grigoriev, M.S.; Grudova, M.V.; Zubkov, F.I. Intramolecular sp^2–sp^3 Disequalization of Chemically Identical Sulfonamide Nitrogen Atoms: Single Crystal X-Ray Diffraction Characterization, Hirshfeld Surface Analysis and DFT Calculations of *N*-Substituted Hexahydro-1,3,5-Triazines. *Crystals* **2020**, *10*, 369. [CrossRef]
27. Zhang, Y.; Wang, J.-G.; Wang, W. Unexpected Sandwiched-Layer Structure of the Cocrystal Formed by Hexamethylbenzene with 1,3-Diiodotetrafluorobenzene: A Combined Theoretical and Crystallographic Study. *Crystals* **2020**, *10*, 379. [CrossRef]
28. Yannacone, S.; Freindorf, M.; Tao, Y.; Zou, W.; Kraka, E. Local Vibrational Mode Analysis of π–Hole Interactions between Aryl Donors and Small Molecule Acceptors. *Crystals* **2020**, *10*, 556. [CrossRef]
29. Novoa-Ramírez, C.S.; Silva-Becerril, A.; Olivera-Venturo, F.L.; García-Ramos, J.C.; Flores-Alamo, M.; Ruiz-Azuara, L. N/N Bridge Type and Substituent Effects on Chemical and Crystallographic Properties of Schiff-Base (*Salen/Salphen*) Ni^{ii} Complexes. *Crystals* **2020**, *10*, 616. [CrossRef]
30. Tiekink, E.R.T. A Survey of Supramolecular Aggregation Based on Main Group Element···Selenium Secondary Bonding Interactions—A Survey of the Crystallographic Literature. *Crystals* **2020**, *10*, 503. [CrossRef]
31. Alkorta, I.; Elguero, J.; Frontera, A. Not Only Hydrogen Bonds: Other Noncovalent Interactions. *Crystals* **2020**, *10*, 180. [CrossRef]

© 2020 by the author. Licensee MDPI, Basel, Switzerland. This article is an open access article distributed under the terms and conditions of the Creative Commons Attribution (CC BY) license (http://creativecommons.org/licenses/by/4.0/).

Review

Not Only Hydrogen Bonds: Other Noncovalent Interactions

Ibon Alkorta [1,*], José Elguero [1,*] and Antonio Frontera [2,*]

1. Instituto de Química Médica, CSIC, Juan de la Cierva, 3, E-28006 Madrid, Spain
2. Departament de Química, Universitat de les Illes Balears, Crta. de Valldemossa km 7.5, 07122 Palma de Mallorca, Spain
* Correspondence: ibon@iqm.csic.es (I.A.); iqmbe17@iqm.csic.es (J.E.); toni.frontera@uib.es (A.F.)

Received: 11 February 2020; Accepted: 3 March 2020; Published: 6 March 2020

Abstract: In this review, we provide a consistent description of noncovalent interactions, covering most groups of the Periodic Table. Different types of bonds are discussed using their trivial names. Moreover, the new name "Spodium bonds" is proposed for group 12 since noncovalent interactions involving this group of elements as electron acceptors have not yet been named. Excluding hydrogen bonds, the following noncovalent interactions will be discussed: alkali, alkaline earth, regium, spodium, triel, tetrel, pnictogen, chalcogen, halogen, and aerogen, which almost covers the Periodic Table entirely. Other interactions, such as orthogonal interactions and π-π stacking, will also be considered. Research and applications of σ-hole and π-hole interactions involving the p-block element is growing exponentially. The important applications include supramolecular chemistry, crystal engineering, catalysis, enzymatic chemistry molecular machines, membrane ion transport, etc. Despite the fact that this review is not intended to be comprehensive, a number of representative works for each type of interaction is provided. The possibility of modeling the dissociation energies of the complexes using different models (HSAB, ECW, Alkorta-Legon) was analyzed. Finally, the extension of Cahn-Ingold-Prelog priority rules to noncovalent is proposed.

Keywords: noncovalent interactions; Lewis acids; Lewis bases; spodium bonds; σ/π-hole interactions

1. Introduction

The aim of this review is to present an original, systematic and prospective view of all noncovalent interactions (NCI). There are several books treating different aspects of NCIs [1–4] but none offers a unified view of the subject, for instance the term Lewis acid/Lewis base does only appear in the most recent one [3]. See on this topic a recent conference paper entitled "Some interesting features of the rich chemistry around electron-deficient systems" [5].

We excluded hydrogen bonds from this survey on NCIs because they are well known and because the bibliography covering HBs is more extensive than the sum of the references on the other NCIs [6–11]. We also excluded anions and cations limiting this review to neutral molecules.

In the modified IUPAC periodic table of the elements reported in Figure 1, we noted in black all the NCIs reported up to now and in blue these not yet discussed. A similar representation was used by Caminati et al. for the front page of their publication [12]. They called the bonds of the groups MB (2), IB (13), TB (14), NB (15), CB (16), and XB (17) following previous authors.

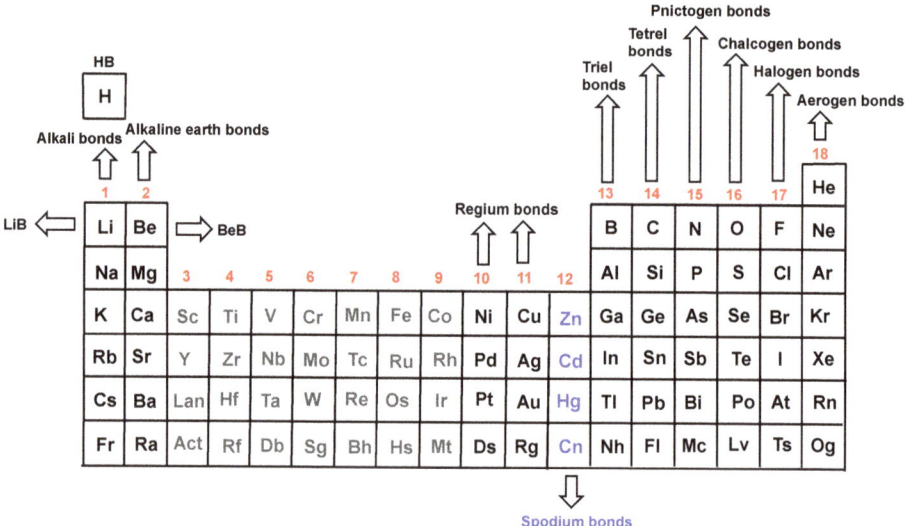

Figure 1. The different noncovalent bonds formed by elements of the Periodic Table. In black are accepted names, and in blue are the proposed new names. Groups 3 to 9 (in grey) are not included in this review.

Usually, the bond is associated with the Lewis acidity of a group, this is the case with groups 11, 13, 14, 15, 16, 17, and 18. For groups 1 and 2, besides HBs, the bond is associated to an element, lithium, sodium and beryllium. We propose to call these bonds Alkali Bonds and Alkaline Earth Bonds (we used this name very recently) [13]. Although Regium Bonds were used for group 11, we propose to use it for both 10 and 11 groups. In grey are the atoms corresponding to groups 3 to 9 that we will not discuss, not that they were unable to form NCIs, but in order not to stretch too much this mini review.

Concerning the rows, we should indicate that Li, Be, B, and C derivatives as Lewis acids have been more studied than Na, Mg, Al, and Si. On the other hand, P, S, and Cl are better representatives of their kind of NCIs than N, O, and F. This observation is related to size and to the softness of the Lewis acid atom that interacts with the Lewis base [14]

Gilbert N. Lewis published his interpretation of acid/base behavior in 1923 [15]; according to him any species with a reactive vacant orbital or available lowest unoccupied molecular orbital is classified as a "Lewis acid" [14,16].

A Lewis base (LB) is associated with a region of the space where there is an excess of negative charge (electron density) in the proximity of an atom or several atoms of a molecule. This happens in anions and in some neutral molecules, such as lone pairs (LP: carbenes, amines, phosphines, N-oxides, ...), multiple bonds (olefins, acetylenes, benzenes, and other aromatic molecules, ...), single bonds (alkanes, dihydrogen, ...), radicals, metals (rare), ...

A Lewis acid (LA) is associated with a region of the space where there is an excess of positive charge (a deficit of negative charge, electron deficiency) in the proximity of an atom or several atoms of a molecule. This happens in cations, σ- and π-holes, metals (frequent), ... The concepts of σ-hole and π-hole were introduced by Politzer et al. [17–19] to describe regions of positive potential along the vector of a covalent bond (σ-hole) or perpendicular to an atom of molecular framework (π-hole).

Some atoms have simultaneously (but in different parts of the space) LB and LA zones due to their anisotropic distribution of electron density. The same happens for molecules, but in this case, they correspond to different parts of the molecule. Note that some Lewis acids when interacting with stronger Lewis acids can behave as Lewis bases [20].

When an LB and an LA containing atoms or molecules are free to interact (i.e., non restrained by some geometrical hindrance), they form complexes being their minima or transition states of different order.

The information on NCIs is mostly based on from crystal structures, microwave (MW) spectroscopy and theoretical calculations; consequently, they are related to gas-phase and solid state. Since chemistry is mainly done in solution there is a consistency problem.

Another aspect that is common to all NCIs is cooperativity. The natural evolution of theoretical studies has been moving from dimer complexes to trimers and longer complexes in search of cooperativity, both augmentative and diminutive, present in crystal structures.

Definition: Noncovalent interactions are complexes formed by two or several LBs and LAs. It is the LA that gives the name to the interaction. Dative bonds are included in this definition.

Why were the complexes not named according to the LB? Historically, because all NCI derive from HBs, i.e., where the H-bond donor is the Lewis acid. More fundamentally, it is because it is not possible to define families of NCIs based on LB. For instance, all anions are LBs, and anions can be found all over the Periodic Table. A classification of LBs is given in Figure 2.

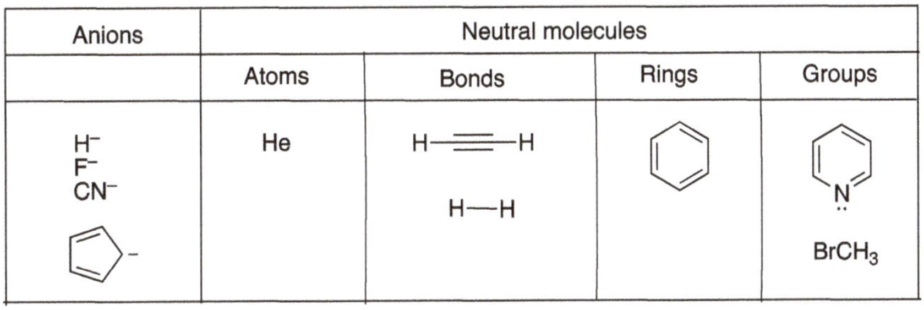

Anions	Neutral molecules			
	Atoms	Bonds	Rings	Groups
H^- F^- CN^- (cyclopentadienyl)$^-$	He	$H\equiv H$ $H-H$	(benzene)	(pyridine) $BrCH_3$

Figure 2. Lewis bases involved in noncovalent interactions.

The proposed definition allows naming immediately the famous $H_3N:BH_3$ complex [21]; since BH_3 is the LA, this is an example of triel bond. The recent controversy Zhou-Frenking/Landis-Weinhold on the $Ca(CO)_8$ complex [22–24] leads us to propose the classify them as alkaline earth bonds, the CO being the Lewis bases.

In a recent paper, it is written: "It is well known that alkynes act as π-acids in the formation of complexes with metals" [25]. If this were correct, then the bond should be a tetrel one; on the other hand, if the alkyne was the base and the metal (in this case Au) the Lewis acid [14], the bond would be a regium bond.

This review does not try to discuss the nature of the bonds [26] we classified as NCIs. This is still a subject not settled [27]. For instance, Mo et al., using the block-localized wave function (BLW), analyzed the halogen bond [28], concluding that it is a charge transfer (CT) interaction, i.e., an intermolecular hyperconjugation consistent with Mulliken proposal [29]. The same authors used the BLW methodology to analyze hydrogen, halogen, chalcogen, and pnictogen bonds, stressing the magnitude of covalency, directionality, and σ-hole concept [30]. A review by Jin et al. [31] compared the σ-hole and π-hole bonds based on halogen bonds. Grabowski et al. [32] discussed halogen, chalcogen, pnictogen, and tetrel bonds as LA-LB complexes.

2. Alkali Bonds

The oldest of NCIs (not including HBs) are the *Halogen Bonds* that, although not named like this, were reported in 1948–1950 by Benasi, Hildebrand, and Mulliken [29,33]. *Lithium Bonds* were introduced by three great chemists: Kollman, Liebman, and Allen in 1970 [34]. We contributed with a paper [35] to this field, where we studied F–Li···N, H–Li···N and H_3C–Li···N lithium bonds. The set of nitrogen Lewis bases consists of two that are sp hybridized (N_2 and HCN); five sp^2-hybridized bases, four of which are aromatic (1,3,5-triazine, 1,2,3-triazine, pyrazine, and pyridine), one nonaromatic (HN=CH_2); and three sp^3-hybridized bases (NH_3, NH_2CH_3, and aziridine).

There have been two theoretical papers reporting *Sodium bonds* [36,37] but, so far, none reporting *Potassium bonds*. For consistency reasons, we propose to call all of them *Alkali bonds*. The paper on sodium bonds reported cooperativity between halogen and sodium bonds in NCX···NCNa···NCY complexes, where Y = F, Cl, Br, I, and Y = H, F, OH. ^{15}N chemical shifts were used to quantify the cooperativity [36].

Although we have excluded cations from this review, we would like to report our studies involving the lithium cation. One characterizing the F–Li^+–F lithium bonds [38]; a number of homo-dimer and hetero-dimer complexes were studied (H_3C–F–Li^+···F_2, H_3C–F–Li^+···F–H, Cl–F···Li^+···F–Cl, F_2···Li^+···F_2, ...) and the spin-spin coupling constants (SSCC) calculated. A different approach was used to study the 1:1 and 2:1 complexes between hydrogen peroxide and its methyl derivatives with lithium cation in order to find if a huge static homogeneous electric field perpendicular to the magnetic field of the NMR spectrometer is able to differentiate enantiomers [39].

3. Alkaline Earth Bonds

Initially, this topic started with *Beryllium bonds* [40,41] and further extended to magnesium and calcium bonds along Group 2. Kollman, Liebman, and Allen suggested, in 1970, studying H_2Be···OH_2, while they explained that HBeF is isoelectronic to HCN [34]. We contributed to this topic starting with a paper of 2009 entitled "Beryllium bonds, do they exist?" [42]. There, we noted that inorganic chemists have described BeX_2L_2 compounds in which X = F, Cl, Br, and L = NH_3 and other Lewis bases (for more recent papers concerning these complexes, see [43,44], and note that they do not call them beryllium bonds).

Beryllium bonds can modulate the strength of HBs (cooperativity) [45], transform azoles into gas-phase superacids [46], create σ-holes in molecules that are devoid of them (like CH_3OF) [47], spontaneous production of radicals [48], beryllium based anion sponges [49], etc.

Magnesium bonds were explored later on. Thus, Q. Li et al. studied the H_2NLi···$HMgX$ complexes where X = H, F, Cl, Br, CH_3, OH and NH_2 that are stabilized though a combination of magnesium and lithium bonds [50]. Scheiner et al. reported the effect of magnesium bonds on the competition between hydrogen and halogen bonds [51]. Montero-Campillo et al. discussed the synergy between tetrel bonds and alkaline earth bonds resulting in weak interactions getting strong [13]. Although NCI are generally studied in intermolecular complexes, there is a paper describing intramolecular magnesium bonds in malonaldehyde-like systems [52].

High-level calculations, using the complete basis set (CBS) extrapolation [CCSD(T)/CBS] of B···BeR_2 and B···MgR_2 complexes were carried out where B is a LB and R = F, H and CH_3 [53]. The Mg series show smaller electrophilicities than the Be series.

Finally, calcium bonds were studied in comparison with beryllium and magnesium bonds at producing huge acidity enhancements [54].

Although some authors have started calling them alkaline earth bonds [13,54], its use has still not become the norm.

4. Regium Bonds

This name (they are also called *Metal Coinage Bonds*) [55–57] is usually given to Group 11; we propose to include also group 10 (Ni, Pd, Pt). We cited Pt (group 10), Co, Rh, and Ir (group 9) in a paper on regium bonds [55], but nobody reports these systems as NCIs.

It is necessary to clearly differentiate clusters (e.g., Au_2 or Ag_{11}) (Figure 3) [58] from molecules (e.g., AuX) [59,60]. Brinck and Stenlid, based on their study of nanoclusters of Cu, Au, Pd, Pt, Rh, ...), proposed a division of σ-holes, depending on the molecular electrostatic potential, into σ_s, σ_p, and σ_d-holes [61,62].

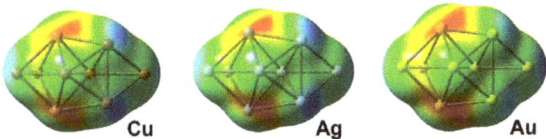

Figure 3. Coinage metal clusters [55].

The higher the oxidation degree (for instance, Au(III) vs. Au(I)) the more acidic the Lewis acid; see, for instance, the complex $(CF_3)_3Au\cdots$pyridine [63]. We cited Legon in a 2014 paper [64] but did not define the Cl–Ag$\cdots C_2H_2$ complex as a regium bond (Figure 4):

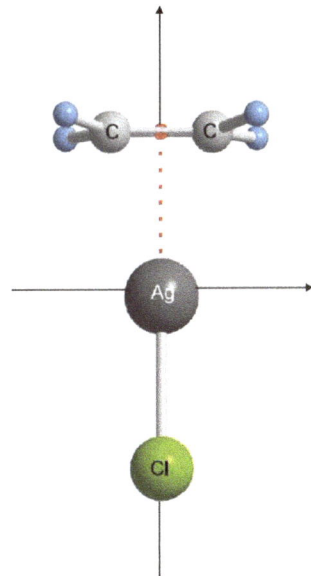

Figure 4. Experimental microwave (MW) structure of complex $C_2H_4\cdots$Ag–Cl.

In 2019, several papers were published on regium bonds, from which we have selected the following four Reference works [65–68].

A comparative study of the regium and hydrogen bonds in Au_2:HX complexes was carried out at CCSD(T) level. In all cases, the regium bond complexes are more stable than HB ones. The binding energies for regium bonds complexes range between −24 and −180 kJ·mol^{-1}, whereas those of the HB complexes are between −6 and −19 kJ·mol^{-1} [65]. Similarly, triel and regium bonds were compared, in particular they augmentative and diminutive interactions; the calculations were carried out at second order Møller-Plesset (MP2) perturbation theory [66]. For Cu, Ag, and Au atoms, the aug-cc-pVDZ-PP pseudopotential was used to account for relativistic effects.

A recent investigation described in detail the synthesis, X-ray characterization, and regium bonding interactions in a trichlorido-(1-hexylcytosine)gold(III) complex [67]. Moreover, this study also included an interesting search in the CSD, revealing that this type of noncovalent interaction is recurrent in X-ray structures and has remained essentially unobserved because of the underestimated van der Waals radius value tabulated for gold. Figure 5 shows the self-assembled dimer that is formed in the solid state of trichlorido-(1-hexylcytosine)gold(III) where two symmetrically equivalent Au···Cl regium bonds are established.

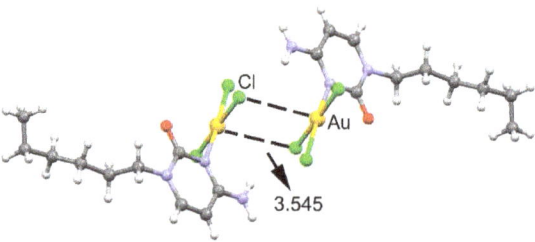

Figure 5. Self-assembled dimer of trichlorido-(1-hexylcytosine)gold(III) complex. Distance in Å.

Finally, regium bonds formed by MX (M = Cu, Ag, Au; X = F, Cl, Br) with phosphine-oxide and phosphinous acid were studied comparing oxygen-shared and phosphine-shared complexes. These complexes were investigated by means of ab initio MP2/aug-cc-pVTZ method [68].

A comparative study of the Lewis acidities of gold(I) and gold (III), specifically ClAu and Cl_3Au, towards different ligands (H, C, N, O, P, S) was carried out at the CCSD(T)/CBS level (an example of N base is given in Figure 6) [69]. The dissociation energies of the complexes are consistent with Yamamoto model. This author, in three fundamental papers [70–72], signaled that $AuCl_3$ behaves preferably as a σ-electrophilic Lewis acid with a η^1 hapticity typically towards heteroatom lone pairs, while AuCl behaves a π-electrophilic Lewis acid with a η^2 hapticity typically towards CC double and triple bonds. Amongst the unexpected findings is that both chlorides open the cyclopropane ring to afford a four-membered metallacycle and that the benzene complexes can show metallotropic shifts. Theoretical [73] and experimental [74] papers related to gold-arene structures have been published. Clearly, this field is one of higher growth in recent times.

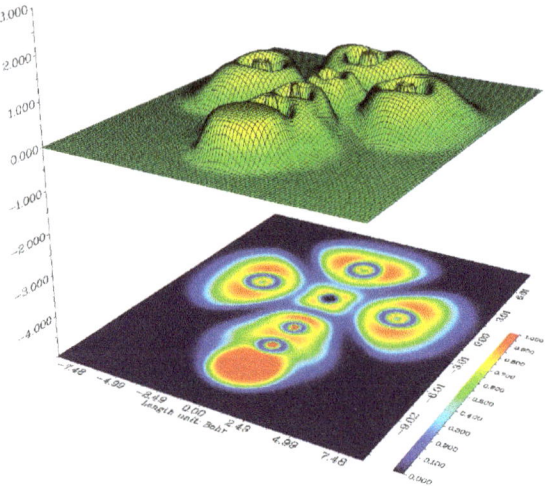

Figure 6. Electron localization function (ELF) analysis of the Cl_3Au···NCH complex.

The nature of the Au–N bond in Au(III) complexes with aromatic heterocycles led Radenkovic et al. to the conclusion that they have higher electrostatic than covalent character [75]. AIM analysis shows that the charge density of the Au–N bond is depleted along the bond path.

5. Spodium Bonds

As aforementioned, for elements of group 11 acting as electron acceptors, the name of regium bonds was proposed to define their interaction with Lewis bases. However, for the adjacent Group 12, the trivial name has not been yet defined. We propose herein to name these bonds "spodium bonds" because a derivative of the first element of the group (ZnO) is called spodium in Latin. It is important to emphasize that the interesting and remarkable work of Joy and Jemmis [76] anticipated that metals of the twelfth group might also participate in noncovalent interactions as Lewis acids. Moreover, these authors also showed that for groups 3–10, this type of interaction (denoted generically as metal bonding) is very scarce. In fact, they searched the Cambridge Structural Database (CSD) [77] and could not find any standard 18-electron transition-metal complexes where the metal participates in a weak interaction of type X–M···:A (A = Lewis Base).

The lack of σ-hole bonding (or metal bonding) in groups 3–10 is due to the fact that the possible σ-hole on the metal center is screened by the core electrons and diminished charge polarization. This is explained by the minimal orbital coefficient on the LUMO in the R–M bond (M belonging to groups 3–10). However, for metal complexes of elements of groups 11 and 12 (fully filled d orbitals), highly diffused valence s and p orbitals can sustain the σ-hole and they are capable to form M–bonds just like the main-group compounds. One of the first manuscripts describing spodium bonds was published by Chieh in 1977 [78]. It corresponds to a dichloro-bis(thiosemicarbazide)-mercury(II) complex that establishes highly directional spodium bonds. It can be clearly observed in Figure 7 that this compound forms in the solid state infinite 1D supramolecular chains where the electron donor (chlorido ligand) is located opposite to the polarized Hg–Cl bond at a distance of 3.25 Å that is slightly shorter than the sum of van der Waals radii (3.30 Å) and significantly longer than the sum of covalent radii (2.39 Å), thus evidencing the noncovalent nature of the interaction.

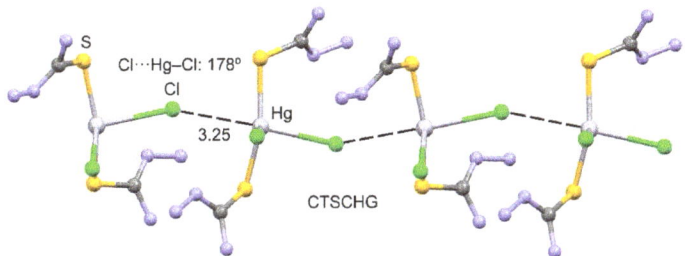

Figure 7. Spodium complexes of $ZnCl_2$. Distances in Å. The CSD reference code is indicated.

The nature of the metal···CO bonds in Group 12 metal carbonyl cations was analyzed by Frenking et al. [79] by studying the geometric and energetic features of their carbonyl complexes, which were also characterized using several computational tools like NBO and distribution of electron density. They showed that in Group 12 the M–CO bond strength in $[M(CO)_n]^{2+}$ complexes exhibits the trend $Zn^{2+} > Hg^{2+} > Cd^{2+}$ and, interestingly, the bond energies are strong for n = 1, 2, moderate for n = 3, 4, and weak for n = 5, 6. Moreover, they showed that Group 12 carbonyls $[M(CO)_n]^{2+}$ exhibit mainly coulombic attraction with quite small covalent contributions apart from $[Hg(CO)]^{2+}$ and $[Hg(CO)_2]^{2+}$ complexes. In contrast, covalent contributions were shown to be significant in the metal carbonyls of Group 11.

It is worthy to highlight the investigation by Vargas et al. where the synthesis and X-ray characterization of unprecedented monomeric 16-electron π-diborene complexes of Zn(II) and Cd(II) are reported, which are good examples of noncovalent spodium bonds [80]. As a matter of fact, stable π-complexes of d^{10} transition metals like copper(I) and nickel(0) with olefins are known. However, such complexes involving d^{10} Zn(II) are not known because the bond is too weak to generate isolable compounds. This fact was explained taking into consideration the limited capacity of elements of Group-12 for π-back-donation. Vargas et al. overcame this drawback by using neutral diborenes because this type of compounds exhibits a high-lying π(B=B) HOMO orbital. In fact, they were able to synthesize in good yields M(II)-diborene (M = Zn, Cd) π-complexes. In addition to their X-ray characterization in the solid state, they were also detected in solution by NMR and UV-visible absorption spectroscopy. The M(II) centers are located over the center of the B=B bond and adopt a trigonal planar geometry almost equidistant to both boron atoms.

6. Triel Bonds

The name of triel bonding was proposed by Grabowski [81] in 2014 to describe the noncovalent interactions between elements of group 13 and electron rich atoms. However, the LA ability of triel atoms has been known for a long time [82–87]. In fact, trivalent triel compounds, such as trihydrides and trihalides, present a strong π-hole due to the empty p orbital, which is perpendicular to the plane of the molecule. This empty p-orbital determines the high directionality of the triel bonding. Since 2014, a number of experimental and theoretical studies have been published devoted to the study of the triel bond and its relation to reactivity [88–93]. As an example, in Figure 8, we show the X-ray structure of the hydrochloride of 4-pyridinylboronic acid, where the anion is located precisely over the B-atom in line with the location of the π-hole, as shown in the MEP surface (see Figure 8).

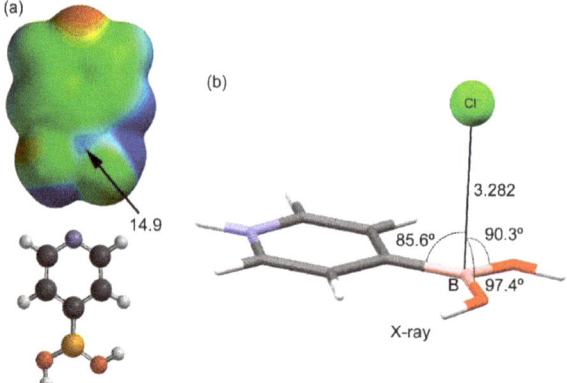

Figure 8. (a) Molecular electrostatic (MEP) surfaces of 4-pyridinylboronic acid with indication of the MEP value at the π-hole in kcal·mol^{-1}. (b) X-ray structure of the hydrochloride of 4-pyridinylboronic acid. The anion is located over the π hole at the boron atom. Distances in Å.

Energetically, the triel bond is very strong and presents highly covalent character. Actually, Leopold et al. [94] have named these type of complexes as "partially bonded complexes" after performing a systematic investigation on the geometric features of triel bonding complexes. The equilibrium distances are intermediate between van der Waals contacts and covalent bonds. It is interesting to highlight the behavior of triel bonds depending on the state. For instance, the triel bonding complex between F$_3$B and acetonitrile exhibits a B···N distances that is 2.01 Å in the gas phase and 1.63 A in the solid state due to cooperativity effects [95].

As a matter of fact, a significant attention has been paid to synergetic effects between triel bonds and a great deal of interactions, including hydrogen bonding [96], and other σ-hole based interactions in elements of group 17 [97], group 16 [98], group 15 [99], group 14 [100], and even regium bonding [66]. In these type of complexes, where two or more interactions coexist, the triel bond is usually the most favored one. Upon formation of the complex, the trivalent triel atom usually suffers a large deformation, changing its planar structure to a pseudo-tetrahedral one thus changing to an sp^3-hybridization. Recently, 'like-like' In(III)···In(III) interactions was studied by Echeverría [101,102] in the crystal of trimethyltriphenyl-phosphine-indium. These unprecedented metallophilic interactions have not been described for the lighter elements of group 13.

7. Tetrel Bonds

A tetrel bond [103] was defined as a noncovalent bond between any electron donating moiety and a LA atom belonging to Group 14 of elements. The initial investigations were basically theoretical; [104–110] however, experimental research on TrB has rapidly grown in the last decade. Actually, there are plenty of examples in the literature reporting experimental [111,112] investigations on tetrel bonding, which was named as such in 2013 [113–116]. A differential feature of tetrel bonding compared to halogen, chalcogen and pnictogen bonding interactions is that the charge density distribution on the tetrel atom is not anisotropic (absence of lone pairs). Moreover, it should be emphasized that the accessibility of the σ-holes is reduced in tetrels because they are located in the middle of three sp^3-hybridized bonds. The behavior of carbon (also named carbon bonding in some studies) [111] is usually different because the rest of tetrels has a strong tendency to expand their valence. Indeed, the heavier tetrels tin and lead, which are commonly seen as metals, have rich coordination chemistry [117–120]. Furthermore, hypervalent species of silicon and germanium are very common [121–131]. Nevertheless, the heavier tetrel atoms (Ge–Pb) participate in noncovalent tetrel bonding interactions when they are in a chemical context avoiding hypervalency, see for instance the SiO$_{12}$(OH)$_8$ cage in Figure 9 [132,133]. In fact, since the atomic polarizability increases in a given group of the periodic table on going from lighter to heavier elements, the stronger interactions in this group are expected for tin and lead [134–136].

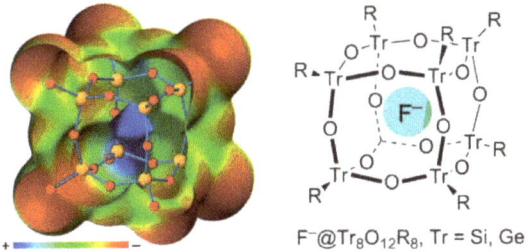

Figure 9. Left: Molecular electrostatic potential open surface of SiO$_{12}$(OH)$_8$ cage. Right: A F$^-$ ion encapsulated inside a Tr$_8$O$_{12}$R$_8$ cage reported by Bauzá et al. [104].

For carbon, tetrel complex can also be understood as the start, [A:···CR$_3$A'] or outcome, [ATrR$_3$···:A'], of an S$_N$2 nucleophilic attack [105] being the transition state an hypervalent specie. Most of the works on tetrel bonding focus on the heavier atoms leaving "carbon bonding" mostly unstudied. In an sp^3 hybridized electron deficient C atom, such as CF$_4$, there is only a limited space available for the LB to interact with C due to its small size. In addition, LB gets very close to negative electrostatic potential of F in CF$_4$. Frontera et al. [107] showed both theoretically and experimentally searching the CSD [77] that a convenient way to expose the σ-hole is to use cyclic X$_2$C–CX$_2$ structures (X = F, CN) where the accessibility of the σ-hole increases as the size of the cycle decreases. In fact, the (CN)$_2$C–C(CN)$_2$ motif was found to be highly directional in 1,1',2,2'-tetracyanocyclopropane/cyclobutane structures.

When sp^2-hybridized electron deficient C-atoms are considered (π-hole instead of σ-hole), the accessibility is not a problem. In this sense, pioneering π-hole interactions were described in 1973 by Bürgi and Dunitz [137,138] in a series of X-ray structural analyses disclosing the trajectory along a LB or nucleophile predominantly attacks the π-hole of a C=O. More than 20 years later, Egli and co-workers described the ability of guanosine to interact with the LBs (O-atom of de-oxiribose) and its importance in the stabilization of Z-DNA form [139].

8. Pnictogen Bonds

These bonds were first described in 2011 in three papers, one experimental [140] and two theoretical [141,142]. An authoritative review was published in a book by some of us (Chapter 8: J. E. Del Bene, I. Alkorta, J. Elguero, The Pnicogen Bond in Review: Structures, Binding Energies, Bonding Properties, and Spin-Spin Coupling Constants of Complexes Stabilized by Pnicogen Bonds, 191-264) [3,143], and another by Scheiner [144]. Although pnictogen bonds are, after halogen bonds, the most studied weak interaction, these bonds have been treated in a reduced number of books and reviews [103,143]. Grabowski classified them as tetrahedral Lewis acid centers [103]. Legon discussed these bonds in an article called " Tetrel, pnictogen and chalcogen bonds identified in the gas phase before they had names: a systematic look at noncovalent interactions" [57].

They are also called "pnicogen bonds" but the pnictogen name should prevail. Similar to halogen bond, pnictogen bond is also a noncovalent interaction. In pnictogen-bond complex, pnictogen atoms (Group VA elements) act as Lewis acid, which can accept electrons from electron donor groups.

Legon pointed out that tetrel, pnictogen, and chalcogen bonds were known in the gas phase (mainly by this author, using rotational spectroscopy) before they had names [56]. Recently, the gas-phase structure of a pnictogen-bonded compound was determined (Figure 10) [145].

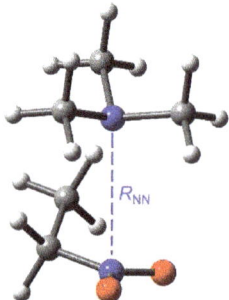

Figure 10. The nitromethane···trimethylamine pnictogen complex [145].

One of our main contributions to pnictogen bonds are the EOM/CCSD calculations, made by J. E. Del Bene, of ^{31}P coupling constants through the pnictogen bond, we called $^{np}J(X-^{31}P)$ [142]. Of our papers concerning pnictogen bonds, we have selected the following eight ones [146–153]. Most of these papers were calculated at the MP2/aug'-cc-pVTZ basis set. We and others have found that FPH_2 and related YPH_2 (Y = H, OH, OCH_3, CH_3, NH_2) and FH_2X (X = P, As) are strong and directional Lewis acid especially suited for theoretical studies [154–156]. Highly acidic heteroboranes yield strong pnictogen bonds [157].

Li, McDowell et al. have shown that upon protonation, the binding distance of the pyridine-(4)-PH_2···NH_3 & PH_3 complexes becomes shorter and the interaction energy is more negative. This shows that the pnictogen bond is strengthened by the protonation of the N atom of pyridine [158]. P···π and π-hole pnictogen bonds have been studied [159,160] and the Cl_3P···C_6H_6 complex studied experimentally by FTIR spectroscopy (Figure 11) [159]. Two important papers have been published, one on the catalysis by pnictogen bonds where there is a distinction between PH_2F σ-hole vs. PO_2F π-hole [161], and the other of supramolecular structures using triple pnictogen bonds [162].

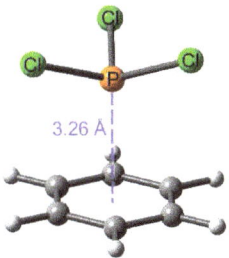

Figure 11. The Cl$_3$P$\cdots$benzene complex [159].

Complexes H$_2$XP$\cdots$NXH$_2$ (X = H, CH$_3$, NH$_2$, OH, F, Cl) presenting P$\cdots$N pnictogen bonds show stabilization energies between 8 and 39 kJ·mol^{-1} [146]. ^{31}P chemical shieldings and ^{1}PJ(N-P) SSCC across the pnictogen interaction were calculated. The last ones exhibit a quadratic dependence on the N–P distance for complexes H$_2$XP$\cdots$NXH$_2$, similar to the dependence of ^{2h}J(X–Y) on the X–Y distance for complexes with X–H$\cdots$Y hydrogen bonds.

The study the influence of F–H$\cdots$F hydrogen bonds on the P$\cdots$P pnictogen bond in complexes nFH$\cdots$(PH$_2$F)$_2$ for n = 1– 3 shows that the formation of F–H$\cdots$F hydrogen bonds leads to a shortening of the P–P distance, a lengthening of the P–F distance involved in the hydrogen bond, a strengthening of the P$\cdots$P interaction, and changes in atomic populations [147]. ^{31}P chemical shieldings, and ^{1}PJ (P–P) coupling constants were calculated.

Pnictogen-bonded cyclic trimers (PH$_2$X)$_3$ with X = F, Cl, OH, NC, CN, CH$_3$, H, and BH$_2$ have been computed (Figure 12) [148]. Most of these complexes have C$_{3h}$ symmetry and binding energies between −17 and −63 kJ·mol^{-1}. The NMR properties of chemical shielding and ^{31}P–^{31}P coupling constants have also been evaluated.

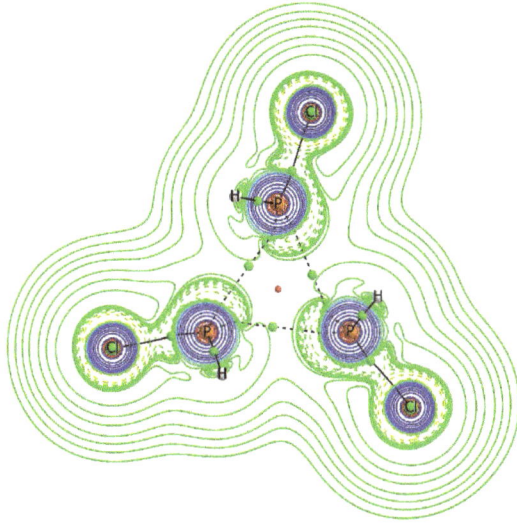

Figure 12. Laplacian of the electron density on the molecular plane of the trimer of (PH$_2$Cl) [148].

Three papers have been reported comparative studies of different NCIs. In the first one [149]. the influence of substituent effects on the formation of P$\cdots$Cl pnictogen bonds or halogen bonds was assessed. There, the potential energy surfaces H$_2$FP$\cdots$ClY for Y = F, NC, Cl, CN, CCH, CH$_3$, and H were explored finding three different types of halogen-bonded complexes with traditional, chlorine-shared, and ion-pair bonds. Two different pnictogen-bonded complexes have also been found on these surfaces. In the second one [153], ab initio calculations were carried out in search of equilibrium dimers on (XCP)$_2$ potential energy surfaces, for X = CN, Cl, F, and H. Five equilibrium dimers with $D_{\infty h}$, $C_{\infty v}$, C_s, C_{2h}, and C_2 symmetries exist on the (ClCP)$_2$ potential energy surface, four on the (FCP)$_2$ and (HCP)$_2$ surfaces, and three on the (NCCP)$_2$ surface. These dimers are stabilized by traditional halogen, pnictogen, and tetrel bonds, and one of them by a hydrogen bond. Finally, Resnati et al. reported an example of a cocrystal where a pnictogen bond prevails over halogen and hydrogen bonds [163].

Another paper reported studies on P(V) complexes [150]. Pnictogen-bonded complexes $H_nF_{5-n}P\cdots$N-Base, for n = 0–5 were studied (two illustrative examples are given in Figure 13). The computed distances and F_{ax}–P–F_{eq} angles in complexes F$_5$P:N-base are consistent with experimental CSD data [77]. All of the complexes with PF$_5$, PHF$_4$, PH$_4$F, and PH$_5$ have C_{4v} symmetry, which is the same symmetry as that of the Berry transition structures of the monomers which lead to the exchange of axial and equatorial atoms.

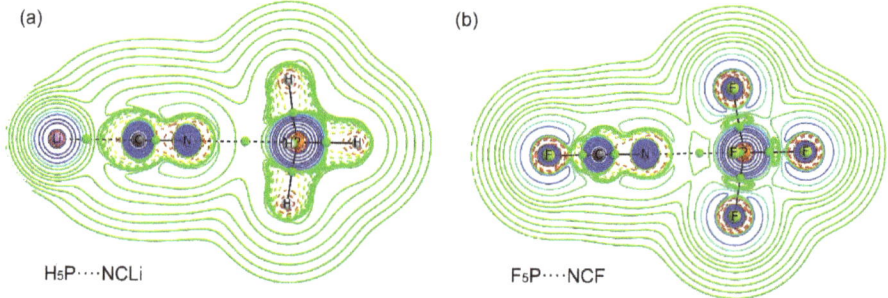

Figure 13. (a) Laplacian of the electron density on the molecular plane of H$_5$P$\cdots$NCLi complex; (b) F$_5$P$\cdots$NCF complex [150].

An ab initio study of the hydration process of metaphosphoric acid shows the importance of the pnictogen interactions [151]. This work was carried out at the MP2/6-31+G(d,p) and MP2/aug-cc-pVTZ computational levels. Up to three explicit water molecules have been considered. The inclusion of more than one water molecule produces important cooperative effects and a shortening of the O$\cdots$P pnictogen interaction simultaneously the reaction barrier drops about 50 kJ mol^{-1}.

A general study of several kinds of NCIs was carried at the MP2/aug-cc-pVTZ computational level. In this paper [152], the dissociation energies D_e of 250 complexes B$\cdots$A composed of 11 Lewis bases B (N$_2$, CO, HC≡CH, CH$_2$=CH$_2$, C$_3$H$_6$, PH$_3$, H$_2$S, HCN, H$_2$O, H$_2$CO, and NH$_3$) and 23 Lewis acids (HF, HCl, HBr, HC≡CH, HCN, H$_2$O, F$_2$, Cl$_2$, Br$_2$, ClF, BrCl, H$_3$SiF, H$_3$GeF, F$_2$CO, CO$_2$, N$_2$O, NO$_2$F, PH$_2$F, AsH$_2$F, SO$_2$, SeO$_2$, SF$_2$, and SeF$_2$) can be represented to good approximation by means of the equation $D_e = c'N_BE_A$, in which N_B is a numerical nucleophilicity assigned to B, E_A is a numerical electrophilicity assigned to A, and c' is a constant, conveniently chosen to have the value 1.00 kJ mol^{-1}. The 250 complexes were chosen to cover a wide range of noncovalent interaction types, namely: (1) the hydrogen bond; (2) the halogen bond; (3) the tetrel bond; (4) the pnictogen bond; and (5) the chalcogen bond.

Diederich orthogonal interactions (N:$\cdots$O$_2$N) are pnictogen bonds when there is a nitrogen lone pair acting as the Lewis base and a nitrogen atom of the nitro group acting as the Lewis acid [164–166]. These interactions have been used by us [167–170] and by others to explain some experimental observations [171]. A theoretical paper entitled "Orthogonal interactions between nitryl derivatives and electron donors: pnictogen bonds"; in this paper complexes from nitryl derivatives (NO$_2$X, X = CN, F, Cl, Br, NO$_2$, OH, CCH, and C$_2$H$_3$) and molecules acting as Lewis bases (H$_2$O, H$_3$N,

CO, HCN, HNC and HCCH) have been obtained at the MP2/aug-cc-pVTZ computational level; a search in the CSD database [77], was carried out, showing a large number of similar interactions in crystallographic structures.

9. Chalcogen Bonds

These bonds have received less attention than the pnictogen bonds, probably due to the fact that P is in chemistry and in biochemistry more important than S. In addition, note that ^{31}P is a very good nucleus for NMR (spin 1/2, natural abundance 100%) and ^{33}S a "bad" one (spin 3/2, natural abundance 0.76%). For books and reviews on chalcogen bonds, see [172–175].

The name "chalcogen bond" was introduced in 2009 by Wang, Ji and Zhang [176]. But papers discussing these NCIs were long known [177–181]. In particular, Gleiter et al. [181] investigated the intermolecular interactions between two molecules containing group 16 elements. The strength of this interaction increases steadily when going from O via S to Se and reaches its maximum for Te. Addition of electron-withdrawing substituents increases the strength of the bond. S···S contacts in thioamides have been studied both experimentally (charge densities) and theoretically [182].

Since most molecules have several kinds of atoms, and since all atoms can be Lewis acids, then, confronted with a Lewis base, several types of NCIs can be formed. For this reason, many papers have been devoted to the competition between some combination of hydrogen, alkaline-earth, tetrel, pnictogen, chalcogen, and halogen bonds [157,183–190]. Curiously, although the nature of the base can change the nature of the most stable acid, none of these publications reported an inversion of acidity. Huynh electronic parameter and its correlation with Hammett σ constants were determined for neutral chalcogen donors [187].

More interesting are the papers reporting cooperative (augmentative) effects where a NCI is reinforced by another NCI, to the point to reach extraordinary values of gas-phase acidity or basicity [191,192].

Although most chalcogen bonds are related to intermolecular situations, a few correspond to intramolecular situations, e.g., to 1,8-disubstituted naphthalenes [193,194]. Other interesting topics related to chalcogen bonds are their use in chiral recognition [195], chalcogen-bonding catalysis [196], and the use by Diederich of benzo[c][1,2,5]thiadiazoles and benzo[c][1,2,5]telluradiazoles to build up capsule dimers [197], followed by a study of "2S-2N" squares formed by benzo[c][1,2,5]thiadiazoles [198].

10. Halogen Bonds

Halogen bonding is a σ-hole interaction of type R–X···:A (X = any element of group 17 including astatine [199]); that is currently experiencing a significant interest in the field of supramolecular chemistry [200–204]. It is the most directional interaction [205] of the σ-hole family, and it can be easily tuned by selecting the type of halogen atom involved (X = I > Br > Cl >> F) [206,207] and nature of the substituent R. This tunability facilitates the rational design of X-bonded catalysts [208,209] and supramolecular synthons to be utilized in crystal engineering [210–212]. The distribution of the electron density in a covalently bonded halogen atom is anisotropic. That is, it shows a region of positive electrostatic potential [213] along the extension of the covalent bond that confers it the ability to act as Lewis acid (i.e., halogen bond donor) [214]. Moreover, it also has a region of negative potential (negative belt) associated to the electron lone pairs conferring it the possibility to act as an electron-rich halogen bond acceptor (Lewis base) [215]. Recently, the X–Bond interaction was used in the field of molecular machines [216–218] providing a new dimension to this interaction. In addition, regarding its counterpart (Lewis base), it was recently demonstrated that transition metal complexes can act as halogen bond acceptors [219–221]. Clark [222,223] and Hobza [224,225] related the strength of halogen bonding to the so-called "polar flattening".

Several excellent reviews [181,201,202,226] and books [33,227] are available in the literature describing most aspects of halogen bonding; therefore, only some general features are commented herein briefly. Halogen bonding is comparable in strength [228] to the ubiquitous hydrogen bond, however, more sensitive to steric effects because the σ-hole is located in a small region of the van der Waals surface along the extension of the R–X bond. A differentiating feature is that the H-bond can be only tuned varying the nature of R and the halogen bond can be tuned varying both R and X. The nature of the X-bond is still under discussion in the literature [229,230]. Note that most theoretical studies propose that an important contribution comes from the stabilization due to donor-acceptor orbital interactions. That is, a filled π or n orbital from the Lewis basic site donates electron density to the antibonding R–X sigma bond orbital [231–233]. Other important contributions are electrostatic effects, polarization in heavy halogens, and dispersion forces that depend upon on the nature of both the Lewis acid and Lewis base [234]. Finally, Kozuch and Martin used halogen bonds as benchmarks for theoretical analyses of wave methods and DFT methods [235].

11. Aerogen Bonds

A noble gas (or aerogen) [236] bond (NgB) was recently defined as: *the interaction between an electron rich atom or group of atoms and any element of Group-18 acting as electron acceptor* [237]. While reports on π,σ-hole interactions involving atoms of groups 14 to 17 as LA have exponentially grown in recent years, investigations on experimental aerogen bonding are scarce. One of those was reported by Schrobilgen's group [238], where they synthesized and X-ray characterized several xenon salts [N(C$_2$H$_5$)$_4$]$_3$ [X$_3$(XeO$_3$)$_3$] X = Cl, Br. These salts form three aerogen bonding interactions with the halides by using the three σ-holes opposite to the O=Xe bonds. Similar behavior was observed by Goettel et al. [239] in their investigation of a series of XeO$_3$ adducts with nitriles since they also form three aerogen bonds in the solid state.

In Figure 14, two X-ray structures are represented where the XeO$_3$ establishes three concurrent aerogen bonds with pyridine N-atoms [232]. These aerogen bonds are shorter for the *p*-dimethylaminopyridine Lewis base due to its stronger basicity compared to pyridine.

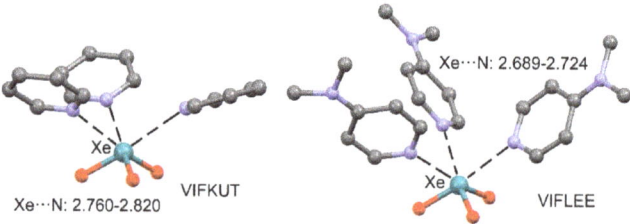

Figure 14. Aerogen bonging interactions in two XeO$_3$ adducts (CSD refcodes VIFKUT and VIFLEE [77]). Aerogen bonds are represented as black dashed lines.

Britvin et al. also demonstrated the tendency of xenon(VI) to form oxide structures synthesizing K$_4$Xe$_3$O$_{12}$, an unprecedented perovskite based on xenon. Its importance is due to the fact that xenon is the only p-block element that forms perovskite frameworks by using a single cation (K$^+$). Remarkably, the authors showed that aerogen bonds are the NCIs that preserve the structural integrity of the perovskite. It is interesting to highlight that these compounds are explosive and the aerogen bonds have been proposed to be the trigger bonds responsible for the detonation [240,241].

Several computational works studied this interaction energetically and geometrically, including its physical insights [242–253]. Interestingly, the effect of increasing the pressure (up to 50 GPa) on the aerogen interactions in XeO$_3$ was also analyzed, resulting in O-hopping along the noncovalent Xe–O···Xe aerogen bonds, resembling H-hopping commonly observed in hydrogen bonds [254]. Moreover, cooperativity effects in aerogen bonding clusters were studied [255] and the interplay with other interactions, as well [256–259].

12. Other Bonds

Cation-π and anion-π (or lone pair-π) [260,261] and even π-π stacking between a π-excessive and π-deficient aromatic rings (Figure 15) can be classified as LA/LB complexes (the Lewis acids being the cation and the hexafluorobenzene and the Lewis bases the anion, the lone pair and hexamethylbenzene) could be classified as tetrel bond since the carbon atoms act like LA (in the case of C^+ it depends on its nature, i.e., C = Na should be an alkaline bond). However, we have decided not to force our systematization running against practices shared by the scientific community.

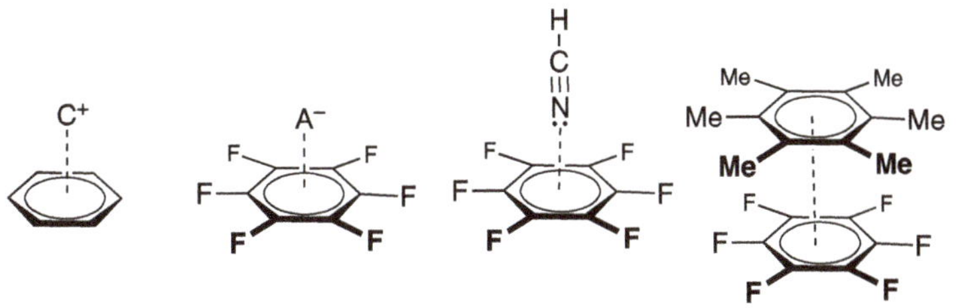

Figure 15. Cation-π, anion-π, lone pair–π and π-π stacking.

13. Modeling

The use of statistical methods to establish extra-thermodynamic relationships [262] for discussing values obtained by quantum methods presents the problem that they have no error, unlike experimental values; note that, without error, no statistical methods can be applied. In spite of this flaw, regression analysis is currently applied to values without error [263].

Two kinds of models are most commonly used: geometrical models like the Hammett, Taft, Grunwald-Winstein equations and the models subjacent to the HSAB (Hard Soft Acid Base)principle. Since we are dealing with Lewis acids and bases, it would be interesting to write a quantitative model that corresponds to hard-hard and soft-soft interactions being strong and hard-soft/soft-hard being weak. We are aware of Mayr et al. criticism of HSAB [264] but note a paper of 2002 by Chandrakumar and Pal entitled "A systematic study on the reactivity of Lewis acid-Base complexes through the local Hard-Soft Acid-Base principle" [265] where they succeed in calculating correctly the interaction energy of complexes using a HSAB model (not cited by Mayr in 2011). A quantitative version of the HSAB principle is Drago's ECW model [266,267].

Alkorta and Legon in two papers, which are (i) "Nucleophilicities of Lewis bases B and electrophilicities of Lewis acids A determined from the dissociation energies of complexes B···A involving hydrogen bonds, tetrel bonds, pnictogen bonds, chalcogen bonds and halogen bonds" and (ii) "Noncovalent interactions involving alkaline-earth atoms and Lewis bases B: An ab initio investigation of beryllium and magnesium bonds, B···MR$_2$ (M = Be or Mg, and R = H, F or CH$_3$") use geometrical models to analyze D_e (equilibrium dissociation energies) in function of k_σ (quadratic force constants) or NB (nucleophilicity of the Lewis base, B) plus E_A (electrophilicity of the Lewis acid): $D_e = a_0 + a_{ij} N_B E_A$ [53,152].

Steric effects are inexistent for protonation in the gas-phase due to the small size of the proton and appear in solution due to solvation, for example, by water molecules [268,269]. For HBs, steric effects have been found, but they are weak or inexistent [270–273]; on the other hand, steric effects are important in NCIs giving yield to a new concept, that of "Frustrated Lewis Pairs" (FLP) [274–278].

14. Application Con Cahn-Ingold-Prelog Rules to Complexes Formed by Weak Interactions (Including Hydrogen Bonds)

For all the situations where the Cahn-Ingold-Prelog priority rules apply for covalent and coordinative structures (ligancy four, axial, planar, ...) [279,280], the priority rules also apply for noncovalent complexes [281,282]. This is particularly useful for crystal structures.

15. A General Definition for Weak Interactions (Including HBs)

A weak interaction between a Lewis acid and a Lewis base is established if the stabilizing forces (electrostatic, dipole-dipole, covalent, ...) overcome the repulsion forces (steric). It is not necessary that the complex should be the lowest minimum; it suffices that there is a barrier between the complex and other minima of lower energy.

16. Summary and Outlook

The number and quality of recent references prove that NCIs are a topic of great and increasing interest. However, as the analysis of the authors of these references show, they belong to a reduced number of groups proving that NCIs are still not part of the large community of chemists. We hope this review will contribute to their diffusion and general acceptance.

A systematic naming resulting from identifying the interaction referring to the Group of the periodic table is very convenient for the sake of unambiguousness. Basically, all donor-acceptor noncovalent interactions can be identified by the element acting as the electrophile. This criterion has been already adopted by the IUPAC for the definition of hydrogen, halogen, and chalcogen bonds. This can be systematically applied to attractive interactions formed by the elements of Groups 1, 2, 10–18 and also to transition metals in a near future. Other names used in the literature like lithium bond, bromine bond or carbon bond can be considered sub-classes of alkali metal bond, halogen bond, and tetrel bond, respectively. Other interactions, like π–π stacking, lp–π, or anion–π interactions involving heteroaromatics, cannot be included in this systematic nomenclature. In contrast, the cation–π interaction could be classified using this nomenclature by using the name of the group to which the cationic element belongs.

It can be predicted that more gas-phase MW structures will be determined in a not so-distant future. Organometallic chemists will report new structures of the regium and spodium classes. Other future developments will be attached to the biological importance of the NCIs.

Author Contributions: Conceptualization, I.A., A.F. and J.E.; methodology, I.A., A.F. and J.E.; validation, I.A., A.F. and J.E.; writing—original draft preparation, I.A., A.F. and J.E.; writing—review and editing, I.A., A.F. and J.E. All authors have read and agreed to the published version of the manuscript.

Funding: This work was carried out with financial support from the Ministerio de Ciencia, Innovación y Universidades of Spain (PGC2018-094644-B-C22) and Comunidad Autónoma de Madrid (P2018/EMT-4329 AIRTEC-CM). Thanks are also given to the CTI (CSIC) for their continued computational support. AF thanks the MICIU/AEI (project CTQ2017-85821-R FEDER funds) for financial support.

Acknowledgments: This is dedicated to Anthony C. Legon, pioneer of studies of noncovalent interactions.

Conflicts of Interest: The authors declare no conflict of interest.

References

1. Hobza, P.; Müller-Dethlefs, K. *Noncovalent interactions. Theory and Experiment*; RSC Theoretical and Computational Chemistry Series; Royal Society of Chemistry: Cambridge, UK, 2010.
2. Sauvage, J.P.; Gaspard, P. (Eds.) *From Noncovalent Assemblies to Molecular Machines*; Wiley-VCH: Weinheim, Germany, 2011.
3. Scheiner, S. (Ed.) *Noncovalent Forces*; Springer: Cham, Switzerland, 2015.
4. Maharramov, A.M.; Mahmudov, K.T.; Kopylovich, M.N.; Pombeiro, A.J.L. (Eds.) *Noncovalent Interactions in the Synthesis and Design of New Compounds*; John and Wiley and Sons: Hoboken, NJ, USA, 2016.

5. Mó, O. Some interesting features of the rich chemistry around electron-deficient systems. *Pure Appl. Chem.* **2019**. [CrossRef]
6. Pimentel, G.C.; McClellan, A.L. *The Hydrogen Bond*; W.H. Freeman & Co.: San Francisco, CA, USA, 1960.
7. Desiraju, G.R.; Steiner, T. The Weak Hydrogen Bond. In *Structural Chemistry and Biology*; International Union of Crystallography, Oxford Science Publications: Oxford, UK, 1999.
8. Gilli, G.; Gilli, P. *The Nature of the Hydrogen Bond: Outline of a Comprehensive Hydrogen Bond Theory*; International Union of Crystallography, Oxford Science Publications: Oxford, UK, 2009.
9. Grabowski, S.J.; Lesczcynski, J. (Eds.) *Hydrogen Bonding—New Insights*; Springer: Dordrecht, The Netherlands, 2006.
10. Li, Z.T.; Wu, L.Z. (Eds.) *Hydrogen Bonded Supramolecular Structures*; Springer: Berlin/Heidelberg, Germany, 2015.
11. Alkorta, I.; Rozas, I.; Elguero, J. Non-conventional hydrogen bonds. *Chem. Soc. Rev.* **1998**, *27*, 163–170. [CrossRef]
12. Juanes, M.; Saragi, R.T.; Caminati, W.; Lesarri, A. The hydrogen bond and beyond: Perspectives for rotational investigations of non-covalent interactions. *Chem. Eur. J.* **2019**, *25*, 11402–11411. [CrossRef] [PubMed]
13. Alkorta, I.; Montero-Campillo, M.M.; Mó, O.; Elguero, J.; Yáñez, M. Weak interactions get strong: Synergy between tetrel and alkaline earth bonds. *J. Phys. Chem. A* **2019**, *123*, 7124–7132. [CrossRef] [PubMed]
14. Fürstner, A.; Davies, P.W. Catalytic carbophilic activation: Catalysis by platinum and gold π acids. *Angew. Chem. Int. Ed.* **2007**, *46*, 3410–3449. [CrossRef]
15. Lewis, G.N. *Valence and the Structure of Atoms and Molecules*; American Chemical Monograph Series; Chemical Catalogue Company: New York, NY, USA, 1923.
16. Jensen, W.B. The Lewis acid-base definitions: A status report. *Chem. Rev.* **1978**, *78*, 1–22. [CrossRef]
17. Politzer, P.; Murray, J.S.; Clark, T. Halogen bonding: An electrostatically-driven highly directional noncovalent interaction. *Phys. Chem. Chem. Phys.* **2010**, *12*, 7748–7757. [CrossRef]
18. Politzer, P.; Murray, J.S. Halogen bonding and other σ-hole interactions: A perspective. *Phys. Chem. Chem. Phys.* **2013**, *15*, 11178–11189. [CrossRef]
19. Bauzá, A.; Mooibroek, T.J.; Frontera, A. The bright future of unconventional σ/π-hole interactions. *ChemPhysChem* **2015**, *16*, 2496–2517.
20. Montero-Campillo, M.M.; Ferrer, M.; Mó, O.; Yáñez, M.; Alkorta, I.; Elguero, J. Insights into the bonding between electron-deficient elements. Systems with X-Y-X (X = B, Al; Y = Be, Mg) bridges. *Phys. Chem. Chem. Phys.* **2020**. submitted.
21. Ott, H.; Matthes, C.; Ringe, A.; Magull, J.; Stalke, D.; Klingebiel, U. On the track of novel triel-stabilized silylaminoiminoborenes. *Chem. Eur. J.* **2009**, *15*, 4602–4609. [CrossRef] [PubMed]
22. Wu, X.; Zhao, L.; Jin, J.; Pan, S.; Li, W.; Jin, X.; Wang, G.; Zhou, M.; Frenking, G. Observation of alkaline earth complexes $M(CO)_8$ (M = Ca, Sr, or Ba) that mimic transition metals. *Science* **2018**, *361*, 912–916. [CrossRef] [PubMed]
23. Landis, C.R.; Hughes, R.P.; Weinhold, F. Comment on "Observation of alkaline earth complexes $M(CO)_8$ (M = Ca, Sr, or Ba) that mimic transition metals". *Science* **2019**, *365*, eaay2355. [CrossRef] [PubMed]
24. Zhao, L.; Pan, S.; Zhou, M.; Frenking, G. Response to comment on "Observation of alkaline earth complexes $M(CO)_8$ (M = Ca, Sr, or Ba) that mimic transition metals". *Science* **2019**, *365*, eaay5021. [CrossRef]
25. Iwasaki, M.; Shichibu, Y.; Konishi, K. Unusual attractive Au-π interactions in small diacetylene-modified gold clusters. *Angew. Chem. Int. Ed.* **2019**, *58*, 2443–2447. [CrossRef] [PubMed]
26. Frenking, G.; Shaik, S. (Eds.) *The Chemical Bond, Chemical Bonding Across the Periodic Table*; Wiley-VCH: Weinheim, Germany, 2014.
27. Brinck, T.; Borrfors, A.N. Electrostatic and polarization determine the strength of the halogen bond: A red card for charge transfer. *J. Mol. Model.* **2019**, *25*, 125. [CrossRef]
28. Wang, C.; Danovich, D.; Mo, Y.; Shaik, S. On the nature of the halogen bond. *J. Chem. Theor. Comput.* **2014**, *10*, 3726–3737. [CrossRef]
29. Mulliken, R.S. Structures of complexes formed by halogen molecules with aromatic and with oxygenated solvents. *J. Am. Chem. Soc.* **1950**, *72*, 600–608. [CrossRef]
30. Wang, C.; Guan, L.; Danovich, D.; Shaik, S.; Mo, Y. The origins of the directionality of noncovalent intermolecular interactions. *J. Comput. Chem.* **2016**, *37*, 34–45. [CrossRef]
31. Wang, H.; Wang, W.; Jin, W.J. σ-Hole vs π-hole bond: A comparison based on halogen bond. *Chem. Rev.* **2016**, *116*, 5072–5104. [CrossRef]

32. Grabowski, S.J.; Sokalski, W.A. Are various -hole bonds steered by the same mechanism? *ChemPhysChem* **2017**, *18*, 1569–1577. [CrossRef] [PubMed]
33. Metrangolo, P.; Resnati, G. (Eds.) Halogen Bonding: Fundamentals and Applications. In *Structure and Bonding*; Springer: Berlin/Heidelberg, Germany, 2008; Volume 126.
34. Kollman, P.A.; Liebman, J.F.; Allen, L.C. The lithium bond. *J. Am. Chem. Soc.* **1970**, *92*, 1142–1150. [CrossRef]
35. Del Bene, J.E.; Alkorta, I.; Elguero, J. Characterizing complexes with F–Li···N, H–Li···N, and CH_3Li···N lithium bonds: Structures, binding energies, and spin-spin coupling constants. *J. Phys. Chem. A* **2009**, *113*, 10327–10334. [CrossRef] [PubMed]
36. Esrafili, M.D.; Mohammadirad, N. Halogen bond interactions enhanced by sodium bonds—Theoretical evidence for cooperative and substitution effects in NCX···NCNa···NCY complexes (X = F, Cl, Br, I; Y = H, F, OH). *Can. J. Chem.* **2014**, *92*, 653–658. [CrossRef]
37. Solimannejad, M.; Rabbani, M.; Ahmadi, A.; Esrafili, M.D. Cooperative and diminutive interplay between the sodium bonding with hydrogen and dihydrogen bondings in ternary complexes of NaC_3N with HMgH and HCN (HNC). *Mol. Phys.* **2014**, *112*, 2017–2022. [CrossRef]
38. Del Bene, J.E.; Alkorta, I.; Elguero, J. Characterizing complexes with $F–Li^+–F$ lithium bonds: Structures, binding energies, and spin-spin coupling constants. *J. Phys. Chem. A* **2009**, *113*, 8359–8365. [CrossRef]
39. Alkorta, I.; Elguero, J.; Provasi, P.F.; Pagola, G.I.; Ferraro, M.B. Electric field effects on nuclear magnetic resonance shielding of the 1.1 and 2:1 (homo and heterochiral) complexes of XOOX (X, X = H, CH_3) with lithium cation and their chiral discrimination. *J. Chem. Phys.* **2011**, *135*, 104116. [CrossRef]
40. Brea, O.; Alkorta, I.; Corral, I.; Mó, O.; Yáñez, M.; Elguero, J. Intramolecular Beryllium Bonds. Further Insights into Resonance Assistance Phenomena. In *Intermolecular Interactions in Crystals*; Novoa, J.J., Ed.; The Royal Society of Chemistry: London, UK, 2018; Chapter 15.
41. Montero-Campillo, M.M.; Mó, O.; Yáñez, M.; Alkorta, I.; Elguero, J. The beryllium bond. *Adv. Inorg. Chem.* **2019**, *73*, 73–121.
42. Yáñez, M.; Sanz, P.; Mó, O.; Alkorta, I.; Elguero, J. Beryllium bonds, do they exists? *J. Chem. Theory Comput.* **2009**, *5*, 2763–2771. [CrossRef]
43. Parameswaran, P.; Frenking, G. Chemical bonding in transition metal complexes with beryllium ligands [$(PMe_3)_2$M-$BeCl_2$], [$(PMe_3)_2$M-BeClMe], and [$(PMe_3)_2$M-$BeMe_2$] (M) Ni, Pd, Pt). *J. Phys. Chem. A* **2010**, *114*, 8529–8535. [CrossRef]
44. Buchner, M.R. Recent contributions to the coordination chemistry of beryllium. *Chem. Eur. J.* **2019**, *25*, 12018–12036. [CrossRef] [PubMed]
45. Mó, O.; Yáñez, M.; Alkorta, I.; Elguero, J. Modulating the strength of hydrogen bonds through beryllium bonds. *J. Chem. Theory Comput.* **2012**, *8*, 2293–2300. [CrossRef] [PubMed]
46. Yáñez, M.; Mó, O.; Alkorta, I.; Elguero, J. Can conventional bases and unsaturated hydrocarbons be converted into gas-phase superacids that are stronger than most of the known oxyacids? The role of beryllium bonds. *Chem. Eur. J.* **2013**, *19*, 11637–11643. [CrossRef] [PubMed]
47. Brea, O.; Mó, O.; Yáñez, M.; Alkorta, I.; Elguero, J. Creating σ-holes through the formation of beryllium bonds. *Chem. Eur. J.* **2015**, *21*, 12676–12682. [CrossRef] [PubMed]
48. Brea, O.; Alkorta, I.; Mó, O.; Yáñez, M.; Elguero, J. Exergonic and spontaneous production of radicals through beryllium bonds. *Angew. Chem. Int. Ed.* **2016**, *55*, 8736–8739. [CrossRef] [PubMed]
49. Brea, O.; Corral, I.; Mó, O.; Yáñez, M.; Alkorta, I.; Elguero, J. Beryllium-based anion sponges: A close relatives of proton sponges. *Chem. Eur. J.* **2016**, *22*, 18322–18325. [CrossRef]
50. Yang, X.; Li, Q.; Cheng, J.; Li, W. A new interaction mechanism of $LiNH_2$ with MgH_2: Magnesium bond. *J. Mol. Model.* **2013**, *19*, 247–253. [CrossRef]
51. Xu, H.L.; Li, Q.Z.; Scheiner, S. Effect of magnesium bond on the competition between hydrogen and halogen bonds and the induction of proton and halogen transfer. *ChemPhysChem* **2018**, *19*, 1456–1464. [CrossRef]
52. Sanz, P.; Montero-Campillo, M.M.; Mó, O.; Yáñez, M.; Alkorta, I.; Elguero, J. Intramolecular magnesium bonds in malonaldehyde-like systems: A critical view of the resonance-assisted phenomena. *Theor. Chem. Acc.* **2018**, *137*, 97. [CrossRef]
53. Alkorta, I.; Legon, A.C. Non-covalent interactions involving alkaline-earth atoms and Lewis bases B: An ab initio investigation of beryllium and magnesium bonds, B···MR_2 (M = Be or Mg, and R = H, F or CH_3). *Inorganics* **2019**, *7*, 35. [CrossRef]

54. Montero-Campillo, M.M.; Sanz, P.; Mó, O.; Yáñez, M.; Alkorta, I.; Elguero, J. Alkaline-earth (Be, Mg and Ca) bonds at the origin of huge acidity enhancements. *Phys. Chem. Chem. Phys.* **2018**, *20*, 2413–2420. [CrossRef] [PubMed]
55. Frontera, A.; Bauzá, A. Regium-π bonds: An unexplored link between noble metal nanoparticles and aromatic surfaces. *Chem. Eur. J.* **2018**, *24*, 7228–7234. [CrossRef] [PubMed]
56. Legon, A.C.; Walker, N.R. What's in a name? 'Coinage-metal' non-covalent bonds and their definition. Tetrel, pnictogen and chalcogen bonds identified in the gas phase before they had names: A systematic look at non-covalent interactions. *Phys. Chem. Chem. Phys.* **2018**, *20*, 19332–19338. [CrossRef] [PubMed]
57. Legon, A.C.; Walker, N.R. Tetrel, pnictogen and chalcogen bonds identified in the gas phase before they had names: A systematic look at non-covalent interactions. *Phys. Chem. Chem. Phys.* **2017**, *19*, 14884–14896. [CrossRef] [PubMed]
58. Stenlid, J.H.; Johansson, A.J.; Brinck, T. σ-Holes and σ-lumps direct the Lewis basic and acidic interactions of noble metal nanoparticles: Introducing regium bonds. *Phys. Chem. Chem. Phys.* **2018**, *20*, 2676–2692. [CrossRef] [PubMed]
59. Li, H.; Li, Q.; Li, R.; Li, W.; Cheng, J. Prediction and characterization of HCCH···AuX (X = OH, F, Cl, Br, CH_3, CCH, CN and NC) complexes: A π Au-bond. *J. Chem. Phys.* **2011**, *135*, 074304. [CrossRef] [PubMed]
60. Zierkiewicz, W.; Michalczyk, M.; Wysokinski, R.; Scheiner, S. Dual geometry schemes in tetrel bonds: Complexes between TF_4 (T = Si, Ge, Sn) and pyridine derivatives. *Molecules* **2019**, *24*, 376. [CrossRef]
61. Stenlid, J.H.; Johansson, A.J.; Brinck, T. σ-Holes on transition metal nanoclusters and their influence on the local Lewis acidity. *Crystals* **2017**, *7*, 222. [CrossRef]
62. Brinck, T.; Stenlid, J.H. The Molecular Surface Property Approach: A Guide to Chemical Interactions in Chemistry, Medicine, and Material Science. *Adv. Theor. Simul.* **2019**, *2*, 1800139. [CrossRef]
63. Pérez-Bitrián, A.; Baya, M.; Casas, J.M.; Falvello, L.R.; Martín, A.; Menjón, B. $(CF_3)_3Au$ as a highly acidic organogold(III) fragment. *Chem. Eur. J.* **2017**, *23*, 14918–14930. [CrossRef]
64. Martín-Somer, A.; Montero-Campiño, M.M.; Mó, O.; Yáñez, M.; Alkorta, I.; Elguero, J. Some interesting features of non-covalent interactions. *Croat. Chim. Acta* **2014**, *87*, 291–396. [CrossRef]
65. Sánchez-Sanz, G.; Trujillo, C.; Alkorta, I.; Elguero, J. Understanding regium bonds and their competition with hydrogen bonds in Au_2:HX complexes. *ChemPhysChem* **2019**, *20*, 1572–1580. [CrossRef] [PubMed]
66. Zhang, J.; Wang, Z.; Liu, S.; Cheng, J.; Li, W.; Li, Q. Synergistic and diminutive effects between triel bond and regium bond: Attractive interactions between π-hole and σ-hole. *Appl. Organomet. Chem.* **2019**, *33*, e4806. [CrossRef]
67. Terrón, A.; Buils, J.; Mooibroek, T.J.; Barceló-Oliver, M.; García-Raso, A.; Fiol, J.J.; Frontera, A. Synthesis, X-ray characterization and regium bonding interactions of a trichlorido-(1-hexylcytosine)gold(III) complex. *Chem. Commun.* **2020**. [CrossRef] [PubMed]
68. Zheng, B.; Liu, Y.; Wang, Z.; Zhou, F.; Liu, Y.; Ding, X.; Lu, T. Regium bonds formed by MX (M = Cu, Ag, Au; X = F, Cl, Br) with phosphine-oxide/phosphinous acid: Comparisons between oxygen-shared and phosphine-shared complexes. *Mol. Phys.* **2019**, *117*, 2443–2455. [CrossRef]
69. Trujillo, C.; Sánchez-Sanz, G.; Elguero, J.; Alkorta, I. The Lewis Acidities of Gold(I) and Gold(III): A Theoretical Study of Complexes of AuCl and $AuCl_3$. *Organometallics* **2020**. submitted.
70. Patil, N.T.; Yamamoto, Y. Gold-catalyzed reactions of oxo-alkynes. *Arkivoc* **2007**, *6*, 19.
71. Yamamoto, Y. From σ- to π-electrophilic Lewis acids. Application to selective organic transforma-tions. *J. Org. Chem.* **2007**, *21*, 7817–7831. [CrossRef]
72. Yamamoto, Y.; Gridnev, I.D.; Patil, N.T.; Jin, T. Alkyne activation with Brønsted acids, iodine, or gold complexes, and its fate leading to synthetic application. *Chem. Commun.* **2009**, 5075–5087. [CrossRef]
73. Bauzá, A.; Frontera, A. Regium–π vs. cation–π interactions in M_2 and MCl (M = Cu, Ag and Au) complexes with small aromatic systems: An ab initio study. *Inorganics* **2018**, *6*, 64. [CrossRef]
74. Lu, P.; Boorman, T.C.; Slawin, A.M.Z.; Larrosa, I. Gold(I)-mediated C–H activation of arenes. *J. Am. Chem. Soc.* **2010**, *132*, 5580–5581. [CrossRef] [PubMed]
75. Radenkovic, S.; Antic, M.; Savic, N.D.; Glisic, B.D. The nature of the Au–N bond in gold(III) complexes with aromatic nitrogen-containing heterocycles: The influence of Au(III) ions on the ligand aromaticity. *New J. Chem.* **2017**, *41*, 12407–12415. [CrossRef]
76. Joy, J.; Jemmis, E.D. Contrasting Behavior of the Z Bonds in X–Z···Y Weak Interactions: Z = Main Group Elements Versus the Transition Metals. *Inorg. Chem.* **2017**, *56*, 1132–1143. [CrossRef] [PubMed]

77. CSD: Allen, F.H. The Cambridge Structural Database: A quarter of a million crystal structures and rising. *Acta Crystallogr. Sect. B* **2002**, *58*, 380–388.
78. Chieh, C. Synthesis and structure of dichlorobis(thiosemicarbazide)mercury(II). *Can. J. Chem.* **1977**, *55*, 1583–1587. [CrossRef]
79. Lupinetti, A.J.; Jonas, V.; Thiel, W.; Strauss, S.H.; Frenking, G. Trends in molecular geometries and bond strengths of the homoleptic d^{10} metal carbonyl cations $[M(CO)_n]^{x+}$ (M^{x+} = Cu^+, Ag^+, Au^+, Zn^{2+}, Cd^{2+}, Hg^{2+}; n = 1–6): A theoretical study. *Chem. Eur. J.* **1999**, *5*, 2573–2583. [CrossRef]
80. Wang, S.R.; Arrowsmith, M.; Braunschweig, H.; Dewhurst, R.D.; Dömling, M.; Mattock, J.D.; Pranckevicius, C.; Vargas, A. Monomeric 16-electron π-diborene complexes of Zn(II) and Cd(II). *J. Am. Chem. Soc.* **2017**, *139*, 10661–10664. [CrossRef]
81. Grabowski, S.J. Boron and other triel Lewis acid centers: From hypovalency to hypervalency. *ChemPhysChem.* **2014**, *15*, 2985–2993. [CrossRef]
82. Hiberty, P.C.; Ohanessian, G. Comparison of minimal and extended basis sets in terms of resonant formulas. Application to 1, 3-dipoles. *J. Am. Chem. Soc.* **1982**, *104*, 66–70. [CrossRef]
83. Brinck, T.; Murray, J.S.; Politzer, P. A computational analysis of the bonding in boron trichloride and their complexes with ammonia. *Inorg. Chem.* **1993**, *32*, 2622–2625. [CrossRef]
84. Kutzelnigg, W. Chemical bonding in higher main group elements. *Angew. Chem. Int. Ed.* **1984**, *23*, 272–295. [CrossRef]
85. Rowsell, B.D.; Gillespie, R.J.; Heard, G.L. Ligand close-packing and the Lewis acidity of BF3 and BC3. *Inorg. Chem.* **1999**, *38*, 4659–4662. [CrossRef] [PubMed]
86. Bessac, F.; Frenking, G. Why is BCl_3 a stronger Lewis acid with respect to strong bases than BF_3? *Inorg. Chem.* **2003**, *42*, 7990–7994. [CrossRef] [PubMed]
87. Fau, S.; Frenking, G. Theoretical investigation of the weakly bonded donor-acceptor complexes H_3B-H_2, X_3B-C_3H_4, and X_3B-C_2H_2 (X = H, F, Cl). *Mol. Phys.* **1999**, *96*, 519–527.
88. Grabowski, S.J. π-Hole bonds: Boron and aluminum Lewis acid centers. *ChemPhysChem* **2015**, *16*, 1470–1479. [CrossRef]
89. Bauzá, A.; Frontera, A. On the versatility of BH_2X (X = F, Cl, Br, and I) compounds as halogen-, hydrogen-, and triel-bond donors: An ab initio study. *ChemPhysChem* **2016**, *17*, 3181–3186. [CrossRef]
90. Bauzá, A.; García-Llinás, X.; Frontera, A. Charge-assisted triel bonding interactions in solid state chemistry: A combined computational and crystallographic study. *Chem. Phys. Lett.* **2016**, *666*, 73–78. [CrossRef]
91. Gao, L.; Zeng, Y.; Zhang, X.; Meng, L. Comparative studies on group III σ-hole and π-hole interactions. *J. Comput. Chem.* **2016**, *37*, 1321–1327. [CrossRef]
92. Bauzá, A.; Frontera, A. Competition between lone pair-π, halogen-π and triel bonding interactions involving BX_3 (X = F, Cl, Br and I) compounds: An ab initio study. *Theor. Chem. Acc.* **2017**, *136*, 37. [CrossRef]
93. Chi, Z.; Li, Q.; Li, H.B. Comparison of triel bonds with different chalcogen electron donors: Its dependence on triel donor and methyl substitution. *Int. J. Quant. Chem.* **2019**, *120*, e26046. [CrossRef]
94. Leopold, K.R.; Canagaratna, M.; Phillips, J.A. Partially Bonded Molecules from the Solid State to the Stratosphere. *Acc. Chem. Res.* **1997**, *30*, 57–64. [CrossRef]
95. Phillips, J.; Giesen, D.; Wells, N.; Halfen, J.; Knutson, C.; Wrass, J. Condensed-phase effects on the structural properties of C_6H_5CN-BF_3 and $(CH_3)_3CCN$-BF_3: IR spectra, crystallography, and computations. *J. Phys. Chem. A* **2005**, *109*, 8199–8208. [CrossRef] [PubMed]
96. Fiacco, D.; Leopold, K. Partially bound systems as sensitive probes of microsolvation: A microwave and ab initio study of HCN$\cdots$HCN–BF_3. *J. Phys. Chem. A* **2003**, *107*, 2808–2814. [CrossRef]
97. Tang, Q.J.; Li, Q.Z. Abnormal synergistic effects between Lewis acid-base interaction and halogen bond in $F_3B\cdots NCX\cdots NCM$. *Mol. Phys.* **2015**, *113*, 3809–3814. [CrossRef]
98. Zhang, J.R.; Li, W.Z.; Cheng, J.B.; Liu, Z.B.; Li, Q.Z. Cooperative effects between π-hole triel and π-hole chalcogen bonds. *RSC Adv.* **2018**, *8*, 26580–26588. [CrossRef]
99. Liu, M.X.; Zhuo, H.Y.; Li, Q.Z.; Li, W.Z.; Cheng, J.B. Theoretical study of the cooperative effects between the triel bond and the pnicogen bond in $BF_3\cdots NCXH_2\cdots Y$ (X = P, As, Sb; Y = H_2O, NH_3) complexes. *J. Mol. Model.* **2016**, *22*, 10. [CrossRef]
100. Yourdkhani, S.; Korona, T.; Hadipour, N.L. Interplay between tetrel and triel bonds in $RC_6H_4CN\cdots MF_3CN\cdots BX_3$ complexes: A combined symmetry-adapted perturbation theory, Møller-Plesset, and Quantum Theory of Atoms-in-Molecules study. *J. Comput. Chem.* **2015**, *36*, 2412–2428. [CrossRef]

101. Echeverría, J. In(III)···In(III) short contacts: An unnoticed metallophilic interaction? *Chem. Commun.* **2018**, *54*, 6312–6315. [CrossRef]
102. Echeverría, J. Intermolecular carbonyl···carbonyl interactions in transition-metal complexes. *Inorg. Chem.* **2018**, *57*, 5429–5437. [CrossRef]
103. Grabowski, S.J. Pnicogan and tetrel bonds—Tetrahedral Lewis acid centers. *Struct. Chem.* **2019**, *30*, 1141–1152. [CrossRef]
104. Bauzá, A.; Mooibroek, T.J.; Frontera, A. Tetrel bonding interactions. *Chem. Rec.* **2016**, *16*, 473–487. [CrossRef] [PubMed]
105. Grabowski, S.J. Tetrel bond-σ-hole as a preliminary stage of the S_N2 reaction. *Phys. Chem. Chem. Phys.* **2014**, *16*, 1824–1834. [CrossRef] [PubMed]
106. Bauzá, A.; Ramis, R.; Frontera, A. Computational study of anion recognition based on tetrel and hydrogen bonding interaction by calix[4]pyrrole derivatives. *Comput. Theor. Chem.* **2014**, *1038*, 67–70. [CrossRef]
107. Bauzá, A.; Mooibroek, T.J.; Frontera, A. Small cycloalkane $(CN)_2C–C(CN)_2$ structures are highly directional non-covalent carbon-bond donors. *Chem. Eur. J.* **2014**, *20*, 10245–10248. [CrossRef]
108. Bauzá, A.; Mooibroek, T.J.; Frontera, A. Tetrel-bonding interaction: Rediscovered supramolecular force? *Angew. Chem. Int. Ed.* **2013**, *52*, 12317–12321. [CrossRef]
109. Mani, D.; Arunan, E. The X–C···Y (X = O/F, Y = O/S/F/Cl/Br/N/P) 'carbon bond' and hydrophobic interactions. *Phys. Chem. Chem. Phys.* **2013**, *15*, 14377–14383. [CrossRef]
110. Naseer, M.M.; Hussain, M.; Bauzá, A.; Lo, K.M.; Frontera, A. Intramolecular noncovalent carbon bonding interaction stabilizes the *cis* conformation in acylhydrazones. *ChemPlusChem* **2018**, *83*, 881–885. [CrossRef]
111. Thomas, S.P.; Pavan, M.S.; Guru Row, T.N. Experimental evidence for 'carbon bonding' in the solid state from charge density analysis. *Chem. Commun.* **2014**, *50*, 49–51. [CrossRef]
112. Mirdya, S.; Frontera, A.; Chattopadhyay, S. Formation of a tetranuclear supramolecule via non-covalent Pb···Cl tetrel bonding interaction in a hemidirected lead(II) complex with a nickel(II) containing metaloligand. *CrystEngComm.* **2019**, *21*, 6859–6868. [CrossRef]
113. Sohail, M.; Panisch, R.; Bowden, A.; Bassindale, A.R.; Taylor, P.G.; Korlyukov, A.A.; Arkhipov, D.E.; Male, L.; Callear, S.; Coles, S.J.; et al. Pentacoordinate silicon complexes with dynamic motion resembling a pendulum on the SN2 reaction pathway. *Dalton. Trans.* **2013**, *42*, 10971–10981. [CrossRef]
114. Mikosch, J.; Trippel, S.; Eichhorn, C.; Otto, R.; Lourderaj, U.; Zhang, J.X.; Hase, W.L.; Weidemüller, M.; Wester, R. Imaging Nucleophilic Substitution Dynamics. *Science* **2008**, *319*, 183–186. [CrossRef] [PubMed]
115. Langer, J.; Matejcik, S.; Illenberger, E. The nucleophilic displacement (S_N2) reaction F^- + $CH_3Cl \rightarrow CH_3F$ + Cl^- induced by resonant electron capture in gas phase clusters. *Phys. Chem. Chem. Phys.* **2000**, *2*, 1001–1005. [CrossRef]
116. Levy, C.J.; Puddephatt, R.J. Rapid reversible oxidative addition of group 14–halide bonds to platinum (II): rates, equilibria, and bond energies. *J. Am. Chem. Soc.* **1997**, *119*, 10127–10136. [CrossRef]
117. The Chemistry of Functional Groups. *The Chemistry of Organic Germanium, Tin and Lead Compounds*; Patai, S., Rappoport, Z., Eds.; John and Wiley and Sons: Hoboken, NJ, USA, 1995; Volume 19.
118. Parr, J. *Comprehensive Coordination Chem. II*; McCleverty, J.A., Meyer, T.J., Eds.; Pergamon Press: Oxford, UK, 2004; Volume 3, p. 545.
119. Sato, T. *Comprehensive Organometallic Chem. II*; Abel, E.W., Stone, F.G.A., Wilkinson, G., Eds.; Pergamon Press: Oxford, UK, 1995; Volume 11, p. 389.
120. Pinhey, J.T. *Comprehensive Organometallic Chem. II*; Abel, E.W., Stone, F.G.A., Wilkinson, G., Eds.; Pergamon Press: Oxford, UK, 1995; Volume 11, p. 461.
121. Greenberg, A.; Wu, G. Structural relationships in silatrane molecules. *Struct. Chem.* **1990**, *1*, 79–85. [CrossRef]
122. Hencsei, P. Evaluation of silatrane structures by correlation relationships. *Struct. Chem.* **1991**, *2*, 21–26. [CrossRef]
123. Voronkov, M.G.; Barishok, V.P.; Petukhov, L.P.; Rahklin, R.G.; Pestunovich, V.A. 1-Halosilatranes. *J. Organomet. Chem.* **1988**, *358*, 39–55. [CrossRef]
124. Lukevics, E.; Dimens, V.; Pokrovska, N.; Zicmane, I.; Popelis, J.; Kemme, A. Addition of nitrile oxides to 2,3-dihydrofurylsilanes. Crystal and molecular structure of tetrahydrofuro [2,3-*d*]-isoxazolylsilanes. *J. Organomet. Chem.* **1999**, *586*, 200–207. [CrossRef]
125. Corriu, R.J.P. Hypervalent species of silicon: Structure and reactivity. *J. Organomet. Chem.* **1990**, *400*, 81–106. [CrossRef]

126. Förgács, G.; Kolonits, M.; Hargittai, I. The gas-phase molecular structure of 1-fluorosilane from electron diffraction. *Struct. Chem.* **1990**, *1*, 245–250. [CrossRef]
127. Eujen, R.; Petrauskas, E.; Roth, A.; Brauer, D.J. The structures of 1-chlorogermatrane and of 1-fluorogermatrane, revisited. *J. Organomet. Chem.* **2000**, *613*, 86–92. [CrossRef]
128. Lukevics, E.; Ignatovich, L.; Beliakov, S. Synthesis and molecular structure of phenyl and tolylgermatranes. *J. Organomet. Chem.* **1999**, *588*, 222–230. [CrossRef]
129. Livant, P.; Northcott, J.; Webb, T.R. Structure of an oxo-bridged germatrane dimer. *J. Organomet. Chem.* **2001**, *620*, 133–138. [CrossRef]
130. Karlov, S.S.; Shutov, P.L.; Churakov, A.V.; Lorberth, J.; Zaitseva, G.S. New approach to 1-(phenylethynyl) germatranes and 1-(phenylethynyl)-3,7,10-trimethylgermatrane. Reactions of 1-(phenylethynyl) germatrane with N-bromosuccinimide and bromine. *J. Organomet. Chem.* **2001**, *627*, 1–5. [CrossRef]
131. Shen, Q.; Hilderbrandt, R.L. The structure of methyl silatrane (1-methyl-2,8,9-trioxa-5-aza-1-silabicyclo(3.3.3)undecane) as determined by gas phase electron diffraction. *J. Mol. Struct.* **1980**, *64*, 257–262. [CrossRef]
132. Scilabra, P.; Kumar, V.; Ursini, M.; Resnati, G. Close contacts involving germanium and tin in crystal structures: Experimental evidence of tetrel bonds. *J. Mol. Model.* **2018**, *24*, 37. [CrossRef]
133. Frontera, A.; Bauzá, A. S···Sn tetrel bonds in the activation of peroxisome proliferator-activated receptors (PPARs) by organotin molecules. *Chem. Eur. J.* **2018**, *24*, 16582–16587. [CrossRef]
134. Southern, S.A.; Errulat, D.; Frost, J.; Gabidullin, B.; Bryce, D.L. Prospects for ^{207}Pb solid-state NMR studies of lead terel bonds. *Faraday Discuss.* **2017**, *203*, 165–186. [CrossRef]
135. Mahmoudi, G.; Bauzá, A.; Amini, M.; Molins, E.; Mague, J.T.; Frontera, A. On the importance of tetrel bonding interactions in lead (II) complexes with (iso) nicotinohydrazide based ligands and several anions. *Dalton Trans.* **2016**, *45*, 10708–10716. [CrossRef]
136. Servati Gargari, M.; Stilinović, V.; Bauzá, A.; Frontera, A.; McArdle, P.; van Derveer, D.; Ng, S.W.; Mahmoudi, G. Design of lead(II) metal-organic frameworks based on covalent and tetrel bonding. *Chem. Eur. J.* **2015**, *21*, 17951–17958. [CrossRef]
137. Burgi, H.B. Chemical reaction coordinates from crystal structure data. I. *Inorg. Chem.* **1973**, *12*, 2321–2325. [CrossRef]
138. Burgi, H.B.; Dunitz, J.D.; Shefter, E. Geometrical reaction coordinates. II. Nucleophilic addition to a carbonyl group. *J. Am. Chem. Soc.* **1973**, *95*, 5065–5067. [CrossRef]
139. Egli, M.; Gessner, R.V. Stereoelectronic effects of deoxyribose O4' on DNA conformation. *Proc. Natl. Acad. Sci. USA* **1995**, *92*, 180–184. [CrossRef] [PubMed]
140. Zahn, S.; Frank, R.; Hey-Hawkins, E.; Kirchner, B. Pnicogen bonds: A new molecular linker? *Chem. Eur. J.* **2011**, *17*, 6034–6038. [CrossRef] [PubMed]
141. Scheiner, S. A new noncovalent force: Comparison of P···N interaction with hydrogen and halogen bonds. *J. Chem. Phys.* **2011**, *134*, 094315. [CrossRef] [PubMed]
142. Del Bene, J.E.; Alkorta, I.; Sánchez-Sanz, G.; Elguero, J. ^{31}P–^{31}P spin-spin coupling constants for pnicogen homodimers. *Chem. Phys. Lett.* **2011**, *512*, 184–187. [CrossRef]
143. Del Bene, J.E.; Alkorta, I.; Elguero, J. The Pnicogen Bond in Review: Structures, Binding Energies, Bonding Properties, and Spin-Spin Coupling Constants of Complexes Stabilized by Pnicogen Bonds. *Chall. Adv. Comput. Chem. Phys.* **2015**, *19*, 191–264.
144. Scheiner, S. The pnicogen bond: Its relation to hydrogen, halogen, and other noncovalent bonds. *Acc. Chem. Res.* **2013**, *46*, 280–288. [CrossRef]
145. Li, W.; Spada, L.; Tasinato, N.; Rampino, S.; Evangelisti, L.; Gualandi, A.; Cozzi, P.G.; Melandri, S.; Barone, V.; Puzzarini, C. Theory meets experiment for noncovalent complexes: The puzzling case of pnicogen interactions. *Angew. Chem. Int. Ed.* **2018**, *57*, 13853–13857. [CrossRef]
146. Del Bene, J.E.; Alkorta, I.; Sánchez-Sanz, G.; Elguero, J. Structures, energies, bonding, and NMR properties of pnicogen complexes H$_2$XP:NXH$_2$ (X = H, CH$_3$, NH$_2$, OH, F, Cl). *J. Phys. Chem. A* **2011**, *115*, 13724–13731. [CrossRef]
147. Alkorta, I.; Sánchez-Sanz, G.; Elguero, J.; Del Bene, J.E. Influence of hydrogen bonds on the P···P pnicogen bond. *J. Chem. Theor. Comput.* **2012**, *8*, 2320–2327. [CrossRef]
148. Alkorta, I.; Elguero, J.; Del Bene, J.E. Pnicogen-bonded cyclic trimers (PH$_2$X)$_3$ with X = F, Cl, OH, CN, NC, CH$_3$, H, and BH$_2$. *J. Phys. Chem. A* **2013**, *117*, 4981–4987. [CrossRef] [PubMed]

149. Del Bene, J.E.; Alkorta, I.; Elguero, J. Influence of substituent effects on the formation of P···Cl pnicogen bonds or halogen bonds. *J. Phys. Chem. A* **2014**, *118*, 2360–2366. [CrossRef] [PubMed]
150. Del Bene, J.E.; Alkorta, I.; Elguero, J. Pnicogen-bonded complexes H_nF_{5-n}P:N-base, for n = 0–5. *J. Phys. Chem. A* **2014**, *118*, 10144–10154. [CrossRef] [PubMed]
151. Alkorta, I.; Azofra, L.M.; Elguero, J. Ab initio study in the hydration process of metaphosphoric acid: The importance of the pnictogen interactions. *Theor. Chem. Acta* **2015**, *134*, 30. [CrossRef]
152. Alkorta, I.; Legon, A.C. Nucleophilicities of Lewis bases B and electrophilicities of Lewis acids A determined from the dissociation energies of complexes B···A involving hydrogen bonds, tetrel bonds, pnictogen bonds, chalcogen bonds and halogen bonds. *Molecules* **2017**, *22*, 1786. [CrossRef] [PubMed]
153. Del Bene, J.E.; Alkorta, I.; Elguero, J. What types of noncovalent bonds stabilize dimers (XCP)$_2$, for X = CN, Cl, F, and H? *J. Phys. Chem. A* **2019**, *123*, 10086–10094. [CrossRef]
154. Guan, L.; Mo, Y. Electron transfer in pnicogen bonds. *J. Phys. Chem. A* **2014**, *118*, 8911–8921. [CrossRef]
155. Esrafili, M.D.; Mohammadian-Sabet, F. Pnicogen– pnicogen interactions in O$_2$XP:PH$_2$Y complexes (X = H, F, CN; Y = H, OH, OCH$_3$, CH$_3$, NH$_2$). *Chem. Phys. Lett.* **2015**, *638*, 122–127. [CrossRef]
156. Zhuo, H.; Li, Q.; Li, W.; Cheng, J. The dual role of pnicogen as Lewis acid and base and the unexpected interplay between the pnicogen bond and coordination interaction in H$_3$N···MCN (X = P and As; M = Cu, Ag, and Au). *New J. Chem.* **2015**, *39*, 2067–2074. [CrossRef]
157. Pecina, A.; Lepšík, M.; Hnyk, D.; Hobza, P.; Fanfrlík, J. Chalcogen and pnicogen bonds in complexes of neutral icosahedral and bicapped square-antiprismatic heteroboranes. *J. Phys. Chem. A* **2015**, *119*, 1388–1395. [CrossRef]
158. Wei, Y.; Li, Q.; Li, W.; Cheng, J.; McDowell, S.A.C. Influence of the protonation of pyridine nitrogen on pnicogen bonding: Competition and cooperativity. *Phys. Chem. Chem. Phys.* **2016**, *18*, 11348–11356. [CrossRef] [PubMed]
159. Ramanathan, N.; Sankaran, K.; Sundararajan, K. PCl$_3$–C$_6$H$_6$ heterodimers: Evidence for P···π phosphorus bonding at low temperatures. *Phys. Chem. Chem. Phys.* **2016**, *18*, 19350–19358. [CrossRef] [PubMed]
160. Wang, Y.; Zeng, Y.; Li, Z.; Meng, L.; Zhang, X. The mutual influence between π-hole pnicogen bonds and σ-hole halogen bonds in complexes of PO$_2$Cl and XCN/C$_6$H$_6$ (X = F, Cl, Br). *Struct. Chem.* **2016**, *27*, 1427–1437. [CrossRef]
161. Schmauck, J.; Breugst, M. The potential of pnicogen bonding for catalysis—A computational study. *Org. Biomol. Chem.* **2017**, *15*, 8037–8045. [CrossRef]
162. Moaven, S.; Andrews, M.C.; Polaske, T.J.; Karl, B.M.; Unruh, D.K.; Bosch, E.; Bowling, N.P.; Cozzolino, A.F. Triple-pnictogen bonding as a tool for supramolecular assembly. *Inorg. Chem.* **2019**, *58*, 16227–16235. [CrossRef]
163. Scilabra, P.; Terraneo, G.; Daolio, A.; Baggioli, A.; Famulari, A.; Leroy, C.; Bryce, D.L.; Resnatti, G. 4,4′-Dipyridyl dioxide·SbF$_3$ co-crystal: Pnictogen bond prevails over halogen and hydrogen bonds in driving self-assembly. *Cryst. Growth Des.* **2020**, *20*, 916–922. [CrossRef]
164. Paulini, R.; Müller, K.; Diederich, F. Orthogonal multipolar interactions in structural chemistry and biology. *Angew. Chem. Int. Ed.* **2005**, *44*, 1788–1805. [CrossRef]
165. Fischer, F.R.; Schweizer, W.B.; Diederich, F. Molecular torsion balances: Evidence for favorable orthogonal dipolar interactions between organic fluorine and amide groups. *Angew. Chem. Int. Ed.* **2007**, *46*, 8270–8273. [CrossRef]
166. Fischer, F.R.; Wood, P.A.; Allen, F.H.; Diederich, F. Orthogonal dipolar interactions between amide carbonyl groups. *Proc. Natl. Acad. Sci. USA* **2008**, *105*, 17290–17294. [CrossRef]
167. Yap, G.P.A.; Jové, F.A.; Claramunt, R.M.; Sanz, D.; Alkorta, I.; Elguero, J. The X-ray molecular structure of 1-(2′,4′-dinitrophenyl)-1,2,3-triazole and the problem of the orthogonal interaction between a 'pyridine-like' nitrogen and a nitro group. *Aust. J. Chem.* **2005**, *58*, 817–822. [CrossRef]
168. Pinilla, E.; Torres, M.R.; Claramunt, R.M.; Sanz, D.; Prakask, R.; Singh, S.P.; Alkorta, I.; Elguero, J. The structure of 2,3-dihydro-3-(2,4-dioxo-6-methylpyran-3-ylidene)-2-(2-nitrobenzyl)-1,4-benzo-thiazine and the problem of orthogonal interactions. *Arkivoc* **2006**, 136–142. [CrossRef]
169. García, M.A.; Claramunt, R.M.; Elguero, J. ^{13}C and ^{15}N NMR chemical shifts of 1-(2,4-dinitro-phenyl) and 1-(2,4,6-trinitrophenyl) pyrazoles in the solid state and in solution. *Magn. Reson. Chem.* **2008**, *46*, 697–700. [CrossRef]

170. Alkorta, I.; Elguero, J.; Roussel, C.; Vanthuyne, N.; Piras, P. Atropisomerism and axial chirality in heteroaromatic compounds. *Adv. Heterocycl. Chem.* **2012**, *105*, 1–188.
171. Triballeau, N.; Van Name, E.; Laslier, G.; Cai, D.; Paillard, G.; Sorensen, P.W.; Hoffmann, R.; Bertrand, H.O.; Ngai, J.; Acher, F.C. High-potency olfactory receptor agonists discovered by virtual high-throughput screening: Molecular probes for receptor structure and olfactory function. *Neuron* **2008**, *60*, 767–774. [CrossRef]
172. Mahmudov, K.T.; Kopylovich, M.N.; Guedes da Silva, M.F.C.; Pombeiro, A.J.L. Chalcogen bonding in synthesis, catalysis and design of materials. *Dalton Trans.* **2016**, *46*, 10121–10138. [CrossRef]
173. Gleiter, R.; Haberhauer, G.; Werz, D.B.; Rominger, F.; Bleiholder, C. From noncovalent chalcogen-chalcogen interactions to supramolecular aggregates: Experiments and calculations. *Chem. Rev.* **2018**, *118*, 2010–2041. [CrossRef]
174. Vogel, L.; Wonner, P.; Huber, S.M. Chalcogen bonding: An overview. *Angew. Chem. Int. Ed.* **2019**, *58*, 1880–1891. [CrossRef] [PubMed]
175. Scilabra, P.; Terraneo, G.; Resnati, G. The chalcogen bond in crystalline solids: A world parallel to halogen bond. *Acc. Chem. Res.* **2019**, *52*, 1313–1324. [CrossRef] [PubMed]
176. Wang, W.; Ji, B.; Zhang, Y. Chalcogen bond: A sister noncovalent bond to halogen bond. *J. Phys. Chem. A* **2009**, *113*, 8132–8135. [CrossRef] [PubMed]
177. Minkin, V.I.; Sadekov, I.D.; Maksimenko, A.A.; Kompan, O.E. Molecular and crystal structure of *ortho*-tellurated azomethines with intramolecular N→Te coordination. *J. Organomet. Chem.* **1991**, *402*, 331–348. [CrossRef]
178. Minyaev, R.M.; Minkin, V.I. Theoretical study of O→X (S, Se, Te) coordination in organic compounds. *Can. J. Chem.* **1998**, *76*, 776–788. [CrossRef]
179. Minkin, V.I.; Minyaev, R.M. Cyclic aromatic systems with hypervalent centers. *Chem. Rev.* **2001**, *101*, 1247–1265. [CrossRef] [PubMed]
180. Sanz, P.; Yáñez, M.; Mó, O. Competition between X···H···Y intramolecular hydrogen bonds and X···Y (X = O, S, and Y = Se, Te) chalcogen–chalcogen interactions. *J. Phys. Chem. A* **2002**, *106*, 4661–4668. [CrossRef]
181. Bleiholder, C.; Werz, D.B.; Köppel, H.; Gleiter, R. Theoretical investigations on chalcogen–chalcogen interactions: What makes these nonbonded interactions bonding? *J. Am. Chem. Soc.* **2006**, *128*, 2666–2674. [CrossRef]
182. Owczarzak, A.; Dutkiewicz, Z.; Kurczab, R.; Pietruś, W.; Kubicki, M.; Grzéskiewicz, A.M. Role of staple molecules in the formation of S···S contact in thioamides: Experimental charge density and theoretical studies. *Cryst. Growth Des.* **2019**, *19*, 7324–7335. [CrossRef]
183. Li, Q.Z.; Li, R.; Guo, P.; Li, H.; Li, W.Z.; Cheng, J.B. Competition of chalcogen bond, halogen bond, and hydrogen bond in SCS–HOX and SeCSe–HOX (X = Cl and Br) complexes. *Comput. Theor. Chem.* **2012**, *980*, 56–61. [CrossRef]
184. Trujillo, C.; Sánchez-Sanz, G.; Alkorta, I.; Elguero, J. Halogen, chalcogen and pnictogen interaction in $(XNO_2)_2$ homodimers (X = F, Cl, Br, I). *New J. Chem.* **2015**, *39*, 6791–6802. [CrossRef]
185. Esrafili, M.D.; Asadollahi, S.; Shaamat, Y.D. Competition between chalcogen bond and halogen bond interactions in $YOX_4:NH_3$ (Y = S, Se; X = F, Cl, Br) complexes: An ab initio investigation. *Struct. Chem.* **2016**, *27*, 1439–1447. [CrossRef]
186. Alkorta, I.; Elguero, J.; Del Bene, J.E. Complexes of O=C=S with nitrogen bases: Chalcogen bonds, tetrel bonds, and other secondary interactions. *ChemPhysChem* **2018**, *19*, 1886–1894. [CrossRef]
187. Teng, Q.; Ng, P.S.; Leung, J.N.; Huynh, H.V. Donor strengths determination of pnictogen and chalcogen ligands by the Huynh electronic parameter and its correlation to sigma Hammett constants. *Chem. Eur. J.* **2019**, *25*, 13956–13963. [CrossRef]
188. Del Bene, J.E.; Alkorta, I.; Elguero, J. Exploring N···C tetrel and O···S chalcogen bonds in HN (CH)SX:OCS systems, for X = F, NC, Cl, CN, CCH, and H. *Chem. Phys. Lett.* **2019**, *730*, 466–471. [CrossRef]
189. Del Bene, J.E.; Alkorta, I.; Elguero, J. Potential energy surfaces of HN (CH)SX:CO_2 systems, for X = F, Cl, NC, CN, CCH, and H: N···C tetrel bonds and O···S chalcogen bonds. *J. Phys. Chem. A* **2019**, *123*, 7270–7277. [CrossRef] [PubMed]
190. Lu, J.; Scheiner, S. Effects of halogen, chalcogen, pnicogen, and tetrel bonds on IR and NMR spectra. *Molecules* **2019**, *24*, 2822. [CrossRef] [PubMed]
191. Esrafili, M.D.; Saeidi, N.; Solimannejad, M. Tuning of chalcogen bonds by cation-π interactions: Cooperative and diminutive effects. *J. Mol. Model.* **2015**, *21*, 300. [CrossRef] [PubMed]

192. Mó, O.; Montero-Campillo, M.M.; Alkorta, I.; Elguero, J.; Yáñez, M. Ternary complexes stabilized by chalcogen and alkaline-earth bonds: Crucial role of cooperativity and secondary noncovalent interactions. *Chem. Eur. J.* **2019**, *25*, 11688–11795. [CrossRef]
193. Sánchez-Sanz, G.; Trujillo, C.; Alkorta, I.; Elguero, J. Enhancing intramolecular chalcogen interactions in 1-hydroxy-YH-naphthalene derivatives. *J. Phys. Chem. A* **2017**, *121*, 8995–9003. [CrossRef]
194. Sánchez-Sanz, G.; Alkorta, I.; Elguero, J. Theoretical study of intramolecular interactions in *peri*-substituted naphthalenes: Chalcogen and hydrogen bonds. *Molecules* **2017**, *22*, 227. [CrossRef]
195. Sánchez-Sanz, G.; Trujillo, C.; Alkorta, I.; Elguero, J. Intermolecular weak interactions in HTeXH Dimers (X = O, S, Se, Te): Hydrogen bonds, chalcogen–chalcogen contacts and chiral discrimination. *ChemPhysChem* **2012**, *13*, 496–503. [CrossRef]
196. Strakova, K.; Assies, L.; Goujon, A.; Piazzolla, F.; Humeniuk, H.V.; Matile, S. Dithienothiophenes at work: Access to mechanosensitive fluorescent probes, chalcogen-bonding catalysis, and beyond. *Chem. Rev.* **2019**, *119*, 10977–11005. [CrossRef]
197. Riwar, L.J.; Trapp, N.; Root, K.; Zenobi, R.; Diederich, F. Supramolecular capsules: Strong versus weak chalcogen bonding. *Angew. Chem. Int. Ed.* **2018**, *57*, 17259–17264. [CrossRef]
198. Ams, M.R.; Trapp, N.; Schwab, A.; Milic, J.V.; Diederich, F. Chalcogen bonding "2S–2N squares" versus competing interactions: Exploring the recognition properties of sulfur. *Chem. Eur. J.* **2019**, *25*, 323–333. [CrossRef] [PubMed]
199. Guo, N.; Maurice, R.; Teze, D.; Graton, J.; Champion, J.; Montavon, G.; Galland, N. Experimental and computational evidence of halogen bonds involving astatine. *Nat. Chem.* **2018**, *10*, 428–434. [CrossRef] [PubMed]
200. Metrangolo, P.; Neukirch, H.; Pilati, T.; Resnati, G. Halogen bonding based recognition processes: A world parallel to hydrogen bonding. *Acc. Chem. Res.* **2005**, *38*, 386–395. [CrossRef] [PubMed]
201. Gilday, L.C.; Robinson, S.W.; Barendt, T.A.; Langton, M.J.; Mullaney, B.R.; Beer, P.D. Halogen bonding in supramolecular chemistrt. *Chem. Rev.* **2015**, *115*, 7118–7195. [CrossRef] [PubMed]
202. Cavallo, G.; Metrangolo, P.; Milani, R.; Pilati, T.; Priimagi, A.; Resnati, G.; Terraneo, G. The halogen bond. *Chem. Rev.* **2016**, *116*, 2478–2601. [CrossRef]
203. Molina, P.; Zapata, F.; Caballero, A. Anion recognition strategies based on combined noncovalent interactions. *Chem. Rev.* **2017**, *117*, 9907–9972. [CrossRef]
204. Scheiner, S. Comparison of halide receptors based on H, halogen, chalcogen, pnicogen, and tetrel bonds. *Faraday Discuss.* **2017**, *203*, 213–226. [CrossRef]
205. Huber, S.M.; Scanlon, J.D.; Jiménez-Izal, E.; Ugalde, J.M.; Infante, I. On the directionality of halogen bonding. *Phys. Chem. Chem. Phys.* **2013**, *15*, 10350–10357. [CrossRef]
206. Riley, K.E.; Murray, J.S.; Fanfrlík, J.; Řezáč, J.; Solá, R.J.; Concha, M.C.; Ramos, F.M.; Politzer, P. Halogen bond tunability I: The effects of aromatic fluorine substitution on the strengths of halogen-bonding interactions involving chlorine, bromine, and iodine. *J. Mol. Model.* **2011**, *17*, 3309–3318. [CrossRef]
207. Riley, K.E.; Murray, J.S.; Fanfrlík, J.; Řezáč, J.; Solá, R.J.; Concha, M.C.; Ramos, F.M.; Politzer, P. Halogen bond tunability II: The varying roles of electrostatic and dispersion contributions to attraction in halogen bonds. *J. Mol. Model.* **2013**, *19*, 4651–4659. [CrossRef]
208. Chan, Y.C.; Yeung, Y.Y. Halogen Bond Catalyzed Bromocarbocyclization. *Angew. Chem. Int. Ed.* **2018**, *57*, 3483–3487. [CrossRef] [PubMed]
209. Carreras, L.; Serrano-Torné, M.; van Leeuwen, P.W.N.M.; Vidal-Ferran, A. XBphos-Rh: A halogen-bond assembled supramolecular catalyst. *Chem. Sci.* **2018**, *9*, 3644–3648. [CrossRef] [PubMed]
210. Lieffrig, J.; Jeannin, O.; Fourmigué, M. Expanded halogen-bonded anion organic networks with star-shaped iodoethynyl-substituted molecules: From corrugated 2d hexagonal lattices to pyrite-type 2-fold interpenetrated cubic lattices. *J. Am. Chem. Soc.* **2013**, *135*, 6200–6210. [CrossRef] [PubMed]
211. Kumar, V.; Pilati, T.; Terraneo, G.; Meyer, F.; Metrangolo, P.; Resnati, G. Halogen bonded Borromean networks by design: Topology invariance and metric tuning in a library of multi-component systems. *Chem. Sci.* **2017**, *8*, 1801–1810. [CrossRef] [PubMed]
212. Edwards, A.J.; Mackenzie, C.F.; Spackman, P.R.; Jayatilaka, D.; Spackman, M.A. Intermolecular interactions in molecular crystals: what's in a name? *Faraday Discuss.* **2017**, *203*, 93–112. [CrossRef] [PubMed]
213. Politzer, P.; Murray, J.S. σ-Hole interactions: Perspectives and misconceptions. *Crystals* **2017**, *7*, 212. [CrossRef]

214. Clark, T.; Hennemann, M.; Murray, J.S.; Politzer, P. Halogen bonding: The σ-hole. *J. Mol. Model.* **2007**, *13*, 291–296. [CrossRef]
215. Desiraju, G.R.; Ho, P.S.; Kloo, L.; Legon, A.C.; Marquardt, R.; Metrangolo, P.; Politzer, P.; Resnati, G.; Rissanen, K. Definition of the halogen bond (IUPAC recommendations 2013). *Pure Appl. Chem.* **2013**, *85*, 1711–1713. [CrossRef]
216. Catalano, L.; Pérez-Estrada, S.; Terraneo, G.; Pilati, T.; Resnati, G.; Metrangolo, P.; García-Garibay, M.A. Dynamic characterization of crystalline supramolecular rotors assembled through halogen bonding. *J. Am. Chem. Soc.* **2015**, *137*, 15386–15389. [CrossRef]
217. Catalano, L.; Pérez-Estrada, S.; Wang, H.H.; Ayitou, A.J.L.; Khan, S.I.; Terraneo, G.; Metrangolo, P.; Brown, S.; García-Garibay, M.A. Rotational dynamics of diazabicyclo[2.2.2]octane in isomorphous halogen-bonded co-crystals: Entropic and enthalpic effects. *J. Am. Chem. Soc.* **2017**, *139*, 843–848. [CrossRef]
218. Lemouchi, C.; Vogelsberg, C.S.; Zorina, L.; Simonov, S.; Batail, P.; Brown, S.; García-Garibay, M.A. Ultra-fast rotors for molecular machines and functional materials via halogen bonding: Crystals of 1,4-bis(iodoethynyl)bicyclo[2.2.2]octane with distinct gigahertz rotation at two sites. *J. Am. Chem. Soc.* **2011**, *133*, 6371–6379. [CrossRef] [PubMed]
219. Ivanov, D.M.; Novikov, A.S.; Ananyev, I.V.; Kirina, Y.V.; Kukushkin, V.Y. Halogen bonding between metal centers and halocarbons. *Chem. Commun.* **2016**, *55*, 5565–5568. [CrossRef] [PubMed]
220. Baykov, S.V.; Dabranskaya, U.; Ivanov, D.M.; Novikov, A.S.; Boyarskiy, V.P. Pt/Pd and I/Br isostructural exchange provides formation of C–I···Pd, C–Br···Pt, and C–Br···Pd metal-involving halogen bonding. *Cryst. Growth Des.* **2018**, *18*, 5973–5980. [CrossRef]
221. Rozhkov, A.V.; Krykova, M.A.; Ivanov, D.M.; Novikov, A.S.; Sinelshchikova, A.A.; Volostnykh, M.V.; Konovalov, M.A.; Grigoriev, M.S.; Gorbunova, Y.G.; Kukushkin, V.Y. Reverse arene sandwich structures based upon π-hole···[M^{II}] (d^8M = Pt, Pd) interactions, where positively charged metal centers play the role of a nucleophile. *Angew. Chem. Int. Ed.* **2019**, *58*, 4164–4168. [CrossRef]
222. El Kerdawy, A.; Murray, J.S.; Politzer, P.; Bleiziffer, P.; Hesselmann, A.; Görling, A.; Clark, T. Advanced corrections of hydrogen bonding and dispersion for semiempirical quantum mechanical methods. *J. Chem. Theory Comput.* **2013**, *9*, 2264–2275. [CrossRef]
223. Clark, T. Halogen bonds and σ-holes. *Faraday Discuss.* **2017**, *203*, 9–27. [CrossRef]
224. Sedlak, R.; Kolár, M.H.; Hobza, P. Polar flattening and the strength of halogen bonding. *J. Chem. Theory Comput.* **2015**, *11*, 4727–4732. [CrossRef]
225. Kolár, M.H.; Hobza, P. Computer modelling of halogen bonds and other σ-hole interactions. *Chem. Rev.* **2016**, *116*, 5155–5187. [CrossRef]
226. Costa, P.J. The halogen bond: Nature and applications. *Phys. Sci. Rev.* **2017**, 20170136. [CrossRef]
227. Metrangolo, P.; Resnati, G. (Eds.) Halogen Bonding I. Impact on Materials Chemistry and Life Sciences. In *Topics in Current Chemistry*; Springer: Cham, Switzerland, 2015; Volume 358.
228. Metrangolo, P.; Meyer, F.; Pilati, T.; Resnati, G.; Terraneo, G. Halogen bonding in supramolecular chemistry. *Angew. Chem. Int. Ed.* **2008**, *47*, 6114–6127. [CrossRef]
229. Reddy, C.M.; Kirchner, M.T.; Gundakaram, R.C.; Padmanabhan, K.A.; Desiraju, G.R. Isostructurality, polymorphism and mechanical properties of some hexahalogenated benzenes: The nature of halogen···halogen interactions. *Chem. Eur. J.* **2006**, *12*, 2222–2234. [CrossRef] [PubMed]
230. Politzer, P.; Riley, K.E.; Bulat, F.A.; Murray, J.S. Perspectives on halogen bonding and other σ-hole interactions: *Lex parsimoniae* (Occam's razor). *Comput. Theor. Chem.* **2012**, *998*, 2–8. [CrossRef]
231. Zou, J.W.; Jiang, Y.J.; Guo, M.; Hu, G.X.; Zhang, B.; Liu, H.C.; Yu, Q.S. Ab initio study of the complexes of halogen-containing molecules RX (X = Cl, Br, and I) and NH_3: Towards understanding the nature of halogen bonding and the electron-accepting propensities of covalently bonded halogen atoms. *Chem. Eur. J.* **2005**, *11*, 740–751. [CrossRef] [PubMed]
232. Lu, Y.X.; Zou, J.W.; Wang, Y.H.; Jiang, Y.J.; Yu, Q.S. Ab initio investigation of the complexes between bromobenzene and several electron donors: Some insights into the magnitude and nature of halogen bonding interactions. *J. Phys. Chem. A* **2007**, *111*, 10781–10788. [CrossRef] [PubMed]
233. An, X.; Jing, B.; Li, Q. Novel halogen-bonded complexes H_3NBH_3···XY (XY = ClF, ClCl, BrF, BrCl, and BrBr): Partially covalent character. *J. Phys. Chem. A* **2010**, *114*, 6438–6443. [CrossRef]

234. Esrafili, M.D.; Vakili, M.; Solimannejad, M. Characterization of halogen···halogen interactions in crystalline dihalomethane compounds (CH_2Cl_2, CH_2Br_2 and CH_2I_2): A theoretical study. *J. Mol. Model.* **2014**, *20*, 2102. [CrossRef]
235. Kozuch, S.; Martin, J.M.L. Halogen bonds: Benchmarks and theoretical analysis. *J. Chem. Theor. Comput.* **2013**, *9*, 1918–1931. [CrossRef]
236. Bauzá, A.; Frontera, A. Aerogen bonding interactions: A new supramolecular force? *Angew. Chem. Int. Ed.* **2015**, *54*, 7340–7343. [CrossRef]
237. Bauzá, A.; Frontera, A. σ/π-Hole noble gas bonding interactions: Insights from theory and experiment. *Coord. Chem. Rev.* **2020**, *404*, 213112. [CrossRef]
238. Goettel, J.T.; Haensch, V.G.; Schrobilgen, G.J. Stable chloro- and bromoxenate cage anions; $[X_3(XeO_3)_3]^{3-}$ and $[X_4(XeO_3)_4]^{4-}$ (X = Cl or Br). *J. Am. Chem. Soc.* **2017**, *139*, 8725–8733. [CrossRef]
239. Goettel, J.T.; Matsumoto, K.; Mercier, H.P.A.; Schrobilgen, G.J. Syntheses and structures of xenon trioxide alkylnitrile adducts. *Angew. Chem. Int. Ed.* **2016**, *55*, 13780–13783. [CrossRef] [PubMed]
240. Britvin, S.N.; Kashtanov, S.A.; Krivovichev, S.V.; Chukanov, N.V. Xenon in rigid oxide frameworks: Structure, bonding and explsive properties of layered perovskite $K_4Xe_3O_{12}$. *J. Am. Chem. Soc.* **2016**, *138*, 13838–13841. [CrossRef] [PubMed]
241. Britvin, S.N.; Kashtanov, S.A.; Krzhizhanovskaya, M.G.; Gurinov, A.A.; Glumov, O.V.; Strekopytov, S.; Kretser, Y.L.; Zaitsev, A.N.; Chukanov, N.V.; Krivovichev, S.V. Perovskites with the framework-forming xenon. *Angew. Chem. Int. Ed.* **2015**, *54*, 14340–14344. [CrossRef] [PubMed]
242. Makarewicz, E.; Lundell, J.; Gordon, A.J.; Berski, S. On the nature of interactions in the $F_2OXe\cdots NCCH_3$ complex: Is there the Xe (IV)-N bond? *J. Comput. Chem.* **2016**, *37*, 1876–1886. [CrossRef] [PubMed]
243. Miao, J.; Xiong, Z.; Gao, Y. The effects of aerogen-bonding on the geometries and spectral properties of several small molecular clusters containing XeO_3. *J. Phys. Condens. Matter.* **2018**, *30*, 44. [CrossRef] [PubMed]
244. Borocci, S.; Grandinetti, F.; Sanna, N.; Antoniotti, P.; Nunzi, F. Noncovalent complexes of the noble-gas atoms: Analyzing the transition from physical to chemical interactions. *J. Comput. Chem.* **2019**, *40*, 2318–2328. [CrossRef] [PubMed]
245. Weinhold, F.; Landis, C.R. *Valency and Bonding: A Natural Bond. Orbital Donor–Acceptor Perspective*; Cambridge University Press: Cambridge, UK, 2005.
246. Bauzá, A.; Frontera, A. Theoretical study on the dual behavior of XeO_3 and XeF_4 toward aromatic rings: Lone pair–π versus aerogen–π interactions. *ChemPhysChem* **2015**, *16*, 3625–3630. [CrossRef]
247. Gao, M.; Cheng, J.; Li, W.; Xiao, B.; Li, Q. The aerogen-π bonds involving π systems. *Chem. Phys. Lett.* **2016**, *651*, 50–55. [CrossRef]
248. Zierkiewicz, W.; Michalczyk, M.; Scheiner, S. Aerogen bonds formed between $AeOF2$ (Ae = Kr, Xe) and diazines: Comparisons between σ-hole and π-hole complexes. *Phys. Chem. Chem. Phys.* **2018**, *20*, 4676–4687. [CrossRef]
249. Esrafili, M.D.; Asadollahi, S.; Vakili, M. Investigation of substituent effects in aerogen-bonding interactions between ZO_3 (Z = Kr, Xe) and nitrogen bases. *Int. J. Quantum Chem.* **2016**, *116*, 1254–1260. [CrossRef]
250. Esrafili, M.D.; Mohammadian-Sabet, F. Exploring "aerogen-hydride" interactions between ZOF_2 (Z = Ke, Xe) and metal hydrides: An ab initio study. *Chem. Phys. Lett.* **2016**, *654*, 23–28. [CrossRef]
251. Esrafili, M.D.; Mohammadian-Sabet, F.; Solimannejad, M. Single-electron aerogen bonds: Do they exist? *Chem. Phys. Lett.* **2016**, *659*, 196–202. [CrossRef]
252. Esrafili, M.D.; Mohammadian-Sabet, F. An ab initio study on anionic aerogen bonds. *Chem. Phys. Lett.* **2017**, *667*, 337–344. [CrossRef]
253. Esrafili, M.D.; Sadr-Mousavi, A. A computational study on the strength and nature of bifurcated aerogen bonds. *Chem. Phys. Lett.* **2018**, *698*, 1–6. [CrossRef]
254. Hou, C.; Wang, X.; Botanab, J.; Miao, M. Noble gas bond and the behaviour of XeO_3 under pressure. *Phys. Chem. Chem. Phys.* **2017**, *19*, 27463–27467. [CrossRef]
255. Esrafili, M.D.; Vessally, E. A theoretical evidence for cooperative enhancement in aerogen-bonding interactions: Open-chain clusters of $KrOF_2$ and $XeOF_2$. *Chem. Phys. Lett.* **2016**, *662*, 80–85. [CrossRef]
256. Esrafili, M.D.; Vessally, E. The strengthening of a hydrogen or lithium bond on the Z···N aerogen bond (Z = Ar, Kr and Xe): A comparative study. *Mol. Phys.* **2016**, *114*, 3265–3276. [CrossRef]

257. Esrafili, M.D.; Kiani, H. Cooperativity between the hydrogen bonding and σ-hole interaction in linear NCX···(NCH)$_{n=2-5}$ and O$_3$Z···(NCH)$_{n=2-5}$ complexes (X = Cl, Br; Z = Ar, Kr): A comparative study. *Can. J. Chem.* **2017**, *95*, 537–546. [CrossRef]
258. Esrafili, M.D.; Mousavian, P.; Mohammadian-Sabet, F. The influence of hydrogen- and lithium-bonding on the cooperativity of chalcogen bonds: A comparative ab initio study. *Mol. Phys.* **2019**, *117*, 58–66. [CrossRef]
259. Esrafili, M.D.; Qasemsolb, S. Tuning aerogen bonds via anion-π or lone pair-π interaction: A comparative ab initio study. *Struct. Chem.* **2017**, *28*, 1255–1264. [CrossRef]
260. Alkorta, I.; Rozas, I.; Elguero, J. An attractive interaction between the π-cloud of C$_6$F$_6$ and electron-donor atoms. *J. Org. Chem.* **1997**, *62*, 4687–4691. [CrossRef]
261. Buglioni, L.; Mastandrea, M.M.; Frontera, A.; Pericás, M.A. Anion-π interactions in light-induced reactions: Role in the amidation of (hetero)aromatic systems with activated N-aryloxyamides. *Chem. Eur. J.* **2019**, *25*, 11785–11790. [CrossRef] [PubMed]
262. Leffler, J.E.; Grunwald, E. *Rates and Equilibria of Organic Reactions: As Treated by Statistical, Thermodynamic and Extrathermodynamic Methods (Dover Books on Chemistry)*; Kindle Edition; Dover Publications: Mignola, NY, USA, 1989.
263. Matheron, G. Principles of geostatistics. *Econ. Geol.* **1963**, *58*, 1246–1266. [CrossRef]
264. Mayr, H.; Breugst, M.; Ofial, A.R. Farewell to the HSAB treatment of ambident reactivity. *Angew. Chem. Int. Ed.* **2011**, *50*, 6470–6505. [CrossRef] [PubMed]
265. Chandrakumar, K.R.S.; Pal, S. A systematic study on the reactivity of Lewis acid-base complexes through the local Hard-Soft Acid-Base principle. *J. Phys. Chem. A* **2002**, *106*, 11775–11781. [CrossRef]
266. Drago, R.S.; Vogel, G.C.; Needham, T.E. A four-parameter equation for predicting enthalpies of adduct formation. *J. Am. Chem. Soc.* **1971**, *93*, 6014–6026. [CrossRef]
267. Drago, R.S.; Dadmun, A.P.; Vogel, G.C. Addition of new donors to the E and C model. *Inorg. Chem.* **1993**, *32*, 2473–2479. [CrossRef]
268. Hancock, R.D.; Nakani, B.S.; Marsicano, F. Relationship between Lewis acid-base behavior in the gas phase and in aqueous solution. 1. Role of inductive, polarizability, and steric effects in amine ligands. *Inorg. Chem.* **1983**, *22*, 2531–2535. [CrossRef]
269. Hosmane, R.S.; Liebman, J.F. Paradoxes and paradigms: Why is quinoline less basic than pyridine or isoquinoline? A classical organic chemical perspective. *Struct. Chem.* **2009**, *20*, 693–697. [CrossRef]
270. Yamaguchi, I. Nuclear Magnetic Resonance study of the steric effect in dimethylphenols. *Bull. Chem. Soc. Jpn.* **1961**, *34*, 744–747. [CrossRef]
271. Chisholm, J.; Pidcock, E.; van de Streek, J.; Infantes, L.; Motherwell, S.; Allen, F.H. Knowledge-based approaches to crystal design. *CrystEngComm* **2006**, *8*, 11–28. [CrossRef]
272. Delori, A.; Galek, P.T.A.; Pidcock, E.; Jones, W. Quantifying homo- and heteromolecular hydrogen bonds as a guide for adduct formation. *Chem. Eur. J.* **2012**, *18*, 6835–6846. [CrossRef] [PubMed]
273. Romero-Fernández, M.P.; Avalos, M.; Babiano, R.; Cintas, P.; Jiménez, J.L.; Palacios, J.C. A further look at π-delocalization and hydrogen bonding in 2-arylmalondialdehydes. *Tetrahedron* **2016**, *72*, 95–104. [CrossRef]
274. Stephan, D.W. "Frustrated Lewis pairs": A concept for new reactivity and catalysis. *Org. Biomol. Chem.* **2008**, *6*, 1535–1539. [CrossRef]
275. Stephan, D.W.; Erker, G. Frustrated Lewis Pairs chemistry: Development and perspectives. *Angew. Chem. Int. Ed.* **2015**, *54*, 6400–6441. [CrossRef] [PubMed]
276. Pu, M.; Privalov, T. Chemistry of intermolecular frustrated Lewis pairs in motion: Emerging perspectives and prospects. *Isr. J. Chem.* **2015**, *55*, 179–195. [CrossRef]
277. Hill, M.S.; Liptrot, D.J.; Weetman, C. Alkaline earths as main group reagents in molecular catalysis. *Chem. Soc. Rev.* **2016**, *45*, 972–988. [CrossRef] [PubMed]
278. Rohman, S.S.; Kashyap, C.; Ullah, S.S.; Guha, A.K. Designing metal-free frustrated Lewis pairs for dihydrogen activation based on a carbene–borane system. *Polyhedron* **2019**, *162*, 1–7. [CrossRef]
279. Cahn, R.S.; Ingold, C.; Prelog, V. Specification of molecular chirality. *Angew. Chem. Int. Ed.* **1966**, *5*, 385–415. [CrossRef]
280. Prelog, V.; Helmchen, G. Basic principles of the CIP-system and proposals for a revision. *Angew. Chem. Int. Ed.* **1982**, *21*, 567–583. [CrossRef]

281. Alkorta, I.; Elguero, J.; Cintas, P. Adding only one priority rule allows extending CIP rules to supramolecular systems. *Chirality* **2015**, *27*, 339–343. [CrossRef] [PubMed]
282. Elguero, J. Is it possible to extend the Cahn-Ingold-Prelog priority rules to supramolecular structures and coordination compounds using lone pairs? *Chem. Int.* **2016**, 30–31. [CrossRef]

© 2020 by the authors. Licensee MDPI, Basel, Switzerland. This article is an open access article distributed under the terms and conditions of the Creative Commons Attribution (CC BY) license (http://creativecommons.org/licenses/by/4.0/).

Review

A Survey of Supramolecular Aggregation Based on Main Group Element··· Selenium Secondary Bonding Interactions—A Survey of the Crystallographic Literature

Edward R. T. Tiekink

Research Centre for Crystalline Materials, School of Science and Technology, Sunway University, 47500 Bandar Sunway, Selangor Darul Ehsan, Malaysia; edwardt@sunway.edu.my; Tel.: +60-3-7491-7181

Received: 30 May 2020; Accepted: 10 June 2020; Published: 12 June 2020

Abstract: The results of a survey of the crystal structures of main group element compounds (M = tin, lead, arsenic, antimony, bismuth, and tellurium) for intermolecular M···Se secondary bonding interactions is presented. The identified M···Se interactions in 58 crystals can operate independent of conventional supramolecular synthons and can sustain zero-, one-, two-, and, rarely, three-dimensional supramolecular architectures, which are shown to adopt a wide variety of topologies. The most popular architecture found in the crystals stabilized by M···Se interactions are one-dimensional chains, found in 50% of the structures, followed by zero-dimensional (38%). In the majority of structures, the metal center forms a single M···Se contact; however, examples having up to three M···Se contacts are evident. Up to about 25% of lead(II)-/selenium-containing crystals exhibit Pb···Se tetrel bonding, a percentage falling off to about 15% in bismuth analogs (that is, pnictogen bonding) and 10% or lower for the other cited elements.

Keywords: secondary bonding; supramolecular; crystal engineering; tetrel bonding; pnictogen bonding; chalcogen bonding; selenium; structural chemistry; main group elements

1. Introduction

The term "chalcogen bonding" has only relatively recently been incorporated in the crystallography lexicon [1] and refers to a non-covalent interaction featuring a Group VI element, for example and relevant to the present bibliographic survey, selenium, functioning as an electrophilic center [2–4]. The current use of the term "chalcogen bonding" notwithstanding, such interactions have long been recognized in the chemical crystallography community [5–7] but under the guise of "secondary bonding" [7]. Secondary bonding encompasses a range of bonding circumstances such as classic Lewis acid/Lewis base interactions occurring between a metal center, such as a main group element (or p-block element), acting as the acid, and a lone-pair of electrons residing on the Lewis base. The non-covalent binding between atoms under these circumstances, being electrostatic in nature, is in keeping with expectation, that is, opposites attract. More perplexing are those contacts occurring between two ostensibly electron-rich species such a low-valent main group element, that is, having a lone-pair of electrons interacting with an electron-rich element such as selenium. In the structural chemistry of selenium, a very early example of the discussion of the latter type of interaction, that is, an intermolecular Se···O contact between electron-rich species, and the description of the supramolecular assembly stabilized by this interaction, was reported in 1972 [8], and is now classified as a chalcogen bond. The rationale for the formation of chalcogen bonds and indeed, for example, allied tetrel, pnictogen and halogen bonding interactions in which a Group XIV, XV, and XVII element, respectively, functions as the electrophilic center, revolves around the concept of a polar cap or σ-hole [9–14]. Very briefly, a σ-hole refers to an electron-deficient region at the extension of a covalent bond or at the tip of a lone-pair of electrons, which is available, being a pseudo Lewis acidic site, for interaction with an electron-rich region, such as a lone-pair of electrons, of a participating species. Examples of both types of interaction scenarios between a main group element and selenium are found herein and, therefore, the generic term "secondary bonding" is employed throughout. However, the purpose of this present review of the relevant structural data is not to evaluate bonding considerations, rather to highlight the prevalence of M···Se secondary bonding and the supramolecular architectures they sustain. The present literature survey was conducted in continuation of a long-held interest in secondary bonding and the supramolecular patterns stabilized by these interactions [15–23], and is aimed at summarizing all of the known M···Se supramolecular contacts operating in the crystals of main group element species with M = tin, lead, arsenic, antimony, bismuth, and tellurium, and to provide comprehensive descriptions of the supramolecular aggregates arising from these in a consistent fashion.

2. Methods

In the present analysis of the crystallographic literature, the Cambridge Structural Database (CSD; version 5.41) [24] was searched employing ConQuest (version 2.0.4) [25] for M···Se contacts in crystals based on a distance criterion, that is, the separation between the respective main group element (M = tin, lead, arsenic, antimony, bismuth, and tellurium) and selenium had to be equal to or less than the sum of the respective van der Waals radii being 1.90 Å for selenium, 2.17 and 2.02 Å for tin and lead, 1.85, 2.00, 2.00 for arsenic, antimony, and bismuth, and 2.06 Å for tellurium [25]. In addition, general criteria were applied; structures with $R > 0.100$ were excluded along with disordered structures and polymeric species. All retrieved structures were manually evaluated to ensure that the putative M···Se interaction was operating independently of other supramolecular synthons, such as conventional hydrogen bonding. All crystallographic diagrams are original, being generated with DIAMOND [26].

3. Results

The following gives an outline of the supramolecular association formed between selenium and, in turn, the main group elements, M = tin, lead, arsenic, antimony, bismuth, and tellurium, as revealed by X-ray crystallography. Traditionally, when searching for structures with secondary bonding interactions [7], such as A-D···M in the present analysis where D = Se, contacts between the two elements occurring at distances longer than the assumed sum of the covalent radii but, less than the sum of the van der Waals radii are identified. In this scenario, the angle at the selenium atom might be expected to be close to 180°. However, this is an oversimplification for two key reasons. Firstly, if the donor is selenium(II), as in the majority of the structures described herein, there are two lone-pairs of electrons available for binding to M; for selenium(IV), there is one lone-pair. In addition, the selenium atom may be bound to two or more other atoms; for example, the interaction might be of the type $A_2Se\cdots M$, $A_3Se\cdots M$, and so on. In these ways, the A-D···M contact is distinct from a conventional hydrogen bonding interaction or an analogous halogen bonding interaction. In instances where the selenium-bound lone-pair of electrons is assumed to interact with the σ-hole of the main group element-bound lone-pair of electrons, as appears to be the case in most of the examples discussed in 3.1–3.6, the lone-pair may not necessarily be diagonally opposite to a covalent bond. It is for these reasons, that is, the influence of the bonding circumstances and the variable coordination geometries of the donor and acceptor atoms, angular information is not included in the descriptions of the structures. The identified M···Se contacts occur independently of other obvious supramolecular association such as hydrogen bonding interactions. The supramolecular

3.1. Tin Compounds Featuring Sn···Se Interactions in their Crystals

There are 13 compounds featuring Sn···Se secondary bonding interactions in their crystals, **1–13**, and the chemical structures for these are shown in Figure 1. The aggregation patterns involve both tin(II) and tin(IV) centers and encompass zero-, one- and, two, and three-dimensional architectures.

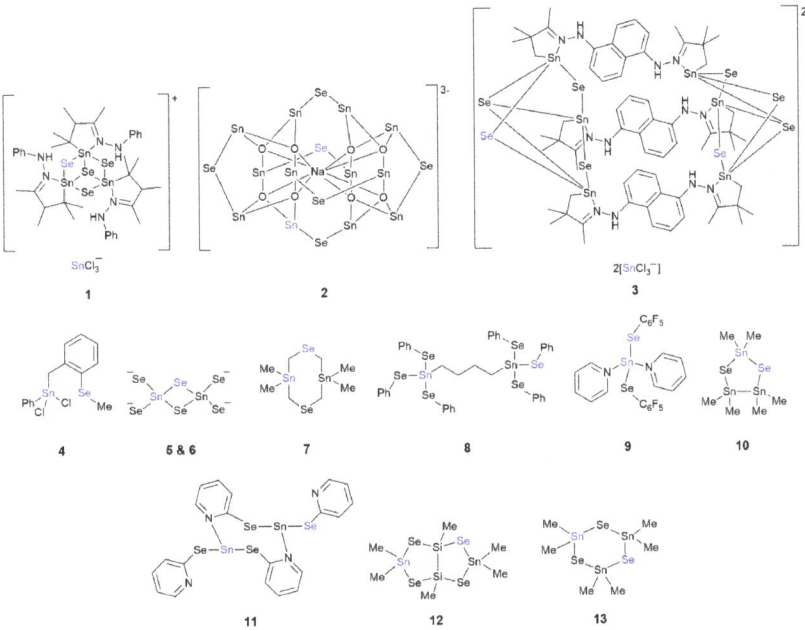

Figure 1. Chemical diagrams for tin compounds **1–13**. The atoms participating in the Sn···Se interactions are highlighted in blue.

The first three structures to be discussed have rather complicated compositions, but the supramolecular association between the interacting species is relatively simple, leading to zero-dimensional aggregates in each case. In **1** [27], comprising interacting cations and anions, the former contains a central Sn_3Se_3 core capped by a μ_3-Se atom forming bonds to the each of the three tin atoms of the core, and the counter-anion is $[SnCl_3]^-$. The tin(II) atom of the latter forms three Sn···Se interactions to the three μ_2-Se atoms of the cyclic core of the cation to form the zero-dimensional aggregate illustrated in Figure 2a. The interacting species in **2** [28] is the $[NaSn_{12}O_8Se_6]^{3-}$ tri-anion and this self-associates about a center of inversion to form a dimeric aggregate mediated by two Sn···Se interactions as shown in Figure 2b. A three-molecule aggregate is observed in ionic **3** [29], Figure 2c. The non-symmetric, di-cation comprises of two bridged Sn_3Se_4 cores, similar to that seen in **1**, and again similar to **1**; one μ_2-Se atom of each core associates via a Sn···Se interaction with a tin(II) atom derived from a $[SnCl_3]^-$ anion.

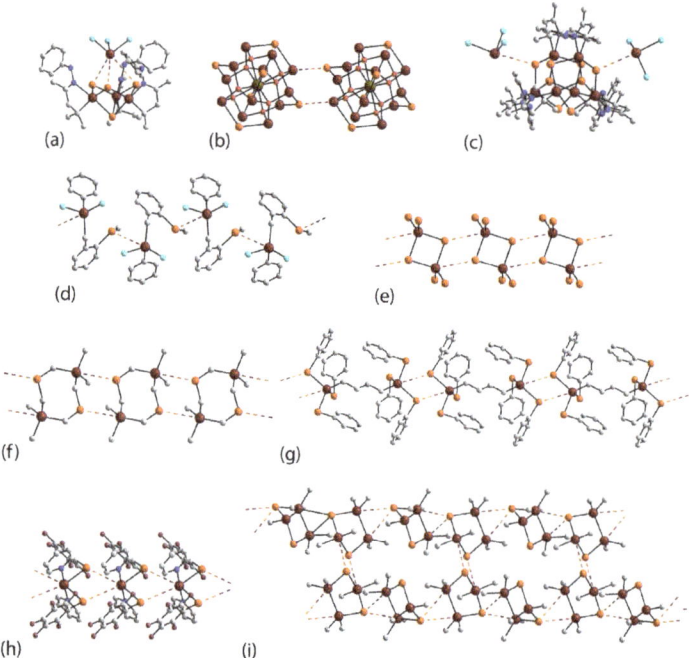

Figure 2. Supramolecular aggregation via Sn···Se secondary bonding in (**a**) **1** {UJAZIQ; d(Sn···Se) = 3.76, 3.79 & 3.90 Å}, (**b**) **2** {PUZCUI; 3.84 Å}, (**c**) **3** {BIFCAX; 3.64 & 3.81 Å}, (**d**) **4** {BELCUS; 3.65 Å}, (**e**) **5** {RESTER; 3.77 Å}, (**f**) **7** {LEVLEE; 3.88 Å}, (**g**) **8** {TORPOG; 3.86 Å}, (**h**) **9** {WUSWOY; 3.79 Å}, and (**i**) **10** {MESESN; 3.77 & 3.91 Å and 3.84, 3.93 & 3.98 Å}. Color code in this and subsequent diagrams: main group element, brown; selenium, orange; chloride, cyan; fluoride, plum; oxygen, red; nitrogen, blue; and carbon, gray.

Seven of the tin compounds self-associate to form one-dimensional chains in their crystals, adopting varying topologies and numbers of Sn···Se interactions sustaining the chains. Compound **4** [30], the first example of a neutral compound and one containing a tin(IV) center, was investigated in terms of systematically varying the substitution pattern in molecules of the formula (2-MeSeC$_6$H$_4$CH$_2$)Sn(Ph)$_{3-n}$Cl$_n$, and ascertaining supramolecular association patterns. In the case of **4**, that is with n = 2, molecules self-associate into a helical chain (2$_1$ screw symmetry) via Sn···Se interactions, as shown in Figure 2d. The next two chains involve the association between tetra-anionic species, [Sn$_2$(µ$_2$-Se)$_2$Se$_4$]$^{4-}$, but the Sn···Se interactions involve the non-charged µ$_2$-Se atoms. The compositions of **5** [31], Figure 2e, and **6** ([32]; SEKYEN) differ in the nature of the counter-cations. The tetra-anion in **5** is disposed about a center of inversion and is connected to centrosymmetrically related aggregates via two Sn···Se interactions and {Sn···Se}$_2$ synthons to form a linear, supramolecular tape. Essentially the same arrangement is observed in **6** where the Sn$_2$Se$_2$ core lies on a mirror plane and is disposed about a center of inversion; the Sn···Se separation is 4.02 Å. The neutral, cyclic compound **7** [33], is disposed about a center of inversion and also connects into a linear, supramolecular tape via Sn···Se interactions involving the µ$_2$-Se atoms, Figure 2f. Binuclear **8** [34], where the tin(IV) atoms are bridged by a butyl chain, is disposed about a center of inversion and associates with inversion related molecules via {Sn···Se}$_2$ synthons, Figure 2g. In **9** [35], designed as a volatile synthetic precursor for SnSe nanomaterials, the tin(IV) atom lies on a 2-fold axis of symmetry, a variation occurs in that the tin atom accepts two Sn···Se interactions from a symmetry related molecule also on the 2-fold axis to form a twisted chain, Figure 2h. The crystallographic asymmetric unit of **10** comprises of two independent five-membered (Me$_2$Sn)$_3$Se$_2$ rings and these form distinct Sn···Se interactions [36]. For the first independent molecule, only two of the constituent tin(IV) atoms, that is, the two tin atoms bonded to each other each forms a Sn···Se interaction and one of the selenium atoms forms two contacts. In the second independent molecule, each of the tin(IV) atoms forms a single Sn···S interaction, one selenium atom forms one contact and the other selenium atom participates in two Sn···Se interactions. In the crystal, alternating independent molecules assemble into a chain, forming comparable Sn···Se interactions involving the bonded tin atoms connecting to the selenium atoms that form two Sn···Se contacts. Centrosymmetrically related chains associate via {Sn···Se}$_2$ synthons involving the second independent molecule only. The resultant double-chain is illustrated in Figure 2i.

The remaining three tin structures assemble into higher-dimensional arrays. In binuclear and centrosymmetric **11** [37], designed as a precursor for the chemical vapor deposition of SnSe nanomaterials, each of the tin(II) atoms forms a single Sn···Se interaction as does one of the two independent selenium atoms. As these extend laterally, a two-dimensional array results with a corrugated topology, as seen in the views of Figure 3a. In binuclear **12** [38], the molecule is disposed about a 2-fold axis of symmetry and has a twisted, U-shape. Each of the tin(IV) and selenium atoms participates in a Sn···Se interaction to form the corrugated array of Figure 3b.

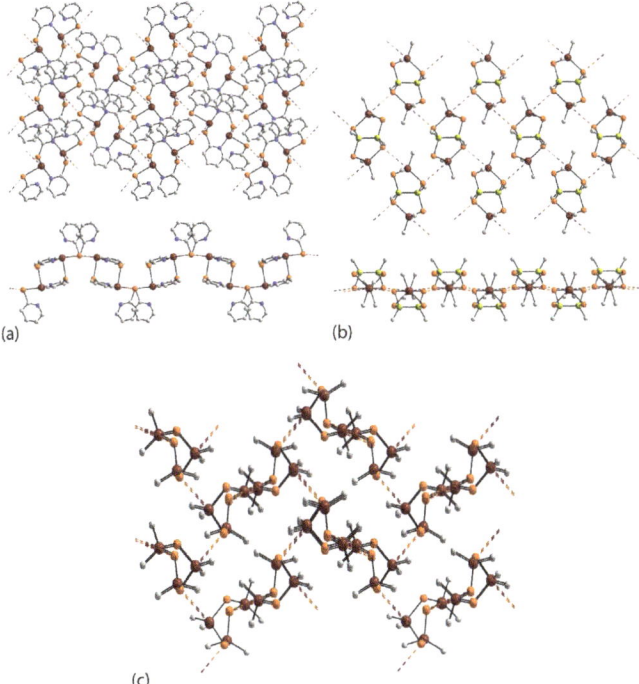

Figure 3. Supramolecular aggregation via Sn···Se secondary bonding in (**a**) **11** {ZOPWIK; d(Sn···Se) = 3.62 Å}, (**b**) **12** {UCOREJ; 4.01 Å}, and (**c**) **13** {HMCTSS; 4.01 Å}. Additional color code: silicon, olive-green.

The cyclic, trinuclear molecule (Me$_2$Sn)$_3$Se$_3$ in **13** [39] has one pair of the diagonally opposite tin(IV) and selenium atoms lying on a 2-fold axis of symmetry, with the ring-atoms not lying on the axis each participating in a single Sn$\cdots$Se interaction. These interactions extend in three-dimensions to consolidate the molecular packing, Figure 3c.

3.2. Lead Compounds Featuring Pb$\cdots$Se Interactions in Their Crystals

There are seven lead compounds satisfying the specified search criteria, **14–20**, and the chemical diagrams for the interacting species in these are shown in Figure 4. The common feature of each structure is the +II oxidation state for the lead atom so all Pb$\cdots$Se interactions can be classified as tetrel bonding interactions. Three of the molecules self-associate to form zero-dimensional aggregates and the remaining examples form one-dimensional chains in their crystals.

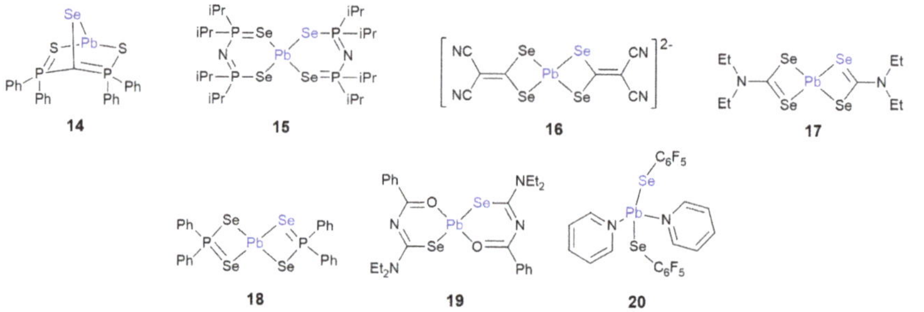

Figure 4. Chemical diagrams for lead compounds **14–20**.

The first aggregate is the centrosymmetric dimer formed by **14** [40] which was developed as a precursor for the chemical vapor deposition (CVD) of PbSe nanoparticles. As shown in Figure 5a, molecules associate through two Pb$\cdots$Se interactions via a {Pb–Se$\cdots$}$_2$ synthon. A similar {Pb–Se$\cdots$}$_2$ synthon is found in **15** [41], Figure 5b, but the dimeric aggregate has crystallographic 2-fold symmetry. The di-anion in **16** [42], which thermally decomposes to PbSe, associates about a center of inversion with each of the selenium atoms of one 2,2-dicyano-ethylene-1,1-diselenolate ligand forming Pb$\cdots$Se interactions, Figure 5c. The remaining molecules in this section associate to form one-dimensional chains.

In **17** [43], developed as a synthetic precursor for PbSe nanomaterials, a selenium atom of each of the asymmetrically chelating diselenocarbamate ligands connects to the same symmetry related lead(II) atom; as a result, a zigzag chain is formed (glide symmetry), Figure 5d. One selenium atom of each of the asymmetrically coordinating diselenophosphinate ligands in **18** [44] also forms a Pb$\cdots$Se interaction but with different centrosymmetrically related molecules, leading to the formation of a twisted supramolecular chain sustained by {Pb–Se$\cdots$}$_2$ synthons, Figure 5e. The compound was prepared in the context of investigating the mechanism of forming quantum dots from tertiary phosphine selenide sources. The lead(II) atom in **19** [45] lies on a 2-fold axis of symmetry and the coordinated selenium atoms associate with the same symmetry related lead(II) atom to form a twisted, supramolecular chain, Figure 5f. The structure of **20** [35] is isostructural with the tin(II) analog, **9**, described as a twisted chain and illustrated in Figure 2h; **9** was investigated for its utility as a single source precursor for PbSe nanoparticles.

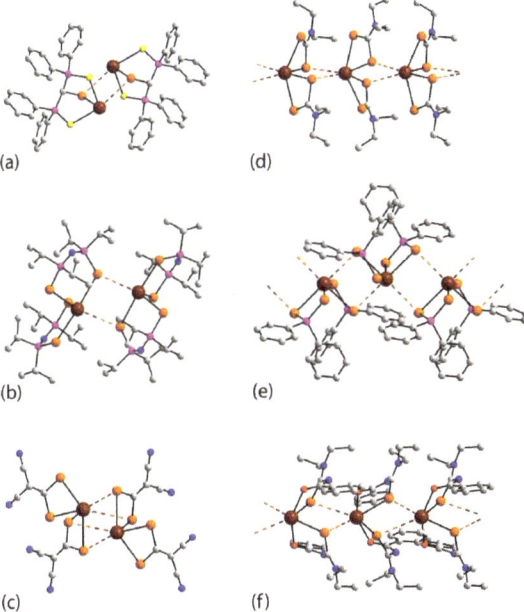

Figure 5. Supramolecular aggregation via Pb···Se secondary bonding in (**a**) **14** {UTAJUV; d(Pb···Se) = 3.41 Å}, (**b**) **15** {TAKLOI; 3.57 Å}, (**c**) **16** {KUHSAH; 3.49 & 3.72 Å}, (**d**) **17** {BOKMUJ; 3.47 & 3.62 Å}, (**e**) **18** {XUZTUI; 3.27 & 3.40 Å}, and (**f**) **19** {YIBHOG; 3.64 Å}. Additional color code: phosphorus, pink.

3.3. Arsenic Compounds Featuring As···Se Interactions in their Crystals

A relatively small number of compounds featuring As···Se interactions in their crystals are known and the chemical structures for the interacting species are shown in Figure 6, that is for **21–27**, and, as demonstrated above, even though there is only a small number of examples, there is a great diversity in supramolecular architectures.

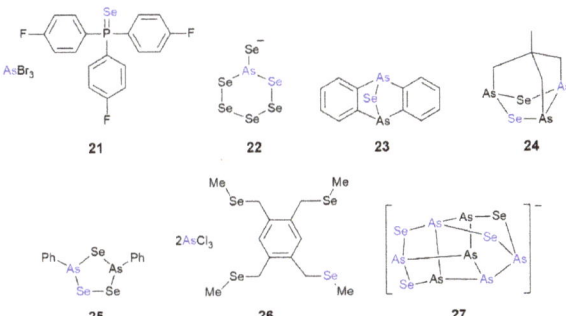

Figure 6. Chemical diagrams for arsenic compounds **21–27**.

The first selenide included in this survey is noted in the crystal of **21** [46], where two distinct molecules associate via As···Se interactions, with the participating atoms being arsenic(III) and selenide-selenium atoms, indicative of a pnictogen interaction. Each of the molecules is located on a crystallographic 3-fold axis of symmetry and associate with a crystallographic site of symmetry 23. It can be noted from the Figure 7a that each phosphaneselenide atom forms three As···Se interactions with three different AsBr$_3$ molecules so that a distorted As$_4$Se$_4$ cube, sustained by eight As···Se interactions, defines the core of the aggregate. The mono-anion in **22** [47] has the charge localized on the exocyclic selenium atom with the dimeric aggregate in the crystal shown in Figure 7b sustained by As···Se interactions between centrosymmetrically related anions.

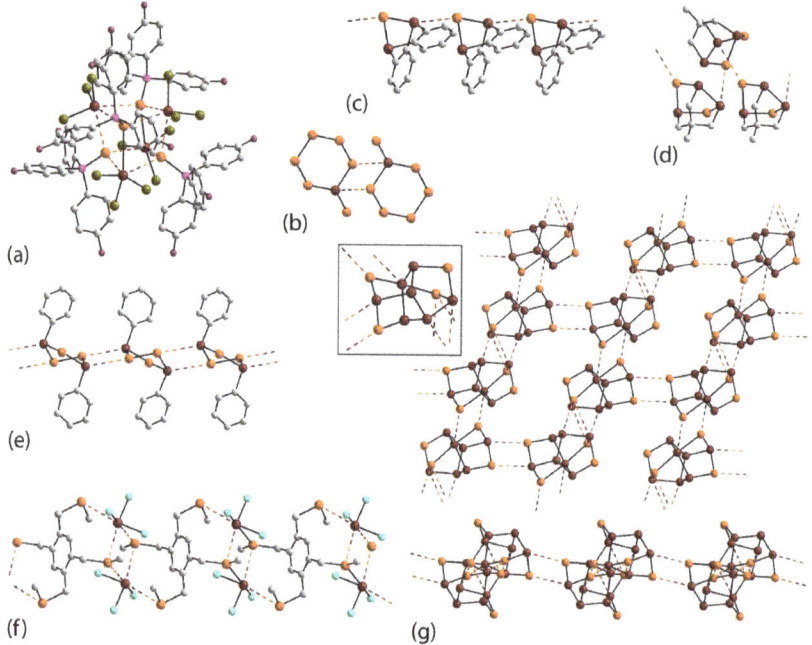

Figure 7. Supramolecular aggregation via As···Se secondary bonding in (**a**) **21** {GEXSOT; d(As···Se) = 3.37 Å}, (**b**) **22** {NEKJIW; 3.48 Å}, (**c**) **23** {ESEARS; 3.63 Å}, (**d**) **24** {COSDIX; 3.64 Å}, (**e**) **25** {SEDMAP; 3.53 Å}, (**f**) **26** {WAMCOE; 3.29, 3.42 & 3.60 Å}, and (**g**) **27** {KAXXUC; 3.60, 3.61, 3.64 & 3.72 Å}. Additional color code: bromide, olive-green.

There are four examples whereby one-dimensional chains are formed through As···Se interactions. In **23** [48], a mirror plane bisects the molecule with the selenium atom lying on the plane. The molecules are assembled into a linear chain via a single As···Se connection per molecule, Figure 7c. Similar connections are noted in the crystal of **24** [49], comprising a five-membered As$_3$Se$_2$ ring, whereby only one of the three potential arsenic(III) atoms and one of the two selenium atoms are engaged in As···Se interactions to form a chain with a helical topology being propagated by 2$_1$-screw symmetry, Figure 7d. A third topology for the chain is seen in the crystal of **25** [50] where the molecule is disposed about a 2-fold axis of symmetry. There are on average two As···Se interactions between the molecules and being propagated by glide symmetry; the chain has a zigzag topology, Figure 7e. The fourth one-dimensional architecture observed for **26** [51] reverts to a helical topology (2_1 screw symmetry), Figure 7f, but exhibits quite distinct features than for **24**. In the crystal, two AsCl$_3$ molecules are bridged by two selenium atoms to form a {As···Se}$_2$ synthon. These are further connected by additional As···Se interactions (3.42 Å) to form the helical, supramolecular chain. In this scheme, the arsenic(III) center participates in three As···Se interactions as seen in **21** and in the next structure to be described, **27**.

A two-dimensional architecture is constructed in the crystal of **27** [52] as a result of three distinct As···Se interactions. As is evident from the inset of Figure 7g, the mono-anion, formulated as As$_7$Se$_4^-$, participates in eight As···Se interactions whereby four arsenic atoms form a single interaction, as do two of the selenium atoms with one selenium atom forming two As···Se contacts. Three of the contacts involve directly bonded arsenic and selenium atoms and occur around a center of inversion in each case; thus, there are three independent {As–Se···}$_2$ synthons. The two remaining interactions occur between bonded arsenic atoms connecting to a single selenium atom, which thereby lead to the formation of a three-membered {···AsAs···Se} synthon. The result is the grid shown in Figure 7g, which define rather large voids that accommodate the tetraphenylphosphonium counter-cations.

3.4. Antimony Compounds Featuring Sb···Se Interactions in their Crystals

Eight crystals feature Sb···Se interactions leading to zero-, one-, and two-dimensional aggregation patterns. The chemical diagrams for the interacting species in these, that is, **28–35**, are shown in Figure 8.

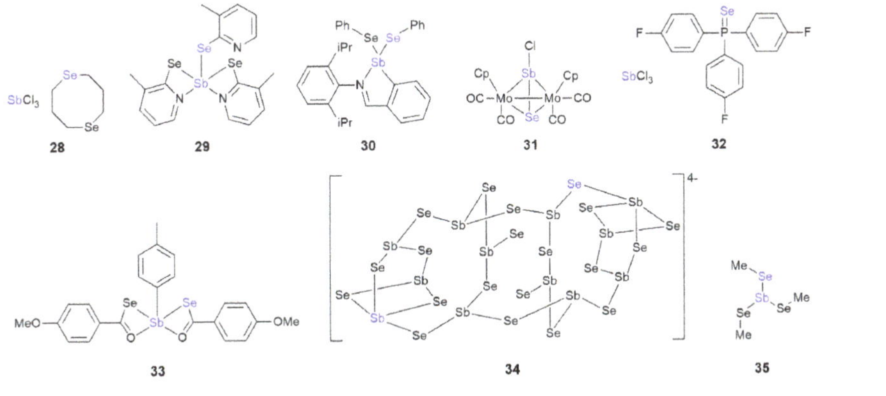

Figure 8. Chemical diagrams for antimony compounds **28–35**; Cp is cyclopentadienyl.

The supramolecular association in the crystal of **28** [53] is an illuminative example of cooperation between Sb···Se and Sb···Cl secondary bonding interactions. As evidenced from Figure 9a, there is a Sb···Se interaction between the SbCl$_3$ molecule and one of the selenium atoms of the eight-membered ring of the 1,5-diselenacyclooctane molecule. These aggregates associate about a center of inversion via Sb···Cl interactions to form a four-molecule aggregate. The molecules in **29** [54], Figure 9b, **30** ([55]; KIMNEB; Sb···Se = 3.69 Å) and **31** ([56]; ISIPEG Sb···Se = 3.88 Å) are centrosymmetric dimers sustained by two Sb···Se interactions; **29** [54] was employed as a precursor for CVD of Sb$_2$Se$_3$ and aerosol-assisted chemical vapor deposition (AACVD) of Sb$_2$Se$_3$ thin films. The last zero-dimensional aggregate is found in the crystal of **32** ([46]; GEXSIN; 3.36 Å). This is centered about a distorted Sb$_4$Se$_4$ cube, sustained by eight Sb···Se interactions, as described above for isostructural **21** [46], Figure 7b.

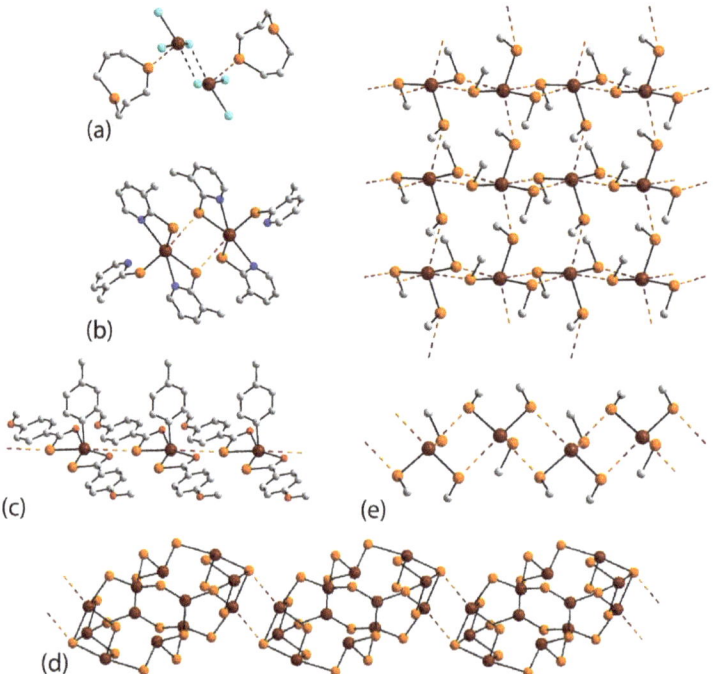

Figure 9. Supramolecular aggregation via Sb···Se secondary bonding in (**a**) **28** {EWIWAI; d(Sb···Se) = 3.29 Å}, (**b**) **29** {TAKSEF; 3.67 Å}, (**c**) **33** {ACUPAQ; 3.87 Å}, (**d**) **34** {HEFCOK; 3.61 Å}, and (**e**) **35** {JAZGIA; 3.55, 3.64 & 3.66 Å}.

Two one-dimensional chains are sustained by Sb···Se interactions. There is an average of one Sb···Se interaction per repeat unit in **33** [57] where the resulting topology is linear and where the interacting selenium atom approaches the antimony atom within the O_2Se_2 skewed-trapezoidal plane in the region between the two oxygen atoms, Figure 9c. The $[Sb_{12}Se_{20}]^{4-}$ Zintl ion in **34** [58] also self-associates into a linear chain whereby centrosymmetrically related tetra-anions are connected by a {Sb···Se}$_2$ synthon, Figure 9d.

The last crystal in this section to be discussed features the smallest molecule in this category, that is, $Sb(SeMe)_3$ in **35** [59]. Similar to that seen in **32**, the antimony atom accepts three Sb···Se interactions as each selenium atom participates in one such contact. To a first approximation, the resultant two-dimensional array has the form of a square grid and displays a corrugated topology, as seen in the views of Figure 9e.

3.5. Bismuth Compounds Featuring Bi···Se Interactions in their Crystals

There are only six bismuth-/selenium-containing crystals featuring Bi···Se interactions and the chemical structures of the interacting species in these, that is, **36–41**, are shown in Figure 10.

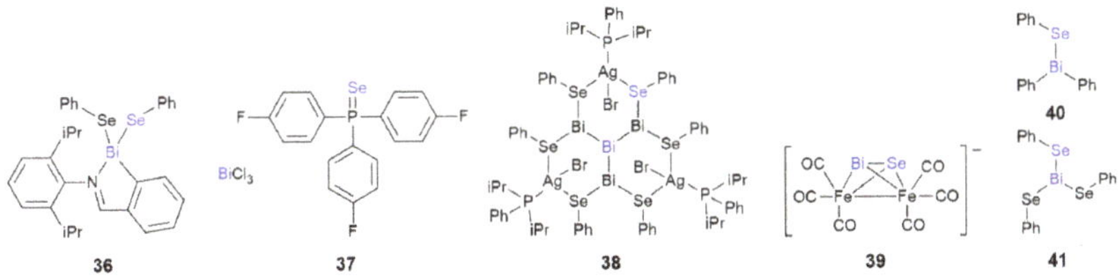

Figure 10. Chemical diagrams for bismuth compounds **36–41**.

A simple dimeric aggregate sustained by two Bi⋯Se interactions and a {Bi–Se⋯}$_2$ synthon is observed in the crystal of **36** [55]. While this has the appearance, at least to a first approximation, of several related species covered above (Figure 11a), the difference here is that the association is not through a center of inversion, as is usually observed. In this case, the contacts occur between two crystallographically independent molecules. The association in **37** ([46]; GEXSEJ; 3.35 Å), with a supramolecular Bi$_4$Se$_4$ core sustained by Bi⋯S interactions, is as described previously for **21**, Figure 7b, and **32**. An aesthetically pleasing Bi$_4$ core is a key feature in the crystal of **38** [60], with each edge of the Bi$_3$ triangle, which encompasses a central bismuth atom, being bridged by a sequence of Se–Ag–Se atoms. Centrosymmetrically related molecules associate through a center of inversion and are sustained by four Bi⋯Se interactions, as is apparent from the two views of Figure 11b.

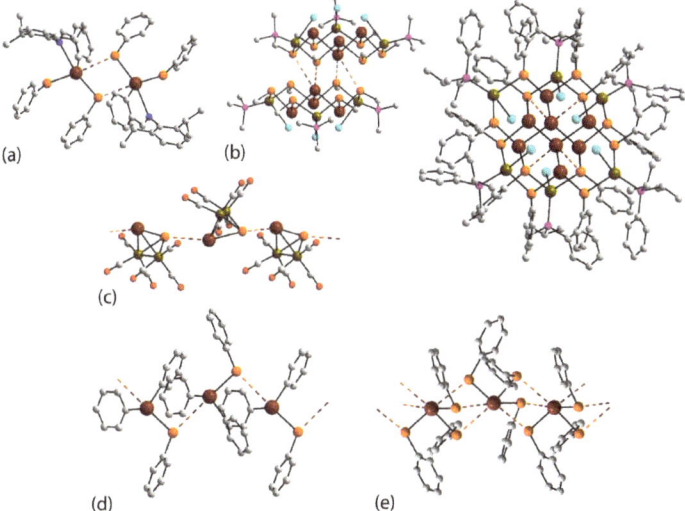

Figure 11. Supramolecular aggregation via Bi⋯Se secondary bonding in (**a**) **36** {KIMMEA; d(Bi⋯Se) = 3.51 & 3.55 Å}, (**b**) **38** {EGEPEM; 3.70 & 3.83 Å}, (**c**) **39** {COFGEM; 3.48 Å}, (**d**) **40** {GIPREC; 3.90 Å}, and (**e**) **41** {MIWFAA; 3.48, 3.50 & 3.57 Å}. Additional color code: silver and iron, dark-green.

The three remaining crystals feature one-dimensional chains. In **39** [61], of interest owing to a semi-conducting character and where a Bi–Se atom pair caps a Fe$_2$(CO)$_6$ unit, the presence of Bi⋯Se interactions lead to a helical, supramolecular chain; Figure 11c. A helical chain is also observed in **40** [62], Figure 11d, again sustained by, on average one Bi⋯Se interaction per repeat unit. When the two bismuth-bound phenyl groups of **40** are replaced by two phenylselenyl groups, leading to **41** [63], significantly more Bi···Se interactions are evident. The asymmetric unit of **41** comprises two independent molecules and each of these self-associates into a helical chain, as for **39** and **40**, but, in this case, there are, on average, three Bi···Se interactions per repeat unit in each of the independent chains formed in the crystal, one of which is illustrated in Figure 11e; the Bi···Se separations for the second independent chain are 3.49 and 2 × 3.59 Å. This propensity to form Bi···Se interactions in Bi(SePh)$_3$ (**40**) is not pervasive as the structure suggests. For example, the molecule highlighted in **38** co-crystallizes with one equivalent of Bi(SePh)$_3$ as well as one equivalent of 1,2-dimethoxyethane (solvate). However, Bi(SePh)$_3$ in **38** (and in the disordered chloride analog of **38**) does not participate in Bi···Se interactions, instead the bismuth atom forms Bi···Br (Bi···Cl) secondary bonding interactions with the other bismuth-containing molecule.

3.6. Tellurium Compounds Featuring Te···Se Interactions in Their Crystals

The most numerous among the main group elements covered in the present survey are those having tellurium, with 17 examples. The chemical diagrams for **42**–**58** are given in Figure 12. As with the earlier series covered, herein a broad range of compounds and supramolecular motifs are noted.

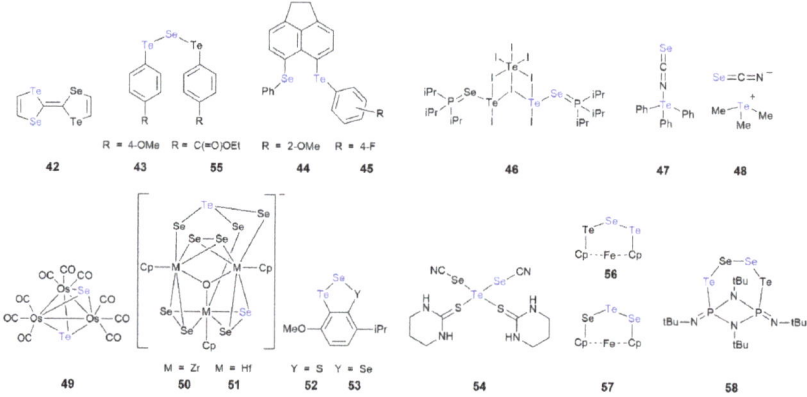

Figure 12. Chemical diagrams for tellurium compounds **42**–**58**; Cp is cyclopentadienyl.

Six of the compounds assemble into zero-dimensional motifs. In **42** [64], the tellurium and selenium atoms of one of the five-membered rings associate about a center of inversion to form the dimeric aggregate shown in Figure 13a. When **42** was cocrystallized with TCNQ (tetracyanoquinodimethane), highly conductive charge-transfer (CT) complexes were formed [64]. Similar centrosymmetric {Te–Se···}$_2$ synthons are observed in each of **43** [65], Figure 13b, **44** ([66], MIVYIB; d(Te···Se) = 3.90 Å), and **45** ([66], MIVZAU; 3.93 Å). Again, a {Te–Se···}$_2$ synthon is noted in **46** [67], Figure 13c, a compound that is particularly noteworthy for the relatively high number of potential iodide donors but, where Te···Se interactions prevail. The ion-pair in **47** [68] is formulated as [Ph$_3$Te][N=C=S] with the closest association between the constituent species being Te···N contacts of 2.81 and 3.12 Å, represented as black dashed lines in Figure 13d, for the two independent ion-pairs comprising the asymmetric unit. In terms of Te···Se interactions, one of the two independent ion-pairs associates with a center of inversion via a {Te–Se···}$_2$ synthon. Associated with this are two of the second independent ion-pairs (each separated by 3.43 Å) so a four-ion-pair aggregate is generated. A related ion-pair, [Me$_3$Te][N=C=S], is seen in **48** [68], where, consistent with the replacement of the tellurium-bound phenyl substituents of **47** with (relatively) electropositive methyl substituents, the Te···N separation is elongated to 3.25 Å. The constituents of the ion-pair are connected into a supramolecular chain with a zigzag topology via Te···Se interactions, Figure 13e. When the weak Te···N interactions are taken into consideration, the aforementioned chains are connected into a two-dimensional array (not illustrated).

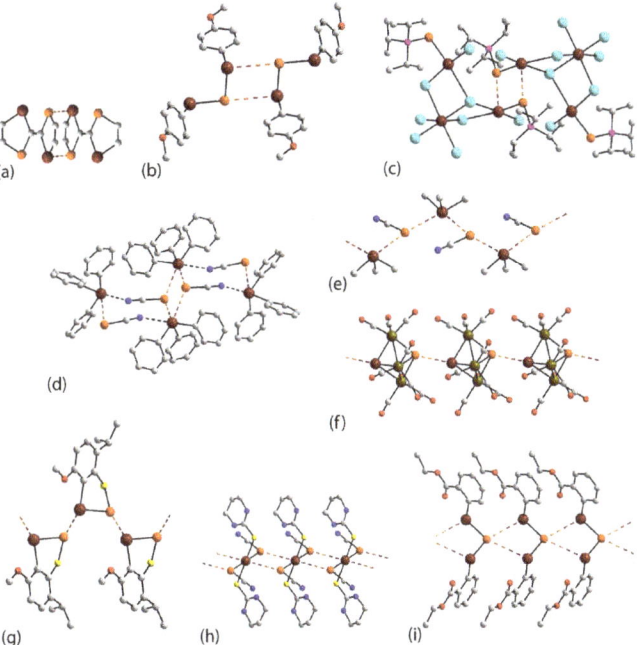

Figure 13. Supramolecular aggregation via Te···Se secondary bonding in (a) **42** {ECITEP; d(Te···Se) = 3.81 Å}, (b) **43** {BAWFUA; 3.68 Å}, (c) **46** {YIKFOO; 3.74 Å}, (d) **47** {ZZZAIJ01; 3.43, 3.44 & 3.54 Å}, (e) **48** {HUHCIW; 3.47 & 3.55 Å}, (f) **49** {QENRIK; 3.95 Å}, (g) **52** {XOTLUN; 3.48 Å}, (h) **54** {TRTUTE; 3.82 Å}, and (i) **55** {BAWGAH; 3.84 Å}. Additional color code: osmium, dark-green; sulfur, yellow.

In cluster compound **49** [69], the osmium atoms of the $Os_3(CO)_9$ core are μ_3-capped on either side by tellurium and selenium atoms, which associate in the crystal to form a linear, supramolecular chain with an average of one Te···Se interaction per repeat unit, Figure 13f. In isostructural **50** and **51** [70], constructed about M_3O cores, M = Zr (**50**) and Hf (**51**), and featuring an unusual $TeSe_3$ capping residue, molecules associate into helical chains (2_1 screw symmetry) via Te···Se interactions. Similar helical chains are observed in **52**, Figure 13g, and **53** [71], which differ in the nature of the atom connecting the aromatic ring to the selenium atom bonded to the tellurium atom, the latter associate to form the chain. On average, there are two Te···Se interactions linking the repeat unit of **54** [72] where the tellurium is located on a center of inversion. The resulting chain has a linear topology, Figure 13h. Compound **55** [65] is closely related to that of **43** in that the methoxy substituents of the latter have replaced by ethoxycarboxyl groups; the central selenium atom in **55** lies in a 2-fold axis of symmetry. Whereas **43** self-associates into a dimer, Figure 13b, **55** self-associates into a linear, supramolecular chain as each selenium atom forms two Te···Se interactions with a translationally related molecule, Figure 13i.

Compound **56** [73] self-associates into a supramolecular chain, Figure 14a. Two independent molecules comprise the asymmetric unit and these differ in the number of Te···Se interactions they form. For the first independent molecule, one tellurium and the selenium atom form a single Te···Se interaction each, whereas for the second molecule, the same situation pertains, except both participating atoms form two Te···Se interactions. The connections between the independent molecules lead to a linear, supramolecular chain. Centrosymmetrically chains are linked into a double-chain via additional Te···Se interactions formed by the second independent molecule. The molecule in **57** [74] is related to that in **56** in that there has been an exchange between selenium and tellurium atoms. This results in a distinct supramolecular assembly. Here, the central tellurium atom forms two Te···Se interactions with each of the selenium atoms forming a single Te···Se interaction. These extend laterally so a two-dimensional array eventuates, Figure 14b. A comparison of the simplified images in Figure 14a,b highlight the different modes of association between molecules. The energies associated with individual Te···Se contacts were calculated for each of **56** and **57**, and for the latter, these were −10.8 and −11.8 kJ mol^{-1} [74]. The molecule in **58** [75] features a seven-membered ring containing a string of Te–Se–Se–Te atoms bridged by a P–N–P link, the latter being a part of a four-membered N_2P_2 ring. Each of the tellurium and selenium atoms forms a Te···Se interaction. Again, these extend laterally to form a two-dimensional array, Figure 14c.

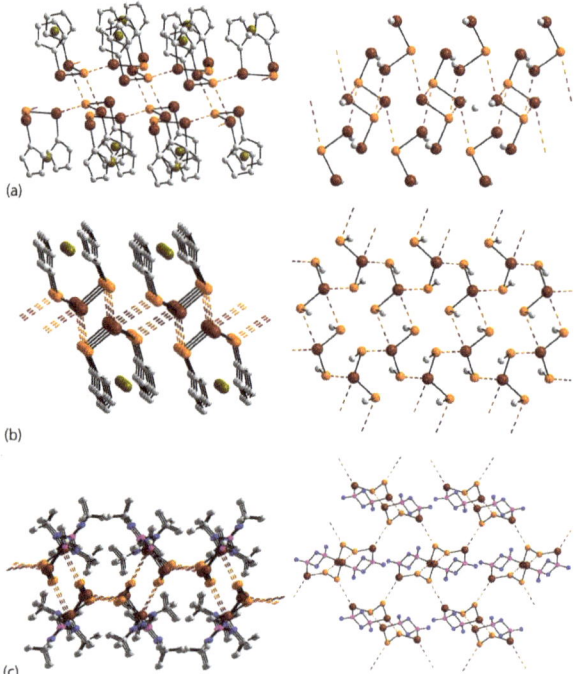

Figure 14. Supramolecular aggregation via Te···Se secondary bonding in (**a**) **56** {QAZGUV; d(Te···Se) = 3.70, 3.79 & 3.92 Å}, (**b**) **57** {OMIHAV; 3.62 & 3.69 Å}, and (**c**) **58** {ONEGIZ; 3.82 & 3.89 Å}. In the simplified views of (**a**) and (**b**), only the carbon atom bound to selenium/tellurium are shown, and in (**c**), the t-butyl groups are omitted.

4. Discussion and Outlook

The foregoing describes 58 crystals featuring M···Se secondary bonding interactions between main group elements (M) and selenium for M = Sn (13 examples), Pb (7), As (7), Sb (8), Bi (6), and Te (17). The percentage adoption of M···Se in the crystals varies considerably. For example, of the 27 crystals containing both lead and selenium, seven feature Pb···Se interactions, giving a percentage adoption of 26%. This falls off to 16% for bismuth to 10% for tellurium and then 6% (arsenic) and 5% (tin and antimony). One reason for the low adoption rates relates to the observation that secondary bonding interactions are extremely sensitive to steric hindrance—bulky groups present on the organometal center and/or ligands bound to the metal can preclude the formation of secondary bonding interactions [15–23]; steric considerations have been exploited for the rational design of coordination polymers in zinc and cadmium dithiolate chemistry [76]. In most of the crystals, the metal center forms a single M···Se contact with few examples of the metal forming two contacts and rarely, three M···Se contacts. With the formation of primarily one M···Se interaction, the supramolecular architectures sustained by these interactions are usually zero- or one-dimensional, being found in 38 and 50% of all crystals, respectively. Two-dimensional architectures sustained by M···Se interactions are found in 10% of the crystals and there is a single example of a three-dimensional architecture. A comment on the likely bonding responsible for the M···Se interactions is appropriate. For the Sn···S contacts, the majority features tin(IV) centers and so the interactions can be considered in terms of classic Lewis Acid/Lewis Base electrostatics. In contrast, all of the Pb···Se contacts can be rationalized in terms of tetrel bonding; the overwhelming majority of M···Se interactions formed by arsenic-triad arise from pnictogen bonding and the tellurium examples in terms of chalcogen bonding where σ-hole considerations come to the fore. Thus far, limited mention has been made of the energy of stabilization provided by M···Se interactions. This is because supporting computational chemistry is largely lacking for M···Se interactions with the exception of **56** and **57** [74]. However, in a recent commentary on supramolecular association involving metal centers, it was concluded that the energies of stabilization provided by various secondary bonding interactions was in the same range and often exceeded the energy of stabilization provided by conventional hydrogen bonding interactions [77]. This conclusion is emphasized in the very recently published analysis of a tetrel, C···O, bond formed between a sp^3-carbon center and the oxygen atom of a tetrahydrofuran molecule, not an interaction that might be expected to be particularly notable, for which an energy of stabilization of about 11 kcal mol^{-1} was calculated [78]. In the context of the foregoing survey of M···Se interactions, with diverse bonding circumstances and supramolecular molecular aggregation patterns, clearly there is enormous scope for further experimental work supported by theoretical analysis.

Funding: Crystallographic research at Sunway University is supported by Sunway University Sdn Bhd (Grant no. STR-RCTR-RCCM-001-2019).

Conflicts of Interest: The author declares no conflict of interest.

References

1. Minyaev, R.M.; Minkin, V.I. Theoretical study of O - > X (S, Se, Te) coordination in organic compounds. *Can. J. Chem.* **1998**, *76*, 776–778. [CrossRef]
2. Wang, W.; Ji, B.; Zhang, Y. Chalcogen bond: A sister noncovalent bond to halogen bond. *J. Phys. Chem. A* **2009**, *113*, 8132–8135. [CrossRef] [PubMed]
3. Aakeröy, C.B.; Bryce, D.L.; Desiraju, G.R.; Frontera, A.; Legon, A.C.; Nicotra, F.; Rissanen, K.; Scheiner, S.; Terraneo, G.; Metrangolo, P.; et al. Definition of the chalcogen bond (IUPAC Recommendations 2019). *Pure Appl. Chem.* **2019**, *91*, 1889–1892. [CrossRef]
4. Alkorta, I.; Elguero, J.; Frontera, A. Not only hydrogen bonds: Other noncovalent interactions. *Crystals* **2020**, *10*, 180. [CrossRef]

5. Bent, H.A. Structural chemistry of donor-acceptor interactions. *Chem. Rev.* **1968**, *68*, 587–648. [CrossRef]
6. Hassel, O. Structural Aspects of interatomic charge-transfer bonding. *Science* **1970**, *170*, 497–502. [CrossRef]
7. Alcock, N.W. Secondary bonding to nonmetallic elements. *Adv. Inorg. Chem. Radiochem.* **1972**, *15*, 1–58. [CrossRef]
8. Llaguno, E.C.; Paul, I.C. Crystal structure of a [1,2,5]oxaselenazolo[2,3-*b*][1,2,5]oxaselenazole-7-SeIV: A molecule with 'short' intramolecular Se ... O distances, or 'long' Se–O bonds. *J. Chem. Soc. Perkin Trans.* **1972**, *2*, 2001–2006. [CrossRef]
9. Murray, J.S.; Lane, P.; Clark, T.; Politzer, P. σ-hole bonding: Molecules containing group VI atoms. *J. Mol. Model.* **2007**, *13*, 1033–1038. [CrossRef]
10. Politzer, P.; Murray, J.S.; Clark, T. Halogen bonding: An electrostatically-driven highly directional noncovalent interaction. *Phys. Chem. Chem. Phys.* **2010**, *12*, 7748–7757. [CrossRef]
11. Politzer, P.; Murray, J.S. Halogen bonding and other σ-hole interactions: A perspective. *Phys. Chem. Chem. Phys.* **2013**, *15*, 11178–11189. [CrossRef] [PubMed]
12. Bauzá, A.; Mooibroek, T.J.; Frontera, A. The bright future of unconventional σ/π-hole interactions. *ChemPhysChem* **2015**, *16*, 2496–2517. [CrossRef] [PubMed]
13. Kolar, M.H.; Hobza, P. Computer modeling of halogen bonds and other σ-hole interactions. *Chem. Rev.* **2016**, *116*, 5155–5187. [CrossRef] [PubMed]
14. Politzer, P.; Murray, J.S. σ-Hole interactions: Perspectives and misconceptions. *Crystals* **2017**, *7*, 212. [CrossRef]
15. Tiekink, E.R.T. Molecular architecture and supramolecular association in the zinc-triad 1,1-dithiolates. Steric control as a design element in crystal engineering? *CrystEngComm* **2003**, *5*, 101–113. [CrossRef]
16. Lai, C.S.; Tiekink, E.R.T. Structural diversity in the mercury(II) bis(*N*,*N*-dialkyldithiocarbamate) compounds: An example of the importance of considering crystal structure when rationalising molecular structure. *Z. Kristallogr. Cryst. Mater.* **2007**, *222*, 532–538. [CrossRef]
17. Tiekink, E.R.T. Exploring the topological landscape exhibited by binary zinc-triad 1,1-dithiolates. *Crystals* **2018**, *8*, 292. [CrossRef]
18. Buntine, M.A.; Kosovel, F.J.; Tiekink, E.R.T. Supramolecular Sn···Cl associations in diorganotin dichlorides and their influence on molecular geometry as studied by *ab initio* molecular orbital calculations. *CrystEngComm* **2003**, *5*, 331–336. [CrossRef]
19. Tiekink, E.R.T. Tin dithiocarbamates: Applications and structures. *Appl. Organomet. Chem.* **2008**, *22*, 533–550. [CrossRef]
20. Liu, Y.; Tiekink, E.R.T. Supramolecular associations in binary antimony(III) dithiocarbamates: Influence of ligand steric bulk, influence on coordination geometry, and competition with hydrogenbonding. *CrystEngComm* **2005**, *7*, 20–27. [CrossRef]
21. Tiekink, E.R.T. Aggregation patterns in the crystal structures of organometallic Group XV 1,1-dithiolates: The influence of the Lewis acidity of the central atom, metal- and ligand-bound steric bulk, and coordination potential of the 1,1-dithiolate ligands upon supramolecular architecture. *CrystEngComm* **2006**, *8*, 104–118. [CrossRef]
22. Lee, S.M.; Heard, P.J.; Tiekink, E.R.T. Molecular and supramolecular chemistry of mono- and di-selenium analogues of metal dithiocarbamates. *Coord. Chem. Rev.* **2018**, *375*, 410–423. [CrossRef]
23. Tiekink, E.R.T.; Zukerman-Schpector, J. Stereochemical activity of lone pairs of electrons and supramolecular aggregation patterns based on secondary interactions involving tellurium in its 1,1-dithiolate structures. *Coord. Chem. Rev.* **2010**, *254*, 46–76. [CrossRef]
24. Taylor, R.; Wood, P.A. A million crystal structures: The whole is greater than the sum of its parts. *Chem. Rev.* **2019**, *119*, 9427–9477. [CrossRef]
25. Bruno, I.J.; Cole, J.C.; Edgington, P.R.; Kessler, M.; Macrae, C.F.; McCabe, P.; Pearson, J.; Taylor, R. New software for searching the Cambridge Structural Database and visualizing crystal structures. *Acta Crystallogr. Sect. B Struct. Sci. Cryst. Eng. Mater.* **2002**, *58*, 389–397. [CrossRef]
26. DIAMOND, Version 3.2k; K. Brandenburg & M. Berndt GbR: Bonn, Germany, 2006.

27. Rinn, N.; Eußner, J.P.; Kaschuba, W.; Xie, X.; Dehnen, S. Formation and reactivity of organo-functionalized tin selenide clusters. *Chem. Eur. J.* **2016**, *22*, 3094–3104. [CrossRef]
28. Krautscheid, H.; Schmidtke, M. [NaSn$_{12}$O$_8$Se$_6$]$^{3-}$—ein chalkogenostannatanion mit schalenförmigem Aufbau. *Z. Anorg. Allg. Chem.* **2002**, *628*, 913–914. [CrossRef]
29. Dehnen, S.; Hanau, K.; Rinn, N.; Argentari, M. Organotin selenide clusters and hybrid capsules. *Chem. Eur. J.* **2018**, *24*, 11711–11716. [CrossRef]
30. Metta-Magaña, A.J.; Lopez-Cardoso, M.; Vargas, G.; Pannell, K.H. Major distinctions in the molecular and supramolecular structures of selenium-containing organotins, (o-MeSe-C$_6$H$_4$CH$_2$)SnPh$_{3-n}$Cl$_n$ (n = 0, 1, 2). *Z. Anorg. Allg. Chem.* **2012**, *638*, 1677–1682. [CrossRef]
31. Santner, S.; Sprenger, J.A.P.; Finze, M.; Dehnen, S. The role of [BF$_4$]$^-$ and [B(CN)$_4$]$^-$ anions in the ionothermal synthesis of chalcogenidometalates. *Chem. Eur. J.* **2018**, *24*, 3474–3480. [CrossRef]
32. Kim, K.-W. DMF Solvothermal synthesis and structural characterization of [dabcoH]$_2$[(CH$_3$)$_2$NH$_2$]$_2$[Sn$_2$Se$_6$] DMF. *J. Korean Chem. Soc.* **2005**, *49*, 603–608. [CrossRef]
33. Block, E.; Dikarev, E.V.; Glass, R.S.; Jin, J.; Li, B.; Li, X.; Zhang, S.-Z. Synthesis, structure, and chemistry of new, mixed group 14 and 16 heterocycles: Nucleophile-induced ring contraction of mesocyclic dications. *J. Am. Chem. Soc.* **2006**, *128*, 14949–14961. [CrossRef] [PubMed]
34. Nayek, H.P.; Niedermeyer, H.; Dehnen, S. Preparation and conformation of organo-bridged bis[tris(arylchalcogenolato)tin] compounds–an experimental and quantum chemical study. *Z. Anorg. Allg. Chem.* **2008**, *634*, 2805–2810. [CrossRef]
35. Holligan, K.; Rogler, P.; Rehe, D.; Pamula, M.; Kornienko, A.Y.; Emge, T.J.; Krogh-Jespersen, K.; Brennan, J.G. Copper, indium, tin, and lead complexes with fluorinated selenolate ligands: Precursors to MSe$_x$. *Inorg. Chem.* **2015**, *54*, 8896–8904. [CrossRef] [PubMed]
36. Dräger, M.; Mathiasch, B. Kristallstrukturbestimmung und schwingungsanalyse von 2,2,4,4,5,5-hexamethyl-1,3-diselena-2,4,5-tristannolan Se$_2$Sn$_3$(CH$_3$)$_6$. *Z. Anorg. Allg. Chem.* **1980**, *470*, 45–58. [CrossRef]
37. Cheng, Y.; Emge, T.J.; Brennan, J.G. Pyridineselenolate Complexes of tin and lead: Sn(2-SeNC$_5$H$_4$)$_2$, Sn(2-SeNC$_5$H$_4$)$_4$, Pb(2-SeNC$_5$H$_4$)$_2$, and Pb(3-Me$_3$Si-2-SeNC$_5$H$_3$)$_2$. Volatile CVD precursors to Group IV–Group VI semiconductors. *Inorg. Chem.* **1996**, *35*, 342–346. [CrossRef]
38. Herzog, U.; Böhme, U.; Brendler, E.; Rheinwald, G. Group 14 chalcogenides featuring a bicyclo[3.3.0]octane skeleton. *J. Organomet. Chem.* **2001**, *630*, 139–148. [CrossRef]
39. Dräger, M.; Blecher, A.; Jacobsen, H.-J.; Krebs, B. Molekül- und kristallstruktur von hexamethylcyclo-tristannaselenan [(CH$_3$)$_2$SnSe]$_3$. *J. Organomet. Chem.* **1978**, *161*, 319–325. [CrossRef]
40. Leung, W.-P.; Wan, C.-L.; Kan, K.-W.; Mak, T.C.W. Synthesis, structure, and reactivity of Group 14 bis(thiophosphinoyl) metal complexes. *Organometallics* **2010**, *29*, 814–820. [CrossRef]
41. Ritch, J.S.; Chivers, T.; Ahmad, K.; Afzaal, M.; O'Brien, P. Synthesis, structures, and multinuclear NMR spectra of tin(II) and lead(II) complexes of tellurium-containing imidodiphosphinate ligands: Preparation of two morphologies of phase-pure PbTe from a single-source precursor. *Inorg. Chem.* **2010**, *49*, 1198–1205. [CrossRef]
42. Hummel, H.-U.; Fischer, E.; Fischer, T.; Gruß, D.; Franke, A.; Dietzsch, W. Synthesen, strukturen und thermische abbaureaktionen von Tl^{I}-, Pb^{II}- und Se^{II}-komplexen mit 2,2-dicyanethylen-1,1-diselenolat. *Chem. Ber.* **1992**, *125*, 1565–1570. [CrossRef]
43. Trindade, T.; Monteiro, O.C.; O'Brien, P.; Motevalli, N. Synthesis of PbSe nanocrystallites using a single-source method. The X-ray crystal structure of lead (II) diethyldiselenocarbamate. *Polyhedron* **1999**, *18*, 1171–1175. [CrossRef]
44. Evans, C.M.; Evans, M.E.; Krauss, T.D. Mysteries of TOPSe revealed: Insights into quantum dot nucleation. *J. Am. Chem. Soc.* **2010**, *132*, 10973–10975. [CrossRef] [PubMed]
45. Schuster, M.; Bensch, W. (Se,O)-Koordinierte komplexe niedervalenter hauptgruppenmetalle: Die kristallstruktur von bis(N,N-diethyl-N'-benzoylselenoureato)blei(II). *Z. Naturforsch. B Chem. Sci.* **1994**, *49*, 1615–1619. [CrossRef]
46. Alhanash, F.B.; Barnes, N.A.; Brisdon, A.K.; Godfrey, S.M.; Pritchard, R.G. Formation of M$_4$Se$_4$ cuboids (M = As, Sb, Bi) via secondary pnictogen–chalcogen interactions in the co-crystals MX$_3$·Se=P(p-FC$_6$H$_4$)$_3$ (M = As, X = Br; M = Sb, X = Cl; M = Bi, X = Cl, Br). *Dalton Trans.* **2012**, *41*, 10211–10218. [CrossRef]

47. Czado, W.; Müller, U. Cyclische polyselenidoarsenate(III) und -antimonate(III): PPh$_4$[Se$_5$AsSe], PPh$_4$[AsSe$_{6-x}$S$_x$], (PPh$_4$)$_2$[As$_2$Se$_6$]·2CH$_3$CN und (PPh$_4$)$_2$[Se$_6$SbSe]$_2$. *Z. Anorg. Allg. Chem.* **1998**, *624*, 239–243. [CrossRef]
48. Kennard, O.; Wampler, D.L.; Coppola, J.C.; Motherwell, W.D.S.; Mann, F.G.; Watson, D.G.; MacGillavry, C.H.; Stam, C.H.; Benci, P. Crystal and molecular structure of 5,10-epoxy-, 5,10-epithio-, 5,10-episeleno-, and 5,10-epitelluro-5,10-dihydroarsanthren. *J. Chem. Soc. C* **1971**, 1511–1515. [CrossRef]
49. Thiele, G.; Rotter, H.W.; Lietz, M.; Ellermann, J. Chemistry of polyfunctional molecules, 80 [1] Crystal and molecular structures of the heteronoradamantanes 5-methyl-2.2.8.8-tetra-ethoxycarbonyl-1.3.7-triarsa-tricyclo[3.3.1.0^{3,7}]nonane and 5-methyl-1.3.7-triarsa-2.8-diselena-[3.3.1.0^{3,7}]nonane. *Z. Naturforsch. B Chem. Sci.* **1984**, *39*, 1344–1349. [CrossRef]
50. Applegate, C.A.; Meyers, E.A.; Zingaro, R.A.; Merijanian, A. reactions of arsinic and arsonic acids with H$_2$S and H$_2$Se: Crystal structure of 1,4-diphenyl-1,4-diarsa-2,3,5-triselenacyclopentane. *Phosphorus Sulfur Rel. Elements* **1988**, *35*, 363–370. [CrossRef]
51. Levason, W.; Maheshwari, S.; Ratnani, R.; Reid, G.; Webster, M.; Zhang, W. Structural diversity in supramolecular complexes of MCl$_3$ (M = As, Sb, Bi) with constrained thio- and seleno-ether ligands. *Inorg. Chem.* **2010**, *49*, 9036–9048. [CrossRef]
52. Angilella, V.; Mercier, H.; Belin, C.J. Heteroatomic polyanions of post-transition elements. Synthesis and structure of a salt containing the novel hybrid hepta-arsenic tetraselenate(1–) anion, As$_7$Se$_4^-$. *Chem. Soc. Chem. Commun* **1989**, 1654–1655. [CrossRef]
53. Hill, N.J.; Levason, W.; Patel, R.; Reid, G.; Webster, M. Unusual structural trends in the [MCl$_3$([8]aneSe$_2$)] (M = As, Sb, Bi) adducts. *Dalton Trans.* **2004**, 980–981. [CrossRef] [PubMed]
54. Sharma, R.K.; Kedarnath, G.; Jain, V.K.; Wadawale, A.; Nalliath, M.; Pillai, C.G.S.; Vishwanadh, B. 2-Pyridyl selenolates of antimony and bismuth: Synthesis, characterization, structures and their use as single source molecular precursor for the preparation of metal selenidenanostructures and thin films. *Dalton Trans.* **2010**, *39*, 8779–8787. [CrossRef] [PubMed]
55. Šimon, P.; Jambor, R.; Růžička, A.; Dostál, L. Oxidative addition of diphenyldichalcogenides PhEEPh (E = S, Se, Te) to low-valent CN- and NCN-chelated organoantimony and organobismuth compounds. *Organometallics* **2013**, *32*, 239–248. [CrossRef]
56. Wagner, C.; Merzweiler, K. Neue [{Cp(CO)$_2$Mo}$_2$ESbCl]-cluster mit tetraedrischem Mo$_2$SbE-Gerüst (E = S, Se). *Z. Anorg. Allg. Chem.* **2011**, *637*, 651–654. [CrossRef]
57. Kimura, M.; Iwata, A.; Itoh, M.; Yamada, K.; Kimura, T.; Sugiura, N.; Ishida, M.; Kato, S. Synthesis, structures, and some reactions of [(thioacyl)thio]- and (acylseleno)antimony and -bismuth derivatives ((RCSS)$_x$MR and (RCOSe)$_x$MR with M = Sb, Bi and x = 1–3). *Helv. Chim. Acta* **2006**, *89*, 747–783. [CrossRef]
58. Martin, T.M.; Wood, P.T.; Kolis, J.W. Synthesis and structure of an [Sb$_{12}$Se$_{20}$]$^{4-}$ salt: The largest molecular Zintl ion. *Inorg. Chem.* **1994**, *33*, 1587–1588. [CrossRef]
59. Breunig, H.J.; Güleç, S.; Krebs, B.; Dartmann, M. Synthese und struktur von (MeSe)$_3$Sb. *Z. Naturforsch. B Chem. Sci.* **1989**, *44*, 1351–1354. [CrossRef]
60. Sommer, H.; Eichhöfer, A.; Drebov, N.; Ahlrichs, R.; Fenske, D. Preparation, Geometric and electronic structures of [Bi$_2$Cu$_4$(SPh)$_8$(PPh$_3$)$_4$] with a Bi$_2$ dumbbell, [Bi$_4$Ag$_3$(SePh)$_6$Cl$_3$(PPh$_3$)$_3$]$_2$ and [Bi$_4$Ag$_3$(SePh)$_6$X$_3$(PPh^{i}Pr$_2$)$_3$]$_2$ (X = Cl, Br) with a Bi$_4$ unit. *Eur. J. Inorg. Chem.* **2008**, 5138–5145. [CrossRef]
61. Shieh, M.; Liu, Y.-H.; Huang, C.-Y.; Chen, S.-W.; Cheng, W.-K.; Chien, L.T. The first naked bismuth–chalcogen metal carbonyl clusters: Extraordinary nucleophilicity of the Bi atom and semiconducting characteristics. *Inorg. Chem.* **2019**, *58*, 6706–6721. [CrossRef]
62. Calderazzo, F.; Morvillo, A.; Pelizzi, G.; Poli, R.; Ungari, F. Reactivity of molecules containing element-element bonds. 1. Nontransition elements. *Inorg. Chem.* **1988**, *27*, 3730–3733. [CrossRef]
63. Sommer, H.; Eichhofer, A.; Fenske, D. Bismutchalkogenolate Bi(SC$_6$H$_5$)$_3$, Bi(SeC$_6$H$_5$)$_3$ und Bi(S-4-CH$_3$C$_6$H$_4$)$_3$. *Z. Anorg. Allg. Chem.* **2008**, *634*, 436–440. [CrossRef]
64. Morikami, A.; Takimiya, K.; Aso, Y.; Otsubo, T. Novel tellurium containing fulvalene-type electron donors, triselenatellurafulvalene (TSTeF) and diselenaditellurafulvalene (DSDTeF); synthesis, conductivities and crystal structures of their TCNQ complexes. *J. Mater. Chem.* **2001**, *11*, 2431–2436. [CrossRef]

65. Dereu, N.L.M.; Zingaro, R.A.; Meyers, E.A. Bis(4-methoxybenzenetellurenyl)selenide, $C_{14}H_{14}O_2SeTe_2$. *Cryst. Struct. Commun.* **1981**, *10*, 1345–1352.
66. Stanford, M.W.; Knight, F.R.; Arachchige, K.S.A.; Camacho, P.S.; Ashbrook, S.E.; Bühl, M.; Slawin, A.M.Z.; Woollins, D.J. Probing interactions through space using spin–spin coupling. *Dalton Trans.* **2014**, *43*, 6548–6560. [CrossRef] [PubMed]
67. Hrib, C.G.; Jeske, J.; Jones, P.G.; du Mont, W.-W. Telluroselenophosphonium ions: Their unusual soft–soft interactions with iodotellurate anions. *Dalton Trans.* **2007**, 3483–3485. [CrossRef] [PubMed]
68. Klapotke, T.M.; Krumm, B.; Mayer, P.; Piotrowski, H.; Schwab, I.; Vogt, M. Synthesis and structures of triorganotelluronium pseudohalides. *Eur. J. Inorg. Chem.* **2002**, 2701–2709. [CrossRef]
69. Mathur, P.; Payra, P.; Ghose, S.; Hossain, M.M.; Satyanarayana, C.V.V.; Chicote, F.O.; Chadha, R.K. Synthesis and characterisation of $[Fe_2M_3(\mu_4\text{-}E)(\mu_3\text{-}E')(CO)_{17}]$ and $[Os_3(\mu_3\text{-}E)(\mu_3\text{-}E')(CO)_9]$ (M=Os or Ru; E=S, Se, Te; E'=Se, Te). *J. Organomet. Chem.* **2000**, *606*, 176–182. [CrossRef]
70. Dibrov, S.M.; Ibers, J.A. $[TeSe_3]^{2-}$ as a tridentate ligand: Syntheses and crystal structures of $[PPh_4][(CpM(\mu_2\text{-}Se_2))_3(\mu_3\text{-}O)(\mu_3\text{-}TeSe_3)]$ (M = Zr, Hf). *Comptes Rendus Chim.* **2005**, *8*, 993–997. [CrossRef]
71. Ogawa, S.; Yoshimura, S.; Nagahora, N.; Kawai, Y.; Mikata, Y.; Sato, R. Novel multi-chalcogen ring systems with three different chalcogen atoms: Synthesis, structure and redox property of five-membered trichalcogenaheterocycles. *Chem. Commun.* **2002**, 1918–1919. [CrossRef]
72. Åse, K.; Foss, O.; Roti, I. The crystal and molecular structures of trans square-planar complexes of tellurium diselenocyanate with trimethylenethiourea and tetramethylthiourea. *Acta Chem. Scand.* **1971**, *25*, 3808–3820. [CrossRef]
73. Karjalainen, M.M.; Oilunkaniemi, R.; Laitinen, R.S. Chalcogen–chalcogen interactions in trichalcogenaferrocenophanes. Crystal structure of 2-selena-1,3-ditellura[3]ferrocenophane $[Fe(C_5H_4Te)_2Se]$. *Inorg. Chim. Acta* **2012**, *390*, 79–82. [CrossRef]
74. Karjalainen, M.M.; Sanchez-Perez, C.; Mikko Rautiainen, J.; Oilunkaniemi, R.; Laitinen, R.S. Chalcogen–chalcogen secondary bonding interactions in trichalcogenaferrocenophanes. *CrystEngComm* **2016**, *18*, 4538–4545. [CrossRef]
75. Nordheider, A.; Hüll, K.; Prentis, J.K.D.; Arachchige, K.S.A.; Slawin, A.M.Z.; Woollins, D.J.; Chivers, T. Main group tellurium heterocycles anchored by a $P_2^V N_2$ scaffold and their sulfur/selenium analogues. *Inorg. Chem.* **2015**, *54*, 3043–3054. [CrossRef] [PubMed]
76. Tiekink, E.R.T. Perplexing coordination behaviour of potentially bridging bipyridyl-type ligands in the coordination chemistry of zinc and cadmium 1,1-dithiolate compounds. *Crystals* **2018**, *8*, 18. [CrossRef]
77. Tiekink, E.R.T. Supramolecular assembly based on "emerging" intermolecular interactions of particular interest to coordination chemists. *Coord. Chem. Rev.* **2017**, *345*, 209–228. [CrossRef]
78. Heywood, V.L.; Alford, T.P.J.; Roeleveld, J.J.; Lekanne Deprez, S.J.; Verhoofstad, A.; van der Vlugt, J.I.; Domingos, S.R.; Melanie Schnell, M.; Davis, A.P.; Mooibroek, T.J. Observations of tetrel bonding between sp^3-carbon and THF. *Chem. Sci.* **2020**, *11*, 5289–5293. [CrossRef]

© 2020 by the author. Licensee MDPI, Basel, Switzerland. This article is an open access article distributed under the terms and conditions of the Creative Commons Attribution (CC BY) license (http://creativecommons.org/licenses/by/4.0/).

Review

Halogen Bonds Fabricate 2D Molecular Self-Assembled Nanostructures by Scanning Tunneling Microscopy

Yi Wang, Xinrui Miao * and Wenli Deng *

College of Materials Science and Engineering, South China University of Technology, Guangzhou 510640, China; 201810103602@mail.scut.edu.cn
* Correspondence: msxrmiao@scut.edu.cn (X.M.); wldeng@scut.edu.cn (W.D.)

Received: 30 September 2020; Accepted: 9 November 2020; Published: 20 November 2020

Abstract: Halogen bonds are currently new noncovalent interactions due to their moderate strength and high directionality, which are widely investigated in crystal engineering. The study about supramolecular two-dimensional architectures on solid surfaces fabricated by halogen bonding has been performed recently. Scanning tunneling microscopy (STM) has the advantages of realizing in situ, real-time, and atomic-level characterization. Our group has carried out molecular self-assembly induced by halogen bonds at the liquid–solid interface for about ten years. In this review, we mainly describe the concept and history of halogen bonding and the progress in the self-assembly of halogen-based organic molecules at the liquid/graphite interface in our laboratory. Our focus is mainly on (1) the effect of position, number, and type of halogen substituent on the formation of nanostructures; (2) the competition and cooperation of the halogen bond and the hydrogen bond; (3) solution concentration and solvent effects on the molecular assembly; and (4) a deep understanding of the self-assembled mechanism by density functional theory (DFT) calculations.

Keywords: halogen bonding; σ-hole interactions; self-assembly; scanning tunneling microscopy

1. Introduction

1.1. The Definition of Halogen Bonds

A halogen bond (XB) is a broader class of noncovalent interaction, which was defined by the International Union of Pure and Applied Chemistry (IUPAC) in 2013 [1]. This definition states that "A halogen bond occurs when there is evidence of a net attractive interaction between an electrophilic region associated with a halogen atom in a molecular entity and a nucleophilic region in another, or the same, molecular entity." A structural scheme for XB is shown in Figure 1.

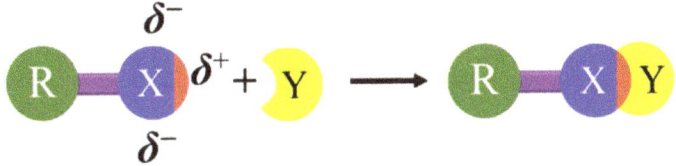

Figure 1. Structural scheme for a halogen bond (XB).

XB exhibits a high directionality, because the electron density in halogen atoms is anisotropically distributed [2,3]. In the halogen atoms, a region of lower electron density (σ-hole) gives rise to a cap of deficient electron density on the extension of the R−X covalent bond, which can act as the attractive electron-rich region. The negative charges form a belt orthogonal to the covalent bond in nearly all cases [4]. In this review, the XBs will be explained by the existence of a positive σ-hole on top of the halogen atom.

1.2. The History Perspective of XB

The investigation of the XB traces back to 200 years ago. The $I_2 \cdots NH_3$ halogen-bonded complex was synthesized by Colin in 1814 after iodine was found by chance and separated in 1812 [5]. However, the precise molecular constituent of the iodine/polyiodide complex was ascertained by Guthrie only 50 years later [6]. The complexes involved Br and Cl as electron acceptor species and were first reported in the late 19th century by Remsen and Norris [7]. However, the first compound, referred to as F_2, was reported nearly 80 years later when the F^{3-} anion could be isolated under very extreme conditions [8–10]. Moreover, the adducts of $F_2 \cdots NH_3$ and $F_2 \cdots OH_2$ did not arise until the 1990s [11]. These early-stage experimental data show that the XB strength is closely related to the polarizability of the XB donor atom, that is, F < Cl < Br < I. In fact, F without polarization cannot easily participate in XB and can serve as an XB donor only when connecting with exceedingly strong electron-withdrawing groups. A critical interest in the stereo electronic evaluation of the XB was supported by computational studies on the electron density distribution of halogen atoms in the early 1990s. The studies by T. Brinck, P. Politzer, and J. S. Murray were particularly noteworthy when they revealed the anisotropic charge distribution on halogen atoms forming one covalent bond [12–14], and supported the theoretical basis of the definition of the "σ-hole": a region of barren and constantly positive electrostatic potential on the surface of halogen atoms. In 2007, a seminal paper by Clark [15] et al. proposed a subtle method of interpreting many properties of XB via the σ-hole. A symposium devoted to the XB was organized by IUPAC in 2009, and some features of the interaction were acknowledged. In 2013, the definition of the XB by IUPAC finally gained the agreement of the scientific community, based on the self-assembly and recognition processes determined by electrophilic halogens. This topic has been well-established, used, and understood up until now.

1.3. The Investigation and Application of XB Focus on Crystal Engineering

Over the last seven decades, supramolecular synthesis has attracted great interest from chemists wanting to fabricate new materials based on noncovalent interactions [16–20]. Different self-assembly strategies have been proposed, and new supramolecular systems will be achieved by considering the relevant scientific literature across the years [18,21–23]. To design and synthesize supramolecular self-assembled materials, molecular building blocks with designed functions can be used to self-assemble ordered structures. The self-assembly process can happen in any dimension, which is the comprehensive outcome of the balanced steric effect, shape complementarity, and specific anisotropic interactions in the assembly process. For example, supramolecular synthesis could be involved in fabricating molecular organic frameworks [24,25], molecular acceptors [26], responsive materials [27], organogels [28,29], polymers [30] and biomimetic systems [31–33], as well as in nanoparticle self-assembly [34,35].

In 1968, Bent [36] published an overall review on the crystal structures of XB systems and discussed at length the structural chemistry of donor–acceptor interactions in bulk crystals, in which the XB and the hydrogen bond (HB) were compared based on their similarities, such as the short interatomic distances and high directionality. Crystal engineering is a self-assembly process in which building blocks can fabricate architectures following the laws of intermolecular interactions and shape complementarity [17,37–40]. Many noncovalent interactions, for example, HB, XB [41], π–π stacking, metal–ligand coordination, dipole–dipole interactions, and hydrophobic interactions, were used to fabricate molecular crystals. Among these weak interactions, HB is commonly used based on its high directionality and moderate strength (25–40 kJ/mol). However, XB similar to HB has not been developed as HB because of the misunderstanding associated with halogen atoms that neutral entities in dihalogens or fully negative elements in halocarbon moieties could not form strong attractive interactions. In general, the strength of XB is closely related to the magnitudes of the positive and negative areas of the electrostatic potential (ESP). Furthermore, the other chemical factors, such as the electronegativity and polarizability of X atoms [42], the hybridization of the carbon atom bound to the XB donor sites [43], and strong electron-withdrawing moieties [44,45], directly affect the magnitudes of the positive and negative areas so as to tune the strength of the XB indirectly.

The directionality of XB refers to the angle between the R–X covalent bond and the X···X or Y noncovalent bonds as θ_2 shown in type-II XB (Figure 2) [46,47]. The great directionality of such XB is attributed to the bond formation. The nucleophile enters the σ-hole of the halogen atom, which is narrowly confined on the elongation of the R–X covalent bond axis. The R–X···Y angle between the covalent and noncovalent bonds around the halogen is approximately 180° [46,47]. Although HB is another simple type of σ-hole interaction, the positive regions on linked hydrogens tend to be almost hemispherical due to hydrogen having only one electron [48–50]. Therefore, the XB is a particularly directional interaction compared to the HB. In 1963, the R–X···X–R bond with different contact geometries was proposed by Sakurai [44] et al., which was classified as type-I and type-II, as illustrated in Figure 2 [51]. There is a clear geometric and chemical distinction between type-I and type-II X···X interactions. The type-I interaction is not XB according to the IUPAC definition. It is a geometry-based contact arising from the close-packing requirement, which is observed for all halogens. The type-II interaction arises from the pairing between the electrophilic area on one halogen atom and the nucleophilic area on the other [52]. Moreover, many investigations reveal that the type-II interaction is most favored by the order of iodinated derivatives > brominated derivatives > chlorinated derivatives.

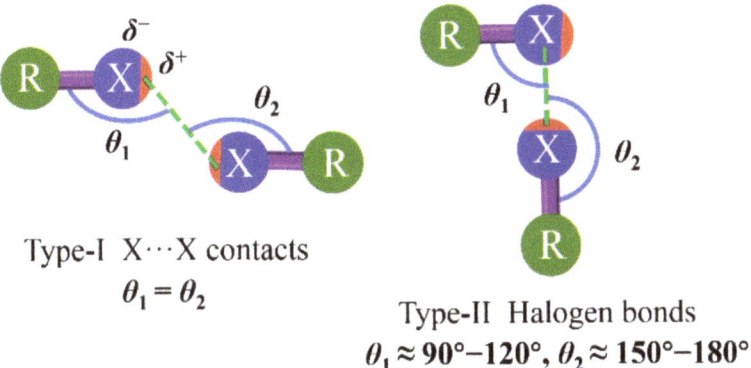

Figure 2. Structural scheme for type-I (left) and type-II (right) halogen···halogen interactions. X = halogen atom; R = C, N, O, halogen atom, etc.; δ is surface potential. Type-II interaction is XB.

In addition, halogen atoms are typically regarded as hydrophobic residues. For example, an I or Br atom is less hydrophilic than a typical HB donor such as an OH or NH group. From the nature of the discrepancy between the two donor sites arise many useful and complementary applications.

1.4. The General Description of XB in 2D Crystal Engineering

Similar to 3D self-assembly processes, the molecular self-assembly process on surfaces can be controlled by balancing molecule–molecule and molecule–substrate interactions [46,47,51–53]. The intermolecular non-covalent bond of XB with high directionality, moderate bond strength (5–30 kJ/mol), and its binding geometry has been used to fabricate 2D nanostructures [40,54–59]. Because of the large size of the halogen atom and the existence of the σ-hole, XB is more sensitive to the steric hindrance and directionality than HB [4,60]. However, as a powerful tool to tune the 2D nanoarchitectures on surfaces, halogens are usually involved in forming HB (total interaction energies up to 140 kJ/mol), where they sever as Lewis bases, and the strength of HBs involving halogens varies with F > Cl > Br > I [61]. Perhaps they can form XBs serving as Lewis acids with an opposite trend, meaning that the strength of the interaction can be tuned by varying halogen substituents. This trend depends on the strength of the σ-hole, which in turn relates to the electron-withdrawing ability of the group. Significantly, XBs are independent from HBs, so both types of interactions can be used to design and adjust supramolecular 2D assemblies on the surface at the same time.

Scanning tunneling microscopy (STM) is a powerful tool that can be used to probe the surface locally and provide structural information in a submolecular resolution whether the nanostructure is crystalline or amorphous [62,63]. The instrument can be used in diverse types of environments such as ultra-high vacuum (UHV), air, and liquid. The temperature of the surface can be precisely controlled (ranging from ~4 K up to a few hundred Kelvin) in a vacuum, which permits both controlled annealing and imaging at low temperatures. UHV provides an ultra-clean environment, which is essential for quantifying supramolecular interactions that are relatively hard to be understood in the solution phase. The solution phase offers a "real-life" view of the assembly process involved in the competition of molecule–solvent and solvent–substrate interactions, except for the molecule–molecule and molecule–substrate interactions [64].

This review will concentrate on our recent research work about 2D supramolecular assembled systems associated with XBs in ambient conditions. Through this review, a comprehensive discussion of supramolecular self-assembled adlayers stabilized by various XBs will be provided. Though it is much more likely to be accepted that the HB is a relatively stronger intermolecular interaction than the XB. In some cases, the XB can be of equivalent strength or even stronger than the typical HB. XB has been widely used to direct and control assembly processes depending on the molecular recognition [40,65–70]. However, the studies of XBs in 2D surface crystal engineering only date back to the last 15 years. Therefore, this is a hot topic in an ongoing research field, and there are many unknown mechanisms that need to be explored.

2. Halogen–Halogen Interactions and Halogen-Bonding in 2D Self-Assembled Networks

To confirm that supramolecular networks are exactly controlled by XBs, a rational design of the building block is essential. Theories of XBs have already been established in 3D crystal engineering, which can be easily applied on 2D assembly, even though differences exist because of the substrate effect. In the following sections, we focus on the research work in our laboratory and discuss the self-assembled patterns fabricated by different halogenated building blocks listed in Scheme 1. The discussion is mainly classified based on the species of π-conjugated cores.

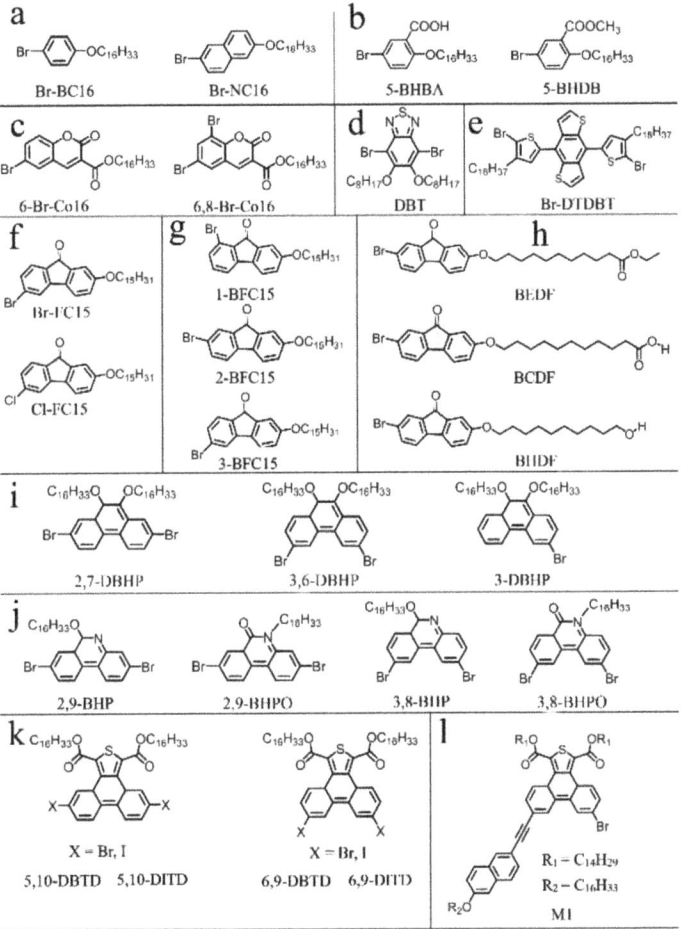

Scheme 1. Chemical structures of the halogenated molecules. (**a**) Molecules with different geometry symmetries of π-conjugated cores along C−Br bond (Br-BC16 and Br-NC16). (**b**) Bifunctional benzene derivatives (5-BHBA and 5-BHDB). (**c**) Bromine substituted coumarin derivatives (6-Br-Co16 and 6,8-Br-Co16). (**d**) Thiadiazole derivative (DBT). (**e**) Dithiophene derivative (Br-BTDBT). (**f**) Different halogen substituted (Br, Cl, and F) fluorenone derivatives (Br-FC15, Cl-FC15, and F-FC15). (**g**) Fluorenone derivatives with different positions of Br substituents (1-BFC15, 2-BFC15, and 3-BFC15). (**h**) Fluorenone derivatives with different terminal groups (BEDF, BCDF, and BHDF). (**i**) Phenanthrene derivatives with different numbers and positions of halogen substituents on the π-conjugated cores (2,7-DBHP, 3,6-DBHP, and 3−DBHP). (**j**) Four regioisomeric phenanthridine derivatives with different positions of halogen substituents and alkoxy chains (2,9-BHP, 2,9-BHPO, 3,8-BHP, and 3,8-BHPO). (**k**) Thienophenanthrene derivatives with different positions of halogen substituents on the π-conjugated cores (6,9-DBTD and 5,10-DBTD). (**l**) An asymmetric thienophenanthrene derivative (M1). All the 28 molecules are used to investigate the 2D supramolecular self-assembled nanostructures based on XBs in our group.

2.1. The Effect of Geometry Symmetry of π-Conjugated Cores Along the C−Br Bond

Varying the geometry of π-conjugated cores can influence the outcome of the self-assembly nanostructures. This aspect can be indicated by the bromine-substituted benzene derivative (Br-BC16) and naphthalene derivative (Br-NC16) (Scheme 1a) [71]. Self-assembled patterns of Br-BC16 and Br-NC16 compounds were studied by STM at the 1-octanoic acid/graphite interface, comparatively (Figure 3). STM results show that the two molecules form different linear fashions stabilized by intermolecular type-I Br···Br contact and H···Br HB (Figure 3b,e) because of their different geometry symmetries of π-conjugated cores along the C−Br covalent bond. The π-conjugated cores and side chains of Br-BC16 absorb with the same direction, while the side chains of Br-NC16 extend into different directions along the lattices of graphite substrate. The π-conjugated cores of two kinds of molecules form a dimer and arrange in a head-to-head mode by a pair of H···Br HB with an antiparallel style.

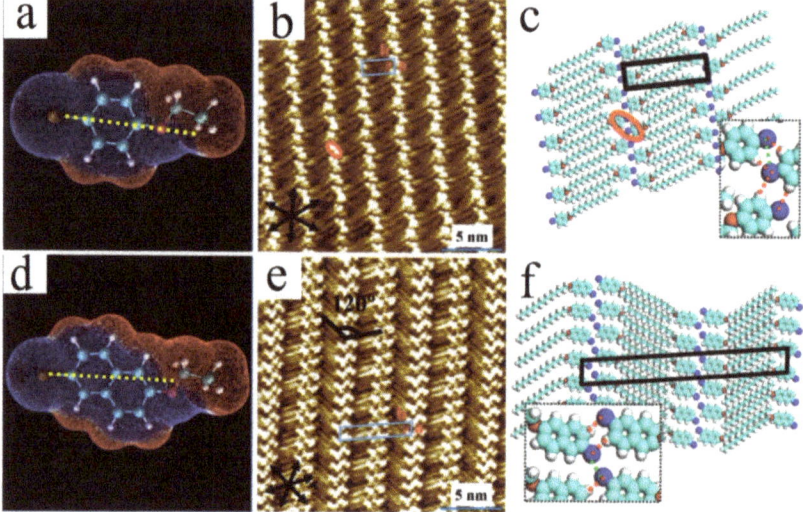

Figure 3. (**a**,**d**) Calculated electrostatic potential (ESP) maps of Br-BC16 and Br-NC16 under vacuum shown by red (positive) and blue (negative) regions. (**b**,**e**) High-resolution STM images of adlayers formed by Br-BC16 and Br-NC16 at the 1-octanoic acid/graphite interface. (**c**,**f**) Proposed molecular models for (**b**,**e**). Insets show the intermolecular interactions. The red lines show the H···Br bonds, and the green lines show the Br···Br contact. Reproduced from [71] with permission from the American Chemical Society.

Calculated 3D ESP maps of π-conjugated cores (Br-BC16 and Br-NC16, Figure 3a,d) and quantum theory of atoms in molecules (QTAIM) of Bader [72,73] obtained by DFT calculations were used to reveal the formation of intermolecular type-I Br···Br contact and HB. The results show that a pair of intermolecular H···Br HBs in each dimer control the structural formation. Furthermore, the type-I Br···Br contact is formed in the dimer of Br-BC16 and the neighboring dimer of Br-NC16, respectively, which is the dominant force to stabilize the two linear nanostructures (Figure 3c,f). It is concluded that the geometry symmetry of π-conjugated cores along the C−Br bond influences the 2D self-assembly. At the same time, the results also indicate that the type-I Br···Br contact is often accompanied by the HB supported by the same Br atom [40].

2.2. Bifunctional Effect of Benzene Derivative

We synthesized a bifunctional molecule (5-BHBA, Scheme 1b), and its 2D self-assembled nanostructures were investigated using STM and DFT calculations [74]. STM experiments were carried out at the 1-octanoic acid/highly oriented pyrolytic graphite (HOPG) interface by varying solution concentrations. Four kinds of patterns (T-like, dislocated, lip-like, and alternating patterns) were observed, as shown in Figure 4a–d. Because of the cooperative and competitive intermolecular XB and HB, these nanostructures consist of dimers, trimers, and tetramers based on rectangular −COOH···HOOC− HB, triangular COO···Br···H−C, Br···O (H), Br···Br, and O···H interactions. At saturated concentration, the T-like pattern is formed (Figure 4a) comprising of two kinds of dimers (the planes of the horizontal and oblique dimer). Every two molecules form a dimer with a head-to-head style. Figure 4e shows that double −Br···OOC− XBs and Br···H HBs are formed in the horizontal dimers, as indicated by the black and blue dashed lines, respectively. In the oblique dimers, two carboxyl groups form −COOH···HOOC− HBs. When decreasing the concentration of 5-BHBA, the molecules self-assemble into a dislocated structure (Figure 4b). There are two different parts (domain 1 and 2). The domain 1 is formed by dimers and trimers alternately, while the domain 2 is composed of pure tetramers. The dimers are still stabilized by intermolecular double −COO···Br− XBs. However, in each trimer, molecules are connected by the type-I Br···Br contact and−COOH···HOOC− HBs (Figure 4f). In each tetramer of domain 2, the arrangement is stabilized by the type-I Br···Br contact and HBs. The type-I Br···Br interaction is formed resulting from the close packing, and thus its binding energy is weaker than that of the type-II XB [75].

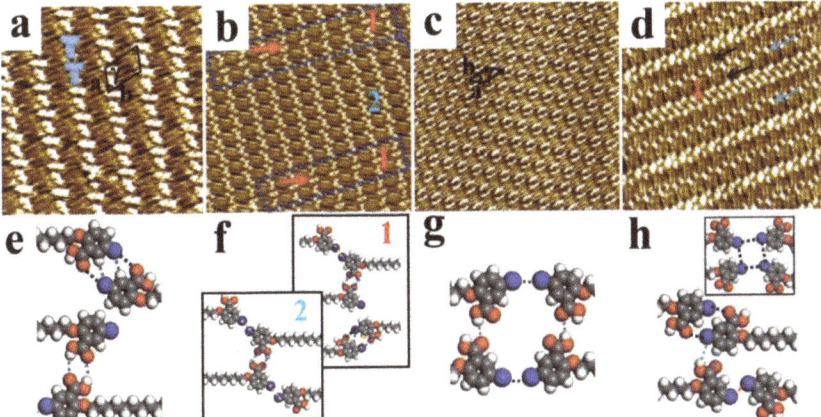

Figure 4. (**a**–**d**) High-resolution STM images show the self-assembled T-like, dislocated, lip-like, and alternate patterns for the 5-BHBA adlayers at the 1-octanoic acid/highly oriented pyrolytic graphite (HOPG) interface by continually diluting concentration. Scanning areas: (**a**) 20 × 20 nm^2; (**b**) 30 × 30 nm^2; (**c**) 35 × 35 nm^2; (**d**) 42 × 42 nm^2. (**e**–**h**) Proposed molecular models for the assembly patterns of (**a**–**d**). Reproduced from [74] with permission from the Royal Society of Chemistry.

When further diluting the solution, the ordered lip-like structure is observed (Figure 4c). The tetramer is found, which is connected by double type-I Br···Br contacts and −OH···OOC− HBs (Figure 4g). Obviously, the type-I Br···Br interaction is weaker in binding energy than the −Br···OOC− XB and the −COOH···HOOC− HB. However, only one HB exists between the carboxyl groups rather than the conventional strong rectangle HBs, which indicates that the lip-like pattern is a metastable phase. At low concentrations, a well-ordered 2D alternated nanoarchitecture is observed (Figure 4d). In one-row lamellae, molecules connect with co-adsorbed 1-octanoic acid molecules by −Br···OOC− XBs [76]. However, in two-row lamellae, the tetramers are formed, which are stabilized by the weak type-I Br···Br contacts (Figure 4h). The results show that the cooperative and competitive intermolecular HBs, XBs, and Br···Br interactions could induce the structural diversity under different concentrations.

Another similar molecule (5-BHDB) mixed with 1-BH (Scheme 1b and Figure 5a) was investigated by dropping the solution on the HOPG surface [77]. The mixture can form a host–guest self-assembled structure. Polar solvent (1-octanoic acid) and nonpolar solvents (1-phenyloctane, n-pentadecane, n-tetradecane, and n-decane) are used to explore the solvent effect. ESP maps (Figure 5b) show the charge distribution of the π-conjugated core. In 1-octanoic acid, the host–guest linear I nanostructure consisting of trimers is observed, which is stabilized by the intermolecular XBs (Figure 5c). This nanostructure is also observed in other solvents at high solution concentrations. When further decreasing the solution concentration, another double-line host–guest nanostructure (linear II, Figure 5d) containing trimers and tetramers is formed in 1-phenyloctane. Co-adsorbed solvent molecules could occupy the gap between side chains in the linear II pattern. Moreover, in n-pentadecane and n-tetradecane, the wavelike structures (Figure 5e) were also observed based on the solvent co-adsorption behavior. Therefore, the van der Waals (vdW) forces of molecule–solvent and the intermolecular XB dominate the formation of co-adsorbed patterns. In n-decane, the linear III pattern (Figure 5f) is observed, driven by the Br···Br type-I contact and the Br···Br type-II XB. In combination with DFT calculations, it is concluded that XBs induce the formation of each structure, and the emergence of relatively stronger host–guest XBs plays a key role in stabilizing these nanostructures and inducing the structural transition.

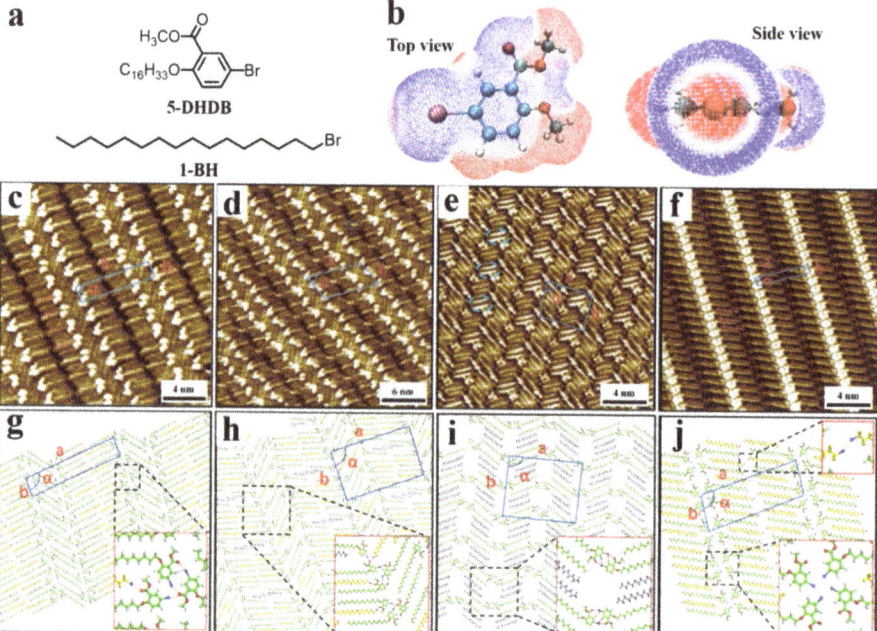

Figure 5. (a) Chemical structures of 5-BHDB and 1-BH. (b) Top and side views of the calculated 3D ESP map of 5-BHDB molecule. (c–f) High-resolution STM images showing the self-assembled linear I, linear II, wavelike, and linear III patterns for the 5-BHBA adlayers in 1-octanoic acid, 1-phenyloctane, n-pentadecane, and n-decane, respectively. (g–j) Proposed molecular models for the assembly (c–f) of the 5-BHBA. Reproduced from [77] with permission from the American Chemical Society.

2.3. Self-Assembled Patterns of Coumarin Derivatives at the 1-Phenyloctane/HOPG Interface

Two coumarin derivatives substituted by one bromine and two bromines (6-Br-Co16 and 6,8-Br-Co16) were synthesized and shown in Figure 6a,d and Scheme 1c [78]. STM results show that 6-Br-Co16 molecules self-assemble into a uniform Z-like linear pattern (Figure 6b). Two 6-Br-Co16 molecules form a dimer adopting an antiparallel orientation through a pair of triangular motifs with the −Br···OOC− XBs and the H···Br HBs (Figure 6c). The 6,8-Br-Co16 molecules fabricate an ordered dislocated linear pattern. In each lamella, the adjacent coumarin cores form a dimer by the same bonding motif as 6-Br-Co16, whereas unlike 6-Br-Co16, the adjacent dimers of 6,8-Br-Co16 align in a tail-to-tail style with type-II Br···Br XB and H···Br HB, leading to their structural difference. The dominant factors in the 2D self-assembled adlayers of the two coumarin derivatives refer to the position and number of Br substituents. Because Br atoms are electron withdrawing groups (strong electronegativity) and highly polarizable, they can induce the rearrangement of the electronic density distribution of the molecules. It is concluded that the Br atom can participant in the formation of the − Br···OOC− XB to induce the formation of different 2D adlayers.

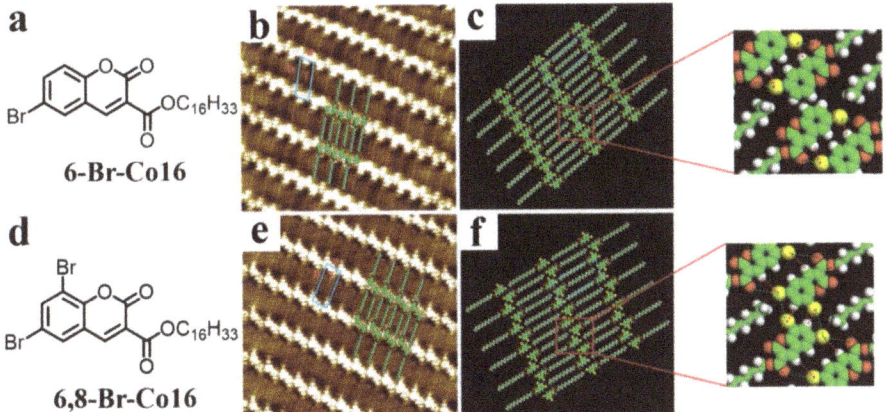

Figure 6. (a,d) Molecular structures of 6-Br-Co16 and 6,8-Br-Co16. (b,e) High-resolution STM images of the 6-Br-Co16 and 6,8-Br-Co16 physiosorbed monolayers at the 1-phenyloctane/HOPG interface. Scanning area: 20 × 20 nm^2. (c,f) Proposed molecular models of (b,e). The inset shows the possible intermolecular bonds. Reproduced from [78] with permission from the Royal Society of Chemistry.

2.4. Self-Assembled Patterns of Thiadiazole Derivatives at the Liquid/HOPG Interface

Self-assembled nanoarchitectures of the DBT molecule (Scheme 1d) were investigated by STM at the liquid/HOPG interface in three kinds of solvents (1-phenyloctane, 1-octanoic acid, and 1-octanol) [79]. Dramatic differences in 2D self-assembly patterns are observed. In 1-phenyloctane, a linear structure (Figure 7a) is stabilized by type-I Br···Br contacts (Figure 7b). However, a lamellar structure is formed at the 1-octanoic acid or 1-octanol/HOPG interface (Figure 7c,e), in which the solvent molecules serve as co-adsorbed components to form the HBs with DBT molecules. The distinct self-assembled nanostructures could be attributed to the solvent polarity. Therefore, the solvent effect plays a significant role in tuning self-assembled nanostructures on solid surfaces.

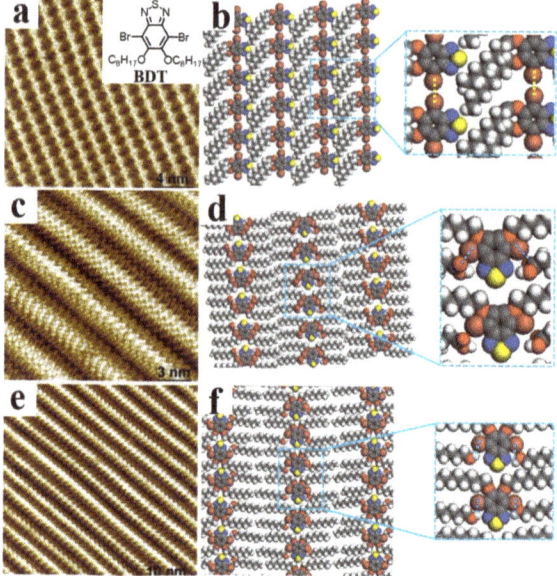

Figure 7. (**a,c,e**) High resolution STM images of DBT adlayers in 1-phenyloctane, 1-octanoic acid, and 1-octanol, respectively. (**b,d,f**) proposed structure models of (**a,c,e**) for the DBT adlayers. Possible interactions are shown in the enlarged insets by yellow and blue dashed lines. Reproduced from [79] with permission from the Hindawi.

2.5. *Self-Assembled Patterns of Dithiophene Derivative at the Liquid/HOPG Interface*

The self-assembly of Br-DTBDT molecule (Scheme 1e) with a cross structure on HOPG surface was studied by STM and DFT calculations (Figure 8a,b) [80]. The self-assembled pattern is shown in the high-resolution STM image (Figure 8c). The proposed model (Figure 8d) implies that the intermolecular Br···S XBs are formed in neighboring molecules, which are the dominant forces to stabilize the well-ordered 2D self-assembled pattern. The bonding mode is also indicated by the ESP map (Figure 8b), in which the charge distribution of Br atom with an electropositive σ-hole and the electronegative area of sulfur atoms provide the possibility to form Br···S XBs.

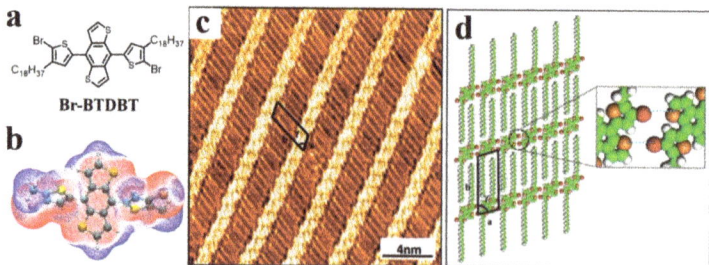

Figure 8. (**a**) Chemical structure of Br-BTDBT. (**b**) Calculated 3D ESP map of Br-BTDBT core, note that the alkyl side chains are replaced by the methyl groups for brevity. (**c**) High-resolution STM image of the adlayer formed by Br-DTBDT at the 1-phenyloctane/HOPG interface. (**d**) Proposed molecular model of the linear structure. Inset shows the intermolecular interactions. Reproduced from [80] with permission from the Royal Society of Chemistry.

2.6. Self-Assembled Patterns of Fluorenone Derivatives at the Liquid/HOPG Interface

Varying the solution concentrations and the halogen substituents are quite significant to adjust the 2D self-assembled pattern. The mechanism is illustrated by comparing self-assembly of Br-FC15 and Cl-FC15 (Scheme 1f and Figure 9a,b) under different concentrations at the 1-phenyloctane/HOPG interface [81]. At high solution concentrations, a lamellar pattern of Br-FC15 is observed (Figure 9c). The molecules in adjoining sides align in antiparallel via C–H···Br HBs and type-I Br···Br contacts (Figure 9d), which is also formed by Cl-FC15 at high solution concentrations (Figure 9e). However, at low solution concentrations of Cl-FC15, twelve Cl-FC15 molecules arrange with two-row and form a dodecamer (Figure 9f). The basic unit of the dumbbell-like pattern is formed by a dodecamer with a neighboring one-row tetramer in each trough, which is stabilized by C–H···Cl and C–H···O–C HBs (Figure 9j). DFT calculations clearly indicate that the binding energies for the Br···Br (−0.19 kcal/mol) and Cl···Cl (−0.19 kcal/mol) are equivalent, so the similar patterns on surface are formed. Therefore, the differences in molecular packing plausibly might arise from the different charge distribution according to ESP maps (Figure 9b) between the Br and Cl atoms, and the nanoarchitectures can be effectively tailored by the introduction of different halogen atoms.

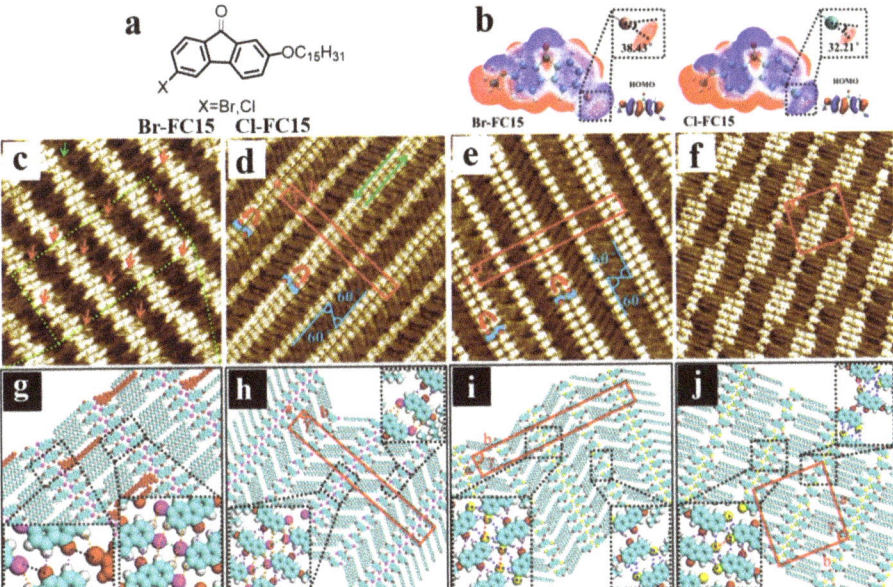

Figure 9. (a,b) Chemical structures of Br-FC15 and Cl-ClC15, and their calculated 3D ESP maps. (c–f) High-resolution STM images of the self-assembled lamellar and alternate-I patterns (Br-FC15), as well as alternate-II and dumbbell-like patterns (Cl-FC15) at the 1-phenyloctane/HOPG interface. Concentration: (c) 3.5×10^{-4} M; (d) 5.4×10^{-5} M; (e) 4.5×10^{-4} M; (f) 6.1×10^{-5} M. (g–j) Tentative structural models of (c–f). Intermolecular interactions are reflected in the enlarged insets. Reproduced from [81] with permission from the American Chemical Society.

Varying the position of Br substituents on the π-conjugated core also affects the self-assembled pattern. This mechanism can be explained by combining STM experiments and DFT calculations. The self-assembly of fluorenone derivatives (1-BFC15, 2-BFC15, and 3-BFC15) with different positions of Br substitution at the 1-octanoic acid/HOPG interface were investigated (Scheme 1g) [82]. The ESP maps of those molecules are shown in Figure 10a–c. Four self-assembled nanostructures: alternate-I pattern (Figure 10d, 1-BFC15), alternate-II pattern (Figure 10e, 2-BFC15), lamellar (Figure 10f, 3-BFC15), and alternate-III patterns (Figure 10g, 3-BFC15) are observed. The alternate-I pattern is stabilized by C−Br···O=C XB and type-I Br···Br contact in the one-row troughs (Figure 10h). In the tetramers of the denser two-row troughs, two diagonal molecules align in an antiparallel fashion bonded with a pair of C−H···Br HBs and connect with the neighboring molecules on the same side through two C−H···Br HBs (Figure 10h). Besides, C−H···O=C HBs are formed between the tetramers. In the looser two-row troughs, there is lack of C−H···Br HB between the diagonal molecules compared with the denser two-row troughs. The carboxyl group of 1-octanoic acid molecules interacts with the Br group to form the COOH···Br HB and C−Br···O=C (COOH) XB (Figure 10h). The 2-BFC15 molecules form the alternate-II pattern (Figure 10e). In the one-row troughs, 2-BFC15 molecules point to the same direction of the trough and align along the trough with a small angle between molecules. Tetramers serve as the elementary unit of the two-row troughs and sequentially align along the troughs (Figure 10i), which contains two dimers bonded by the Br···Br type-I contacts, C−Br···O=C XBs, and C−H···Br HBs. Besides, C−H···O=C and C−H···Br HBs are formed between the two dimers.

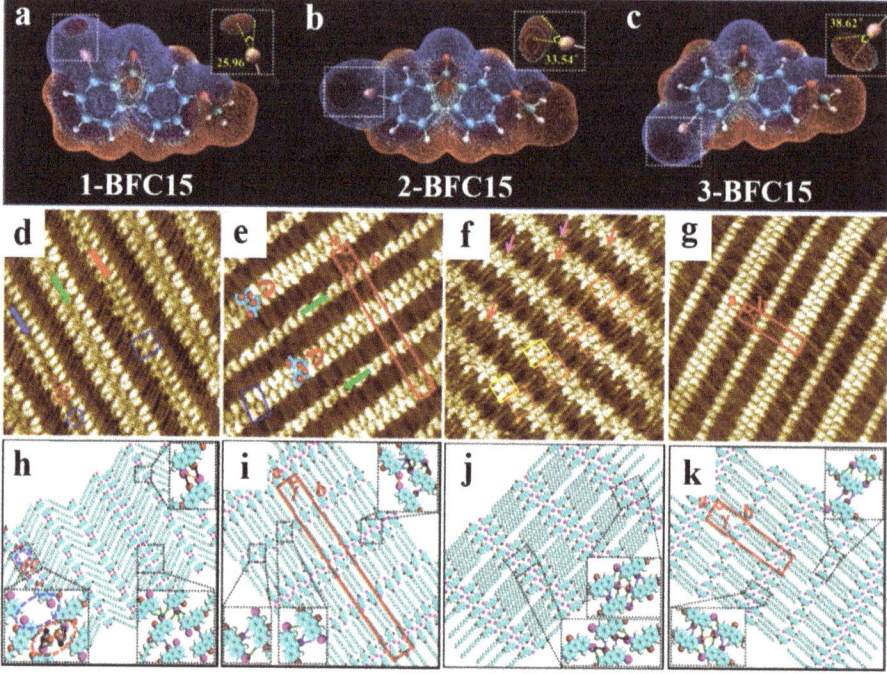

Figure 10. (a–c) Calculated 3D ESP maps of 1-BFC15, 2-BFC15, and 3-BFC15 molecules. (d–g) High-resolution STM images show the self-assembled alternate-I (d, 1-BFC15), alternate-II (e, 2-BFC15), lamellar (f, 3-BFC15), and alternate-III patterns (g, 3-BFC15) at the 1-octanoic acid/HOPG interface, respectively. Scanning areas: 20×20 nm^2. (h–k) Proposed molecular models of (d–g). Intermolecular interactions are reflected in the enlarged insets. Reproduced from [82] with permission from the American Chemical Society.

The lamellar and alternate-III (Figure 10f,g) patterns for 3-BFC15 are observed at high and low concentrations, respectively. In lamellar nanostructure, tetramers and hexamers (indicated by the yellow and orange rectangles in Figure 10f, respectively) are formed, which alternately align along each trough randomly. As the middle-inset shown in Figure 10j, 3-BFC15 molecules on the same side take a parallel style linked by C−H···O=C and C−H···Br HBs, which connect with the antiparallel molecules on the neighboring side through C−H···Br HBs and type-I Br···Br contacts. Whereas, neighboring complexes along the troughs are linked by C–Br⋯O=C XBs, C–H⋯O=C, and C–H⋯Br HBs (Figure 10j). In the alternate-III pattern (Figure 10g), 3-BFC15 molecules in neighboring rows of the two-row troughs take a head-to-head fashion bonded by C–H⋯Br HBs and Br⋯Br type-I contacts. In the one-row troughs, the C–H⋯O=C and C–H⋯Br HBs are formed in and between the dimers, respectively (Figure 10k).

The diverse nanostructures might arise from different positions of bromine substituent in fluorenone derivatives, which can cause various charge distributions of the fluorenone π-core and adjust the positive charge distribution of the σ-hole on the Br atom along the C–Br axis. In addition, the distinct arrangements of the two-row troughs are ascribed to the directionality of XBs and the closest packing principle.

Various terminal groups of side chains can influence the self-assembly patterns governed by XBs at the 1-phenyloctane/HOPG interface. Three bromine substituted fluorenone derivatives bearing an alkoxy chain terminated by ethoxycarbonyl, carboxyl, and hydroxyl groups (Scheme 1h and Figure 11a) are used to systematically investigate the terminal group effect on the molecular assembly [83]. C–H⋯O–C HBs are formed within the π-conjugated cores of adjacent molecules in the assembled nanostructure of BEDF. Besides, the cores form dimers by a pair of C–H⋯Br HBs (Figure 11c,g). Each molecule in dimers interacts with the $COOC_2H_5$–terminated groups of the side chains in the adjacent dimers along the troughs to form C–H⋅⋅O–C ($COOC_2H_5$) HBs (Figure 11g).

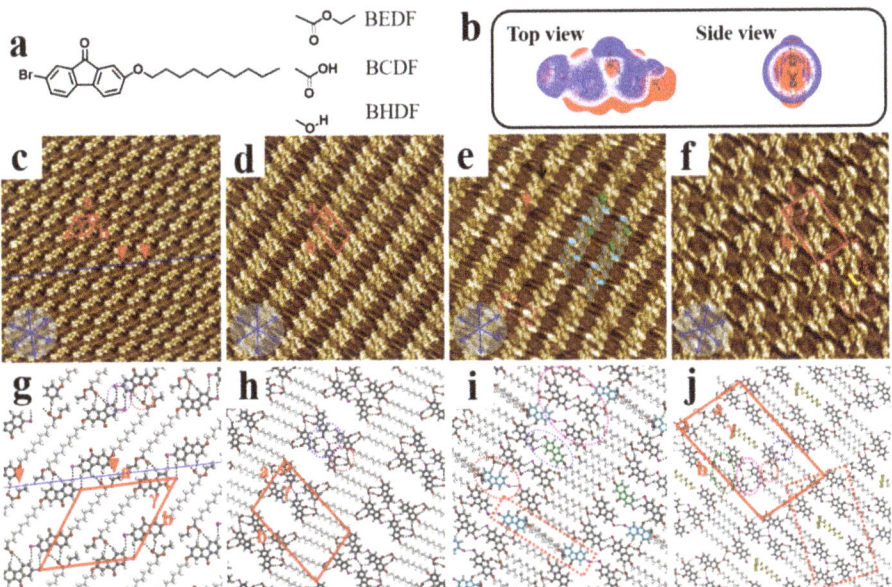

Figure 11. (a) Chemical structures of BEDF, BCDF, and BHDF. (b) Top and side views of the calculated 3D ESP map of Br substituted fluorenone core. (c–f) High-resolution STM images show the self-assembled dimer pattern (BEDF), tetramer pattern (BCDF), lamellar pattern (BHDF) and the octamer pattern (BHDF) at the 1-phenyloctane/HOPG interface, respectively. Scanning areas: 20 × 20 nm². (g–j) Proposed molecular models of (c–f). Reproduced from [83] with permission from the Elsevier B. V.

For BCDF molecule, the linear pattern is composed of tetramers (Figure 11d). Four BCDF molecules form an aggregate, serving as the fundamental unit to arrange orderly along the bright troughs. BCDF cores in diagonal within the tetramers are antiparallel and bonded by a type-I Br⋯Br contact, as well as interact with two adjacent antiparallel cores by C–H⋅⋅Br HBs (Figure 11h).

For BHDF molecule, the lamellar and octamer patterns are observed at high concentrations. After continuous scanning, the octamer pattern will transform into the lamellar pattern. Whereas, at low concentrations, only the octamer pattern is observed. The lamellar pattern (Figure 11e) shows that BHDF molecules self-assemble into two-row troughs. Along the two-row troughs, four molecules form a tetramer (Figure 11e), which are adjacent to a dimer marked by the blue and green ellipses. The tetramers are formed by a pair of C–H···Br HBs, while the other antiparallel cores in top-left and bottom-right make no contribution. Neighboring tetramers along the direction of the side chains are interconnected by the interdigitated chains, leading to form four O–H··O–C HBs and two additional C–H···O–H (OH) HBs. The results reveal that the terminal functional groups play key roles in the regulation of 2D self-assembly nanostructures and serve as the triggers of XBs.

2.7. Self-Assembled Patterns of Phenanthrene Derivatives at the Liquid/HOPG Interface

Tuning the number and position of halogen substituents on the π-conjugated cores also affects the molecular arrangement on surfaces. The adjustment mechanism is illustrated via alkoxy substituted phenanthrene derivatives [84]. Self-assembled nanostructures of 2,7-DBHP, 3,6-DBHP, and 3–DBHP (Scheme 1i and Figure 12a–c) were observed at the 1-octanoic acid/HOPG interface. Both 2,7-DBHP and 3,6-DBHP could form densely packed columnar networks with edge-on orientation on the surface at relatively higher concentrations. This nanostructure is stabilized by interchain vdW interactions and π–π stacking between the π-conjugated cores. At low solution concentrations, the π-conjugated cores of 2,7–DBHP and 3,6–DBHP molecules adsorb on the HOPG surface with a flat-on fashion (Figure 12d,e). Molecular models for the adlayers of 2,7–DBHP and 3,6–DBHP are proposed, as shown in Figure 12g,h. For 2,7–DBHP, Br atoms tend to form Br···O (COO) XB with co-absorbated 1-octanoic acid molecules [41], whereas for 3,6-DBHP, three molecules arrange in a head-to-head fashion through type-I Br···Br contacts and H···Br bonds, which induce the formation of zigzag packing. At moderate concentrations, 2,7-DBHP can form an intermediate pattern, which consists of rows with flat-on and edge-on styled π-conjugated cores.

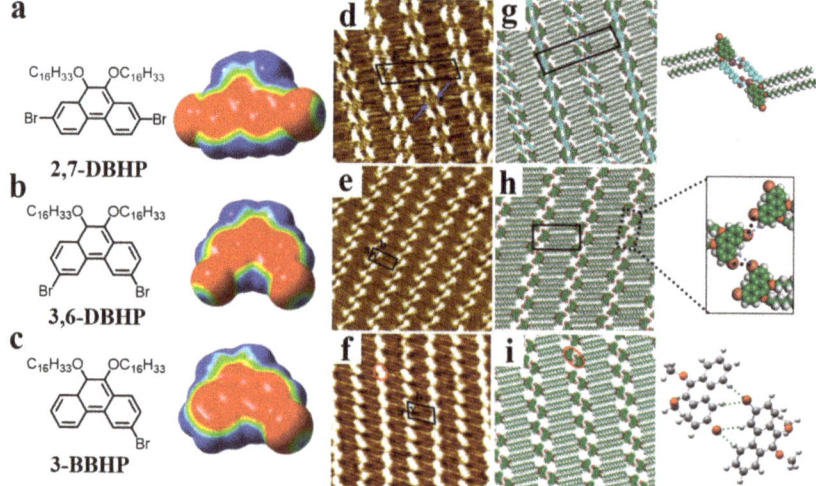

Figure 12. (a–c) Molecular structures and their calculated 3D ESP maps of 2,7-DBHP, 3,6-DBHP, and 3-BBHP, respectively. (d–f) Self-assembled networks formed by 2,7-DBHP, 3,6-DBHP, and 3-BBHP at the 1-octanoic acid/HOPG interface, respectively. Scanning area: 15 × 15 nm². (g–i) Proposed molecular models of (d–f). Reproduced from [84] with permission from the Royal Society of Chemistry.

For 3-BBHP molecules, no solvent and concentration effects are observed in the self-assembly process. The molecules form a dimer with their bromine atoms in an antiparallel style, leading to the formation of two-fold nodes marked by a red circle in Figure 12f,i. The dislocated linear pattern is stabilized by triangular C–H···Br HBs. These results demonstrate that the different positions and numbers of Br atoms on the phenanthrene cores significantly affect intermolecular interactions and determine the outcome of supramolecular architectures.

2.8. Self-Assembled Patterns of Phenanthridine Derivatives at the Liquid/HOPG Interface

Four regioisomeric phenanthridine derivatives with different positions of halogen substituents and alkoxy chains (2,9-BHP, 2,9-BHPO, 3,8-BHP, and 3,8-BHPO) (Scheme 1j and Figure 13a–d) were synthesized in order to investigate the XBs in 2D self-assembled nanostructures by STM at the 1-phenyloctane/HOPG interfaces [85]. For 2,9-BHP and 3,8-BHP molecules, these two molecules form linear structures with the π-conjugated cores parallel to each other in each lamella (Figure 13i,k). The π-conjugated cores of 2,9-BHP molecules in neighboring rows pack in a head-to-head style and form a helical arrangement (Figure 13i). The molecular model (Figure 13m) displays that the intermolecular Br···N XBs and H···Br HBs stabilize the pattern. The arrangement of the π-conjugated cores for 3,8-BHP molecules has a right-handed rotation (Figure 13k). The stronger intermolecular XBs and HBs are the dominated forces to govern the arrangement of 2,9-BHP and 3,8-BHP molecules in 2D self-assembled adlayers.

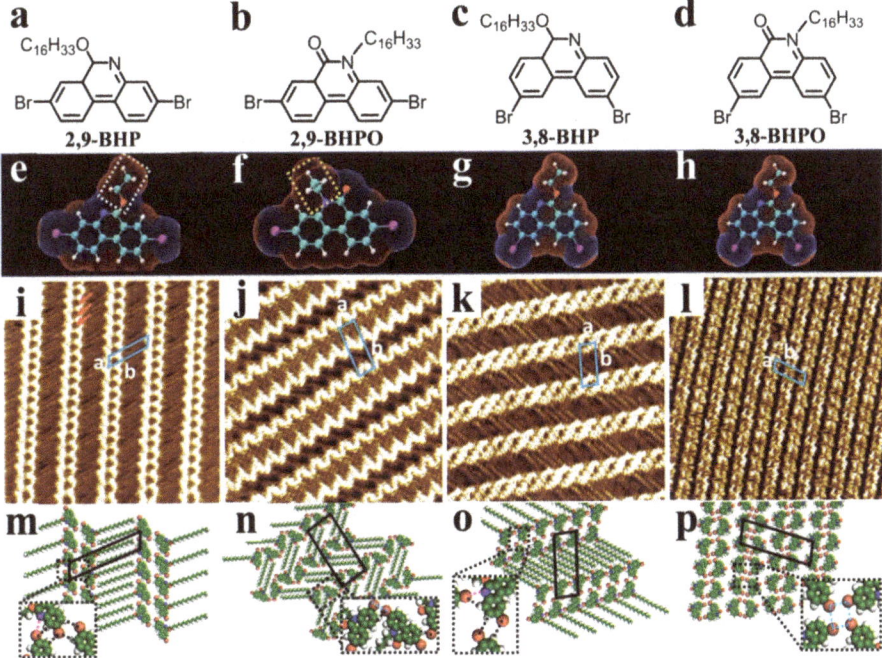

Figure 13. (a–d) Chemical structures of 2,9-BHP, 2,9-BHPO, 3,8-BHP, and 3,8-BHPO and their calculated 3D ESP maps (e–h). (i–l) High-resolution STM images of 2,9-BHP, 2,9-BHPO, 3,8-BHP, and 3,8-BHPO adlayers at the 1-phenyloctane/HOPG interface. (m–p) Proposed molecular models of (i–l). Image area: (i,j,l) 20 × 20 nm^2; (k) 15 × 15 nm^2. The insets in each model show intermolecular interactions. Pink: Br···N bonds; black: Br···H bonds; blue: type-II Br···Br XBs. Reproduced from [85] with permission from the American Chemical Society.

In contrast, 2,9-BHPO forms a zigzag linear pattern (Figure 13j) stabilized by the H···Br and C−H···C=O HBs (Figure 13n); 3,8-BHPO can self-assemble into a lamellar arrangement with all the side chains stretching into the solution (Figure 13l). Four π-conjugated cores of 3,8-BHPO form a tetramer by four type-II Br···Br XBs, which induce the structural formation (Figure 13p). Due to the desorbed side chains in this pattern, the adlayer is not stable on the surface.

The position of the side chain attached to the π-conjugated cores of molecules also plays a significant role in determining the molecular arrangements. The alkoxy chain can rotate randomly to form dense packing. Whereas, the alkyl chain is rigid, and the C−N bond cannot rotate. Therefore, the chain−chain and chain−substrate vdW interactions for four molecules should not be ignored.

2.9. Self-Assembled Patterns of Thienophenanthrene Derivatives at the Liquid/HOPG Interface

Self-assembly of thienophenanthrene (TP) derivatives of 6,9-DBTD and 5,10-DBTD (Scheme 1k and Figure 14a) with triangle π-conjugated cores were investigated at the n-tridecane/HOPG interface by varying the solution concentration (from 10^{-3} to 10^{-6} M) [86,87]. A well-ordered lamellar structure of 6,9-DBTD (Figure 14e) is observed. It is obviously displaying that two TP π-cores form a dimer and arrange in a head-to-head style through two pairs of type-II Br···Br XBs and H···Br bonds, while adjacent dimers are bonded by a pair of C−H···O−C and H···Br HBs. Besides, the side chains arrange in a tail-to-tail style and are vertical to the rows (Figure 14j). When decreasing the solution concentration (1.6×10^{-5} M), the co-adsorbed linear pattern of 6,9-DBTD (Figure 14f) is observed. In one row, two TP π-cores form a dimer via a pair of C−H···O−C HBs with an antiparallel arrangement, while adjacent dimers form type-I Br···Br interactions (Figure 14k).

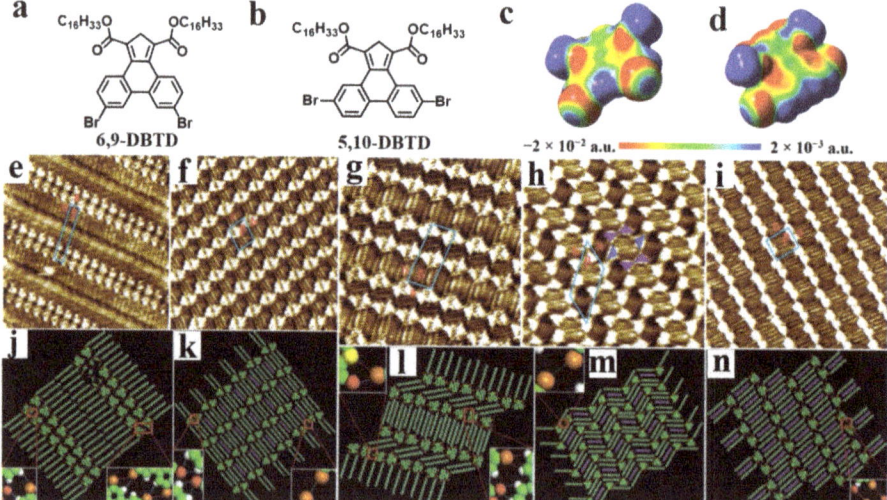

Figure 14. (a–d) Chemical structures of 6,9-DBTD and 5,10-DBTD and their calculated 3D ESP maps. (e,f) High-resolution STM images show the lamellar structure (1.2×10^{-4} M) and co-adsorbed linear pattern (6.2×10^{-5} M) of 6,9-DBTD adlayers at the n-tridecane/HOPG interface. (g–i) High-resolution STM images showing vertebra-like structure (2.8×10^{-4} M), hexagonal network pattern (5.7×10^{-5} M), and linear pattern (5.3×10^{-6} M)) of 5,10-DBTD adlayers at the n-tridecane/HOPG interface. Image area: (e–g,i) 20 × 20 nm^2, (h) 17 × 17 nm^2. (j–n) Proposed structural models of (e–i). Reproduced from [86,87] with permission from the Royal Society of Chemistry.

The 5,10-DBTD molecules (Figure 14b) self-assemble into a large organized domain with a vertebra-like pattern (Figure 14g) at a saturation concentration. Two TP π-cores arrange in a joint-like motif with an angle-to-angle style by intermolecular bifurcated Br···O–C and Br···S XBs. The adjacent 5,10-DBTD units are connected via C–H···Br and C–H···O–C HBs, which further strengthen this motif. Two rows of molecules are packed in an anti-parallel fashion and form the vertebra-like pattern, leading to the minimized polarity of the adlayer [88]. Therefore, 5,10-DBTD exhibits a distinct and nonidentical XB. With decreases in concentration, the hexagonal network of 5,10-DBTD with a 6-fold ring elementary structural motif is encompassed by two pairs of molecules and two single molecules (indicated by triangles) (Figure 14h). Two TP π-conjugated cores appear in pairs with anti-parallel orientations in dimer via type-I Br···Br interactions, marked by triangles with the same color. Besides, the TP π-cores marked by blue triangles form type-I Br···Br contacts with the molecules in the adjacent hexagonal ring (Figure 14 m), respectively. When further diluting the concentration (8.2×10^{-5} M), the typical STM image of the 5,10-DBTD adlayer shows the formation of a linear pattern (Figure 14i). The molecules take an anti-aligned arrangement, and their side chains stretch into one side of rows and vertical to the rows. Therefore, all the carboxyl group orientations might pack toward the conjugated TP core formed Br···O–C XBs, which play key roles in determining the structural formation. Besides, the adjacent TP cores are connected via a single type-I Br···Br interaction.

The halogen–heteroatom interactions can also tune the formation of 2D supramolecular networks at the liquid/solid interface. The 5,10-DBTD and 5,10-DITD molecules are investigated by combinations of STM observations and DFT calculations [89]. This experiment focuses on how the orientations of the ester substituent for 5,10-DBTD and 5,10-DITD affect positive charge distribution of halogens by DFT, which plays a key role in determining the formation of intermolecular XB and various self-assembled arrangements.

To confirm the formation of XB, different concentrations and substituted halogen atoms (Br and I) are adopted. Under saturated solution, a large organized domain with a vertebra-like motif of 5,10-DBTD (Figure 15c) is observed by STM at the 1-phenyloctane/HOPG interface. Two π-cores of 5,10-DBTD arrange in an infinite joint-like motif through bifurcated Br···O–C and Br···S XBs, as the red rectangle indicates in Figure 15c. The 5,10-DBTD molecules are connected by C–H···Br and C–H···O–C HB interactions, which further reinforce this ribbon pattern (Figure 15c). Because of the alkyl substitution in aromatic compounds with halogen atoms, the XBs and vdW forces dominate the formation of a self-assembled network on the HOPG surface. In this arrangement, the orientation of the side chains directly determines the orientation of carboxyl, leading to the formation of intermolecular Br···O–C XBs. The direction of Br···S XB is along the C–Br bond toward the positive charge region of the sulfur atom [70,90]. Thus, 5,10-DBTD exhibits distinct and nonidentical XBs that vary as a function of halogen identity. With decreasing solution concentration, an alternate vertebra-like structure is formed (Figure 15d), in which all the carboxyl groups orient to the π-conjugated cores (Figure 15h). The Br···O–C XBs play significant roles in controlling the formation of this pattern, except for the interchain vdW interactions.

When substituted halogen atoms are changed from a Br atom to an I atom, 5,10-DITD forms a honeycomb-like network composed of two zigzag lines (Figure 15e) at a saturated concentration. In each line, all the π-conjugated cores of 5,10-DITD arrange in an angle-to-angle style by successive intermolecular bifurcated I···O–C and I···S XBs (Figure 15i). All the side chains stretch into different directions. With further decreases in the solution concentration, an alternate-I pattern is formed (Figure 15f). The molecular packing in the zigzag line is the same as that in the honeycomb-like pattern. In the dimers, the side chains have the same stretching direction as those in the zigzag structure (Figure 15j), indicating that only the type-I I···I contact bond is formed. This work gives a deep insight into the role of ester orientation and concentration on the formation of halogen-heteroatom contacts, which proves relevant for identification of multiple XBs in 2D crystal engineering.

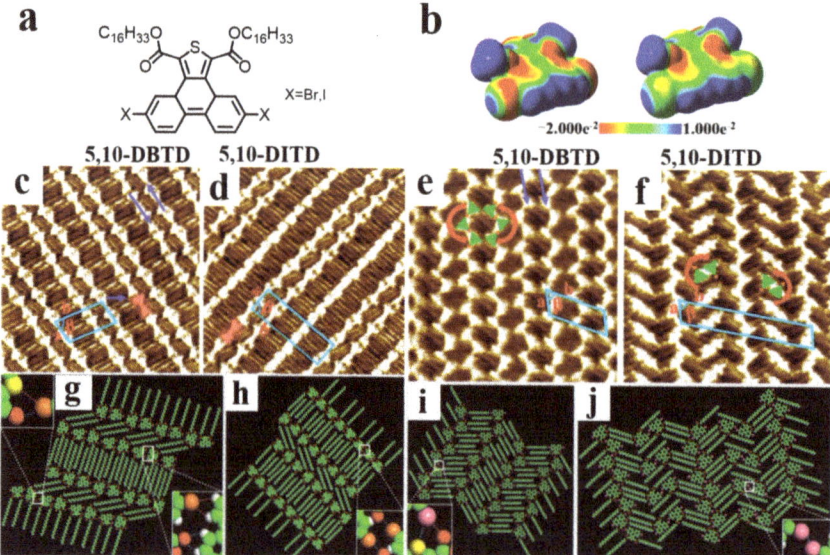

Figure 15. (**a,b**) Chemical structures of 5,10-DBTD and 5,10-DITD, and their calculated 3D ESP maps. (**c,d**) High-resolution STM images of vertebra-like (2.5×10^{-4} M) and alternate vertebra-like (9.7×10^{-5} M) patterns of 5,10-DBTD adlayers at the 1-phenyloctane/HOPG interface, respectively. (**e,f**) High-resolution STM images of honeycomb-like (6.6×10^{-4} M) and alternate chiral (8.2×10^{-5} M) patterns of 5,10-DBTD adlayers at the 1-phenyloctane/HOPG interface, respectively. Scanning area: 20×20 nm^2. (**g–j**) Proposed structural models of (**c–f**). Insets show the intermolecular interactions. Reproduced from [89] with permission from the American Chemical Society.

In addition, the concentration-dependent self-assembly of 5,10-DITD was investigated at the 1-octanic acid/graphite interface by STM [91]. Three chiral arrangements and 2D assembled structural transformation mainly controlled by XBs are clearly revealed. At high concentrations, the molecules self-assemble into a honeycomb-like chiral network (Figure 16c,d). Except for the interchain vdW forces, this nanostructure is stabilized by intermolecular continuous C–O⋯I⋯S XBs in each zigzag line. At a moderate concentration, a chiral kitelike motif (Figure 16e,f) is observed, in which the C–O⋯I⋯S and I⋯O–C XBs, along with the molecule–solvent C–O⋯I⋯H XBs are the dominated forces to determine the structural formation. At low concentrations, the molecules form a chiral cyclic network (Figure 16g,h) by molecule–molecule C–O⋯I⋯S XBs and molecule–solvent C–O⋯I⋯H XBs. The above results reveal that the type of intermolecular XBs and the number of the co-adsorbed 1-octanic acids determine the formation and transformation of chirality.

Based on the TP core, an asymmetric M1 molecule (Figure 17a) was designed and synthesized to investigate C–Br⋯π XB-induced molecular self-assembly at the *n*-hexadecane/solid interface [92]. According to the self-adaption principle of molecule geometry in 2D supramolecular self-assembly, three types of self-assembly dimers are predicted (Figure 17c), which are confirmed by STM experiment and DFT calculations. Three nanostructures ("N" type, "tail-to-tail", and discrete structures) are observed by continually decreasing concentrations (Figure 17d,f,h). Among the three nanostructures, the relative location of two molecular cores in dimer I is maintained, and the difference is only in the arrangement of alkoxy chains. The results demonstrate that dimer I not only has a high binding energy derived from the collaboration of C–H⋯Br⋯π bonds and the rigid directionality of C–Br⋯π XBs, but also has support from vdW forces generated from the suitable lengths of the side chains and *n*-hexadecane molecules.

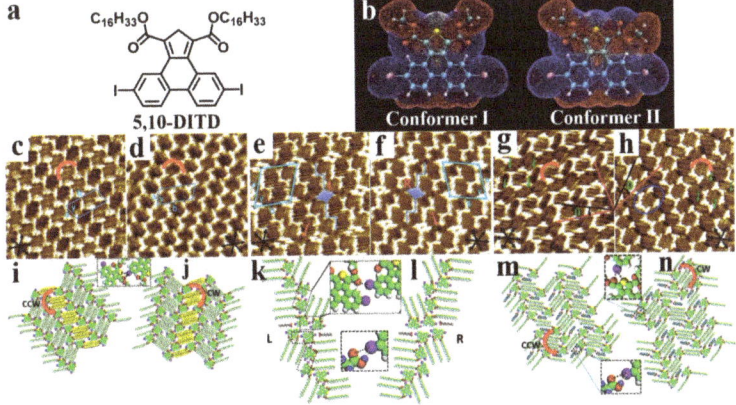

Figure 16. (a) Chemical structure of 5,10-BITD. (b) Calculated 3D charge density maps of 5,10-DITD with different configurations of ester groups. (c–h) High-resolution STM images of CCW and CW honeycomb-like, kite-like, and cyclic networks formed by 5,10-DITD at the 1-octanic acid/HOPG interface (2.0×10^{-3} M; 5.0×10^{-4} M; 5.0×10^{-5} M; scanning area: 20×20 nm^2). (i–n) Proposed molecular models of (c–h) showing R and L kite-like, kite-like, and cyclic networks. Reproduced from [91] with permission from Elsevier B.V.

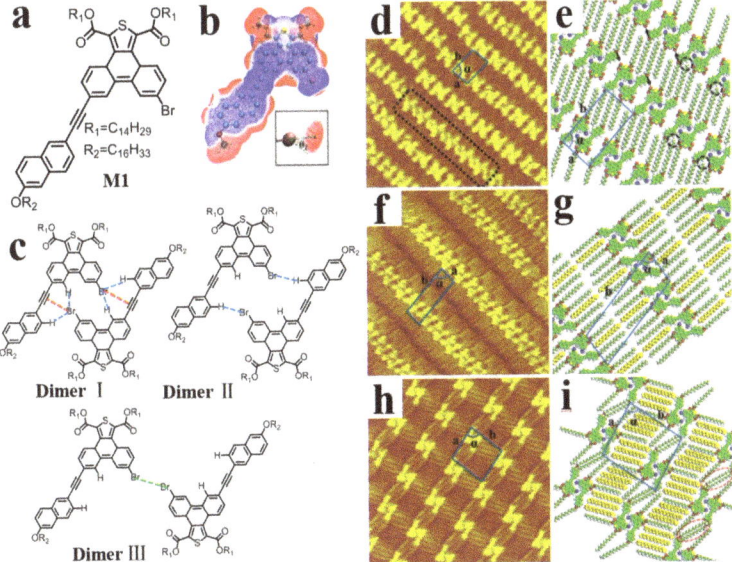

Figure 17. (a) Chemical structure of M1. (b) Calculated 3D ESP map of the M1 conjugated core. (c) Possible self-assembly dimers of M1 containing Br⋯π XB (dimer I), C–H⋯Br HBs (dimer II), and type-I Br⋯Br interactions (dimer III) depicted by red, blue, and green dotted lines, respectively. (d) A high-resolution STM image of the linear structure consisted of dimer I at the n-hexadecane/HOPG interface. Some dimer II structures are excluded by the black dotted rectangle. (f,h) High-resolution STM images showing the "tail-to-tail" and discrete structures, respectively. (e–i) Proposed structural models of (d,f,h). Image size: 20×20 nm^2. Reproduced from [92] with permission from the Royal Society of Chemistry.

Generally, changing the shape of the molecular π-conjugate core and varying the position, type, and number of halogen substituents of the molecules could modify the intermolecular interactions. The results indicate that these methods are efficient to further investigate XBs on surfaces, and these factors play significant roles in controlling the arrangement of nanostructures.

3. Conclusions

This review briefly documented the definition, research history, and essential properties of the XBs. The XB, as a "specific supramolecular interaction" has been widely used in the preparation of complex 2D self-assembly motifs. This review includes 28 molecules that were designed and home-synthesized carefully, and which were grouped by the similarity of π-conjugated cores. Those molecules within the same family are designed by changing the position and number of the halogen substituents on the aromatic conjugated cores and are used to explore the formation of XBs under different solvents and concentrations at the liquid/solid interface. STM observation and DFT calculations show that an X–X contact often is accompanied by a concomitant HB, and the two bonds act as the collaborative forces to stabilize the 2D adlayers. Moreover, varying the type, number and position of halogen substitutes on the π-conjugated cores can induce the rearrangement of the electronic density distribution of the molecules, which can give rise to new molecular arrangements on surfaces. However, to confirm the fabrication of networks driven by XBs, significant efforts need to be involved in the design of 2D crystal engineering. Furthermore, the dominant force of molecule–substrate interactions on the formation of self-assembled nanostructures is difficult to quantify in terms of the real contribution of XBs for stabilizing the supramolecular networks. Fortunately, the high-resolution STM images could support precise determination of interatomic distances (bond length) and angles (bond angle), which thus allows better insight into XBs and X–X interactions.

Author Contributions: Writing—original draft preparation, Y.W.; writing—review and editing, X.M.; supervision, W.D. All authors have read and agreed to the published version of the manuscript.

Funding: This research was funded by the Natural Science Foundation of Guangdong Province grant number 2018A030313452, the Science and Technology Program of Guangzhou grant number 202002030083, and the Fundamental Research Funds for the Central Universities (SCUT).

Acknowledgments: Financial support from the Natural Science Foundation of Guangdong Province (2018A030313452), the Science and Technology Program of Guangzhou (202002030083), and the Fundamental Research Funds for the Central Universities (SCUT) is gratefully acknowledged.

Conflicts of Interest: The authors declare no conflict of interest.

References

1. Desiraju, G.R.; Ho, P.S.; Kloo, L.; Legon, A.C.; Marquardt, R.; Metrangolo, P.; Politzer, P.; Resnati, G.; Rissanen, K. Definition of the Halogen Bond. *Pure Appl. Chem.* **2013**, *85*, 1711–1713. [CrossRef]
2. Murray, J.S.; Lane, P.; Politzer, P. Expansion of the σ-hole concept. *J. Mol. Model.* **2009**, *15*, 723–729. [CrossRef] [PubMed]
3. Politzer, P.; Murray, J.S.; Clark, T. Halogen bonding and other σ-hole interactions: A perspective. *Phys. Chem. Chem. Phys.* **2013**, *15*, 11178–11189. [CrossRef] [PubMed]
4. Cavallo, G.; Metrangolo, P.; Milani, R.; Pilati, T.; Priimagi, A.; Resnati, G.; Terraneo, G. The Halogen Bond. *Chem. Rev.* **2016**, *116*, 2478–2601. [CrossRef]
5. Colin, M.M.; Gaultier de Claubry, H. Sur Le Combinaisons de L'iode Avec Les Substances Végétales et Animales. *Ann. Chim.* **1814**, *90*, 87–100.
6. Guthrie, F. Xxviii.—On the Iodide of Iodammonium. *J. Chem. Soc.* **1863**, *16*, 239–244. [CrossRef]
7. Remsen, I.; Norris, J.F. Action of the Halogens on the Methylamines. *Am. Chem. J.* **1896**, *18*, 90–95.
8. Ault, B.S.; Andrews, L. Infrared and Raman spectra of the M + F^{3-} ion pairs and their mixed chlorine-fluorine counterparts in solid argon. *Inorg. Chem.* **1977**, *8*, 2024–2028. [CrossRef]
9. Ault, B.S.; Andrews, L. Matrix reactions of alkali metal fluoride molecules with fluorine. infrared and raman spectra of the trifluoride ion in the M + F^{3-} species. *J. Am. Chem. Soc.* **1976**, *98*, 1591–1593. [CrossRef]

10. Riedel, S.; Köchner, T.; Andrews, P.L.; Wang, X. Polyfluoride Anions, a Matrix-Isolation and Quantum-Chemical Investigation. *Inorg. Chem.* **2010**, *41*, 7156–7164. [CrossRef]
11. Legon, A.C. Prereactive complexes of dihalogens xy with lewis bases b in the gas phase: A systematic case for the halogen analogue b small middle dot small middle dot small middle dotxy of the hydrogen bond b small middle dot small middle dot small middle dotHX. *Angew. Chem. Int. Ed.* **1999**, *38*, 2686–2714. [CrossRef]
12. Brinck, T.; Jane, S.; Politzer, P. Surface electrostatic potentials of halogenated methanes as indicators of directional intermolecular interactions. *Int. J. Quantum Chem.* **1992**, *44*, 57–64. [CrossRef]
13. Brinck, T.; Murray, J.S.; Politzer, P. Molecular surface electrostatic potentials and local ionization energies of group v–vii hydrides and their anions: Relationships for aqueous and gas-phase acidities. *Int. J. Quantum Chem.* **1993**, *48*, 73–88. [CrossRef]
14. Murray, J.S.; Paulsen, K.; Politzer, P. Molecular surface electrostatic potentials in the analysis of non-hydrogen-Bonding noncovalent Interactions. *Proc. Indian Acad. Sci. Chem. Sci.* **1994**, *106*, 267–275.
15. Clark, T.; Hennemann, M.; Murray, J.S.; Politzer, P. Halogen bonding: The σ-hole. *J. Mol. Model.* **2007**, *13*, 291–296. [CrossRef] [PubMed]
16. Yamamoto, H.M.; Yamaura, J.I.; Kato, R. Structural and electrical properties of (BEDT-TTF)2X(diiodoacetylene) (X = Cl, Br): The novel self-assembly of neutral Lewis-acidic molecules and halide anions in a molecular metal. *J. Mater. Chem.* **1998**, *8*, 15–16. [CrossRef]
17. Maginn, S.J. Crystal engineering: The design of organic solids by G. R. Desiraju. *J. Appl. Crystallogr.* **1991**, *24*, 265. [CrossRef]
18. Desiraju, G.R. Supramolecular synthons in crystal engineering-a new organic synthesis. *Angew. Chem. Int. Ed.* **2010**, *34*, 2311–2327. [CrossRef]
19. Moulton, B.; Zaworotko, M.J. From molecules to crystal engineering: Supramolecular isomerism and polymorphism in network solids. *Chem. Rev.* **2001**, *101*, 1629–1658. [CrossRef] [PubMed]
20. Subramanian, S.; Zaworotko, M.J. Exploitation of the hydrogen bond: Recent developments in the context of crystal engineering. *Coord. Chem. Rev.* **1994**, *137*, 357–401. [CrossRef]
21. Atwood, J.L.; Lehn, J.M. Comprehensive supramolecular chemistry. solid-state supramolecular chemistry. *Cryst. Eng.* **1996**, *6*, 199–206.
22. Tabellion, F.M.; Seidel, S.R.; Arif, A.M.; Stang, P.J. Discrete supramolecular architecture vs. crystal engineering: The rational design of a platinum-based bimetallic assembly with a chair like structure and its infinite, copper analogue. *J. Am. Chem. Soc.* **2001**, *123*, 7740–7741. [CrossRef] [PubMed]
23. Xia, C.; Fan, X.; Locklin, J.; Advincula, R.C.; Gies, A.; Nonidez, W. Characterization, supramolecular assembly, and nanostructures of thiophene dendrimers. *J. Am. Chem. Soc.* **2004**, *126*, 8735–8743. [CrossRef] [PubMed]
24. Cook, T.R.; Zheng, Y.R.; Stang, P.J. Metal–organic frameworks and self-assembled supramolecular coordination complexes: Comparing and contrasting the design, synthesis, and functionality of metal–organic materials. *Chem. Rev.* **2013**, *113*, 734–777. [CrossRef]
25. Li, H.L.; Eddaoudi, M.M.; O'Keeffe, M.; Yaghi, O.M. Design and synthesis of an exceptionally stable and highly porous metal-organic framework. *Nature* **1999**, *402*, 276–279. [CrossRef]
26. Kovbasyuk, L.; Krämer, R. Allosteric supramolecular receptors and catalysts. *Chem. Rev.* **2004**, *104*, 3161–3187. [CrossRef]
27. Yan, X.; Wang, F.; Zheng, B.; Huang, F. Stimuli-responsive supramolecular polymeric materials. *Chem. Soc. Rev.* **2012**, *41*, 6042–6065. [CrossRef]
28. Sangeetha, N.M.; Maitra, U. Supramolecular gels: Functions and uses. *Chem. Soc. Rev.* **2005**, *34*, 821–836. [CrossRef]
29. Buerkle, L.E.; Rowan, S.J. Supramolecular gels formed from multi-component low molecular weight species. *Chem. Soc. Rev.* **2012**, *41*, 6089–6102. [CrossRef]
30. Aida, T.; Meijer, E.W.; Stupp, S.I. Functional supramolecular polymers. *Science* **2012**, *335*, 813–817. [CrossRef]
31. Stupp, S.I.; Hartgerink, J.D.; Beniash, E. Self-assembly and mineralization of peptide-amphiphile nanofibers. *Science* **2001**, *294*, 1684–1688.
32. Gui, S.; Huang, Y.; Zhu, Y.; Jin, Y.; Rui, Z. Biomimetic sensing system for tracing pb^{2+} distribution in living cells based on the metal-peptide supramolecular assembly. *ACS Appl. Mater. Interfaces* **2019**, *11*, 5804–5811. [CrossRef] [PubMed]

33. Shah, V.B.; Ferris, C.; Orf, G.; Kavadiya, S.; Ray, J.R.; Jun, Y.S.; Lee, B.; Blankenship, R.E.; Biswas, P. Supramolecular self-assembly of bacteriochlorophyll c molecules in aerosolized droplets to synthesize biomimetic chlorosomes. *J. Photochem. Photobiol. B* **2018**, *185*, 161–168. [CrossRef] [PubMed]
34. Sanjeeva, K.B.; Pigliacelli, C.; Gazzera, L.; Dichiarante, V.; Bombelli, F.B.; Metrangolo, P. Halogen bond-assisted self-assembly of gold nanoparticles in solution and on a planar surface. *Nanoscale* **2019**, *11*, 18407–18415. [CrossRef] [PubMed]
35. Shirman, T.; Arad, T.; van der Boom, M.E. Halogen bonding: A supramolecular entry for assembling nanoparticles. *Angew. Chem. Int. Ed.* **2010**, *49*, 926–929. [CrossRef] [PubMed]
36. Bent, H.A. Structural chemistry of donor-acceptor interactions. *Chem. Rev.* **1968**, *68*, 587–648. [CrossRef]
37. Braga, D.; Grepioni, F.; Desiraju, G.R. Crystal engineering and organometallic architecture. *Chem. Rev.* **1998**, *98*, 1375–1406. [CrossRef]
38. Aakery, C.B.; Champness, N.R.; Janiak, C. Recent advances in crystal engineering. *CrystEngComm* **2009**, *12*, 22–43. [CrossRef]
39. Cannon, A.S.; Warner, J.C. Noncovalent derivatization: Green chemistry applications of crystal engineering. *Cryst. Growth Des.* **2002**, *2*, 255–257. [CrossRef]
40. Teyssandier, J.; Mali, K.S.; De Feyter, S. Halogen bonding in two-dimensional crystal engineering. *ChemistryOpen* **2020**, *9*, 225–241. [CrossRef]
41. Aakery, C.B.; Panikkattu, S.; Chopade, P.D.; Desper, J. Competing hydrogen-bond and halogen-bond donors in crystal engineering. *CrystEngComm* **2013**, *15*, 3125–3136. [CrossRef]
42. Messina, M.T.; Metrangolo, P.; Panzeri, W.; Ragg, E.; Resnati, G. Perfluorocarbon-hydrocarbon self-assembly. part 3. liquid phase interactions between perfluoroalkylhalides and heteroatom containing hydrocarbons. *Tetrahedron Lett.* **1998**, *39*, 9069–9072. [CrossRef]
43. Metrangolo, P.; Meyer, F.; Pilati, T.; Resnati, G.; Terraneo, G. Halogen bonding in supramolecular chemistry. *Angew. Chem. Int. Ed.* **2010**, *47*, 6114–6127. [CrossRef] [PubMed]
44. Metrangolo, P.; Murray, J.S.; Pilati, T.; Politzer, P.; Resnati, G.; Terraneo, G. Fluorine-centered halogen bonding: A factor in recognition phenomena and reactivity. *Cryst. Growth Des.* **2011**, *11*, 4238–4246. [CrossRef]
45. Metrangolo, P.; Murray, J.S.; Pilati, T.; Politzer, P.; Resnati, G.; Terraneo, G. The fluorine atom as a halogen bond donor, viz. a positive site. *CrystEngComm* **2011**, *13*, 6593–6896. [CrossRef]
46. Politzer, P.; Murray, J.S.; Clark, T. Halogen bonding: An electrostatically-driven highly directional noncovalent interaction. *Phys. Chem. Chem. Phys.* **2010**, *12*, 7748–7757. [CrossRef]
47. Mitzel, N.W.; Blake, A.J.; Rankin, D.W.H. Beta-donor bonds in sion units: An inherent structure-determining property leading to (4+4)-coordination in tetrakis-(n,n-dimethylhydroxylamido)silane. *J. Am. Chem. Soc.* **1997**, *119*, 4143–4148. [CrossRef]
48. Hennemann, M.; Murray, J.S.; Politzer, P.; Riley, K.E.; Clark, T. Polarization-induced σ-holes and hydrogen bonding. *J. Mol. Model.* **2012**, *18*, 2461–2469. [CrossRef]
49. Politzer, P.; Murray, J.S. A unified view of halogen bonding, hydrogen bonding and other σ-hole interactions. In *Noncovalent Forces*; Scheiner, S., Ed.; Springer: Cham, Switzerland, 2015; Volume 19, pp. 291–321.
50. Murray, J.S.; Politzer, P. Hydrogen bonding: A coulombic σ-hole interaction. *J. Indian Inst. Sci.* **2020**, *100*, 21–30. [CrossRef]
51. Sakurai, T.; Sundaralingam, M.; Jeffrey, G.A. A nuclear quadrupole resonance and X-ray study of the crystal structure of 2,5-dichloroaniline. *Acta Crystallogr.* **2010**, *16*, 354–363. [CrossRef]
52. Desiraju, G.R.; Parthasarathy, R. The nature of halogen⋯halogen interactions: Are short halogen contacts due to specific attractive forces or due to close packing of non-spherical atoms? *J. Am. Chem. Soc.* **1989**, *111*, 8725–8726. [CrossRef]
53. Silly, F.; Viala, C.; Bonvoisin, J. Two-dimensional halogen-bonded porous self-assembled nanoarchitectures of copper β-diketonato complexes. *J. Phys. Chem. C* **2018**, *122*, 17143–17148. [CrossRef]
54. Chang, M.H.; Jang, W.J.; Lee, M.W.; Jeon, U.S.; Han, S.; Kahng, S.J. Networks of non-planar molecules with halogen bonds studied using scanning tunneling microscopy on Au (111). *Appl. Surf. Sci.* **2018**, *432*, 110–114. [CrossRef]
55. Silly, F.; Shaw, A.Q.; Castell, M.R.; Briggs, G.A.D. A chiral pinwheel supramolecular network driven by the assembly of ptcdi and melamine. *Chem. Commun.* **2008**, 1907–1909. [CrossRef] [PubMed]

56. Dai, H.; Wang, S.; Hisaki, I.; Nakagawa, S.; Ikenaka, N.; Deng, K.; Xiao, X.; Zeng, Q.D. On-surface self-assembly of a c 3-symmetric π-conjugated molecule family studied by STM: Two-dimensional nanoporous frameworks. *Chem. Asian J.* **2017**, *5*, 2558–2564. [CrossRef] [PubMed]
57. Qiu, X.; Chen, W.; Zeng, Q.; Bo, X.; Bai, C. Alkane-assisted adsorption and assembly of phthalocyanines and porphyrins. *J. Am. Chem. Soc.* **2000**, *122*, 5550–5556. [CrossRef]
58. Silly, F. Selecting two-dimensional halogen-halogen bonded self-assembled 1,3,5-tris(4-iodophenyl)benzene porous nanoarchitectures at the solid–liquid interface. *J. Phys. Chem. C* **2013**, *117*, 20244–20249. [CrossRef]
59. Peyrot, D.; Silly, F. On-surface synthesis of two-dimensional covalent organic structures versus halogen-bonded self-assembly: Competing formation of organic nanoarchitectures. *ACS Nano* **2016**, *10*, 5490–5498. [CrossRef]
60. Gatti, R.; Macleod, J.M.; Lipton-Duffin, J.A.; Moiseev, A.G.; Perepichka, D.F.; Rosei, F. Substrate, Molecular structure, and solvent effects in 2d self-assembly via hydrogen and halogen bonding. *J. Phys. Chem. C* **2014**, *118*, 25505–25516. [CrossRef]
61. Cavallo, G.; Metrangolo, P.; Pilati, T.; Resnati, G.; Terraneo, G. Halogen bond: A long overlooked interaction. In *Halogen Bonding I. Topics in Current Chemistry*; Metrangolo, P., Resnati, G., Eds.; Springer: Cham, Switzerland, 2014; Volume 358, pp. 1–17.
62. Binnig, G.; Rohrer, H. The scanning tunneling microscope. *Sci. Am.* **1985**, *253*, 50–56. [CrossRef]
63. Park, S.I.; Quate, C.F. Scanning tunneling microscope. *Rev. Sci. Instrum.* **1987**, *58*, 2010–2017. [CrossRef]
64. Goronzy, D.P.; Ebrahimi, M.; Rosei, F.; Arramel; Fang, Y.; De Feyter, S.; Tait, S.L.; Perepichka, D.F. Supramolecular assemblies on surfaces: Nanopatterning, functionality, and reactivity. *ACS Nano* **2018**, *12*, 7445–7481. [CrossRef] [PubMed]
65. Mate, E. Scientific Conferences: A big hello to halogen bonding. *Nat. Chem.* **2014**, *6*, 762–764.
66. Xing, L.; Jiang, W.; Huang, Z.; Liu, J.; Song, H.; Zhao, W.; Dai, J.; Zhu, H.; Wang, Z.; Weiss, P.S. Steering two-dimensional porous networks with σ-hole interactions of Br···S and Br···Br. *Chem. Mater.* **2019**, *31*, 3041–3048. [CrossRef]
67. Mukherjee, A.; Teyssandier, J.; Hennrich, G.; De Feyter, S.; Mali, K.S. Two-dimensional crystal engineering using halogen and hydrogen bonds: Towards structural landscapes. *Chem. Sci.* **2017**, *8*, 3759–3769. [CrossRef]
68. Zheng, Q.N.; Liu, X.H.; Chen, T.; Yan, H.J.; Cook, T. Formation of halogen bond-based 2d supramolecular assemblies by electric manipulation. *J. Am. Chem. Soc.* **2015**, *137*, 6128–6131. [CrossRef]
69. Yasuda, S.; Furuya, A.; Murakoshi, K. Control of a two-dimensional molecular structure by cooperative halogen and hydrogen bonds. *RSC Adv.* **2014**, *4*, 58567–58572. [CrossRef]
70. Gutzler, R.; Fu, C.; Dadvand, A.; Hua, Y.; Macleod, J.M.; Rosei, F.; Perepichka, D.F. Halogen Bonds in 2d Supramolecular self-assembly of organic semiconductors. *Nanoscale* **2012**, *4*, 5965–5971. [CrossRef]
71. Li, J.X.; Wu, J.T.; Chen, S.W.; Miao, X.R.; Fabien, S.; Deng, W.L. Geometry symmetry of conjugated cores along C–Br bond effect on the 2d self-assembly by intermolecular H···Br and Br···Br Bonds. *J. Phys. Chem. C* **2018**, *433*, 1075–1082. [CrossRef]
72. Blanco, M.A.; Martín, P.A.; Francisco, E. Interacting quantum atoms: A correlated energy decomposition scheme based on the quantum theory of atoms in molecules. J. Chem. *Theory Comput.* **2005**, *1*, 1096–1109. [CrossRef]
73. Popelier, P.L.A. Molecular similarity and complementarity based on the theory of atoms in molecules. In *Molecular Similarity in Drug Design*; Dean, P.M., Ed.; Springer: Dordrecht, The Netherlands, 1995; Volume 358, pp. 215–240.
74. Wu, Y.C.; Li, J.X.; Yuan, Y.; Dong, M.Q.; Zha, B.; Miao, X.R.; Hu, Y.; Deng, W.L. Halogen bonding versus hydrogen bonding induced 2d self-assembled nanostructures at the liquid-solid interface revealed by STM. *Phys. Chem. Chem. Phys.* **2016**, *19*, 3143–3150. [CrossRef] [PubMed]
75. Li, Y.; Lei, L.; Subramani, R.; Pan, Y.; Bo, L.; Yang, Y.; Chen, W.; Mamdouh, W.; Besenbacher, F.; Dong, M. Building layer-by-layer 3d supramolecular nanostructures at the terephthalic acid/stearic acid interface. *Chem. Commun.* **2011**, *47*, 9155–9157. [CrossRef] [PubMed]
76. Nagula, R.G.; Bolton, O.; Burgess, E.C.; Matzger, A.J. The unprecedented size of the σ-holes on 1,3,5-triiodo-2,4,6-trinitrobenzene begets unprecedented intermolecular interactions. *Cryst. Growth Des.* **2016**, *16*, 1765–1771.

77. Wu, J.T.; Li, J.X.; Dong, M.Q.; Miao, K.; Miao, X.R.; Wu, Y.C.; Deng, W.L. Solvent effect on host-guest two-dimensional self-assembly mediated by halogen bonding. *J. Phys. Chem. C* **2018**, *122*, 22597–22604. [CrossRef]
78. Zha, B.; Li, J.X.; Wu, J.T.; Miao, X.R.; Zhang, M. Cooperation and competition of hydrogen and halogen bonds in 2d self-assembled nanostructures based on bromine substituted coumarins. *New J. Chem.* **2019**, *43*, 17182–17187. [CrossRef]
79. Zha, B.; Miao, X.R.; Li, Y.; Liu, P.; Deng, W.L. Solvent-dependent self-assembly of 4,7-dibromo-5,6-bis(octyloxy)benzo[c][1,2,5] thiadiazole on graphite surface by scanning tunneling microscopy. *J. Nanomater.* **2013**, *2013*, 1–7. [CrossRef]
80. Hu, T.Z.; Wang, Y.J.; Dong, M.Q.; Wu, J.T.; Pang, P.; Miao, X.R.; Deng, W.L. Ordering self-assembly structures via intermolecular Brs interactions. *Phys. Chem. Chem. Phys.* **2020**, *22*, 1437–1443. [CrossRef]
81. Dong, M.Q.; Miao, K.; Wu, J.T.; Miao, X.R.; Li, J.X.; Pang, P.; Deng, W.L. Halogen substituent effects on concentration-controlled self-assembly of fluorenone derivatives: Halogen bond versus hydrogen bond. *J. Phys. Chem. C* **2019**, *123*, 4349–4359. [CrossRef]
82. Dong, M.Q.; Wu, J.T.; Miao, X.R.; Li, J.X.; Deng, W.L. Bromine substituent position triggered halogen versus hydrogen bond in 2d self-assembly of fluorenone derivatives. *J. Phys. Chem. C* **2019**, *123*, 26191–26200. [CrossRef]
83. Dong, M.Q.; Hu, T.Z.; Wang, Y.; Pang, P.; Wang, Y.J.; Miao, X.R.; Deng, W.L. Halogen-bonded building block for 2D self-assembly: Triggered by hydrogen-bonding motifs relative to the terminal functions of the side chains. *Appl. Surf. Sci.* **2020**, *515*, 145983–145992. [CrossRef]
84. Hu, X.Y.; Zha, B.; Wu, Y.C.; Miao, X.R.; Deng, W.L. Effects of the position and number of bromine substituents on the concentration-mediated 2d self-assembly of phenanthrene derivatives. *Phys. Chem. Chem. Phys.* **2016**, *18*, 7208–7215. [CrossRef] [PubMed]
85. Pang, P.; Miao, X.R.; Ying, L.; Kong, G.; Deng, W.L. Halogen-bond-controlled self-assembly of regioisomeric phenanthridine deriatives into nanowires and nanosheets. *J. Phys. Chem. C* **2020**, *124*, 5665–5671. [CrossRef]
86. Zha, B.; Dong, M.Q.; Miao, X.R.; Peng, S.; Wu, Y.C.; Miao, K.; Hu, Y.; Deng, W.L. Cooperation and competition between halogen bonding and van der waals forces in supramolecular engineering at the aliphatic hydrocarbon/graphite interface: Position and number of bromine group effects. *Nanoscale* **2016**, *9*, 237–250. [CrossRef] [PubMed]
87. Zha, B.; Miao, X.R.; Liu, P.; Wu, Y.C.; Deng, W.L. Concentration dependent halogen-bond density in the 2d self-assembly of a thienophenanthrene derivative at the aliphatic acid/graphite interface. *Chem. Commun.* **2014**, *50*, 9003–9006. [CrossRef]
88. Zhang, Y.; Luo, Y.; Zhang, Y.; Yu, Y.J.; Kuang, Y.M.; Zhang, L.; Meng, Q.S.; Luo, Y.; Yang, J.L.; Dong, Z.C. Visualizing coherent intermolecular dipole–dipole coupling in real space. *Nature* **2016**, *531*, 623–627. [CrossRef]
89. Zha, B.; Dong, M.Q.; Miao, X.; Miao, K.; Hu, Y.; Wu, Y.C.; Xu, L.; Deng, W.L. Controllable orientation of ester-group-induced intermolecular halogen bonding in a 2d self-assembly. *J. Phys. Chem. Lett.* **2016**, *7*, 3164–3170. [CrossRef]
90. Gutzler, R.; Ivasenko, O.; Fu, C.; Brusso, J.L.; Rosei, F.; Perepichka, D.F. Halogen bonds as stabilizing interactions in a chiral self-assembled molecular monolayer. *Chem. Commun.* **2011**, *47*, 9453–9455. [CrossRef]
91. Miao, X.R.; Li, J.X.; Zha, B.; Miao, K.; Dong, M.Q.; Wu, J.T.; Deng, W.L. Concentration-dependent multiple chirality transition in halogen-bond-driven 2d self-assembly process. *Appl. Surf. Sci.* **2018**, *433*, 1075–1082. [CrossRef]
92. Wu, J.T.; Li, J.X.; Miao, X.R.; Ying, L.; Dong, M.Q.; Deng, W.L. The brmidline horizontal ellipsis π-halogen bond assisted self-assembly of an asymmetric molecule regulated by concentration. *Chem. Commun.* **2020**, *56*, 2727–2730. [CrossRef]

Publisher's Note: MDPI stays neutral with regard to jurisdictional claims in published maps and institutional affiliations.

© 2020 by the authors. Licensee MDPI, Basel, Switzerland. This article is an open access article distributed under the terms and conditions of the Creative Commons Attribution (CC BY) license (http://creativecommons.org/licenses/by/4.0/).

Article

Regium Bonds between Silver(I) Pyrazolates Dinuclear Complexes and Lewis Bases (N_2, OH_2, NCH, SH_2, NH_3, PH_3, CO and CNH)

Ibon Alkorta [1,*], Cristina Trujillo [2], Goar Sánchez-Sanz [3,4] and José Elguero [1]

1. Instituto de Química Médica, CSIC, Juan de la Cierva, 3, E-28006 Madrid, Spain; iqmbe@iqm.csic.es
2. School of Chemistry, Trinity Biomedical Sciences Institute, Trinity College, Dublin, 152–160 Pearse St, Dublin 2, Ireland; trujillc@tcd.ie
3. Irish Centre of High-End Computing, Grand Canal Quay, Dublin 2, Ireland; goar.sanchez@ichec.ie
4. School of Chemistry, University College Dublin, Belfield, Dublin 4, Ireland
* Correspondence: ibon@iqm.csic.es

Received: 28 January 2020; Accepted: 19 February 2020; Published: 24 February 2020

Abstract: A theoretical study and Cambridge Structural Database (CSD) search of dinuclear Ag(I) pyrazolates interactions with Lewis bases were carried out and the effect of the substituents and ligands on the structure and on the aromaticity were analyzed. A relationship between the intramolecular Ag–Ag distance and stability was found in the unsubstituted system, which indicates a destabilization at longer distances compensated by ligands upon complexation. It was also observed that the asymmetrical interaction with phosphines as ligands increases the Ag–Ag distance. This increase is dramatically higher when two simultaneous PH_3 ligands are taken into account. The calculated ^{109}Ag chemical shielding shows variation up to 1200 ppm due to the complexation. Calculations showed that six-membered rings possessed non-aromatic character while pyrazole rings do not change their aromatic character significantly upon complexation.

Keywords: non-covalent interactions; regium bonds; silver(I); coinage metals; pyrazolates; phosphines

1. Introduction

Non-covalent interactions (NCIs) are present in complexes formed between two or more Lewis acids and Lewis bases. In fact, those interactions are commonly named by the Lewis acid [1]. On the one hand, a Lewis base (LB) or a LB motif is associated with a region of space where there exists an excess of negative charge (i.e., electron density) in the proximity of an atom or atoms within a molecule. This is predominant in anions and in some neutral molecules such as those that exhibit lone pairs (LP: carbenes, amines, phosphines, N-oxides, etc.), multiple bonds (olefins, acetylenes, benzenes and other aromatic molecules), single bonds (alkanes, dihydrogen, etc.), radicals, metals (rarely), etc. On the other hand, a Lewis acid (LA) is associated with a region of space where there is an excess of positive charge (in other words a deficit of negative charge or electron deficiency) in the proximity of an atom or atoms in a molecule. This can be found in cations, molecules or atoms exhibiting σ- and π-holes and metals (frequently). The concept of a σ-hole was introduced by Politzer et al. [2–5] to describe regions of positive potential along the vector of a covalent bond. It was latter extended to other situations, for example, the π-hole [6–11] (positive electrostatic potential perpendicular to an atom of a molecular framework) and lone pair hole (similar to σ-hole but along the atom–lone pair direction) [12–14]. The maximum value of the molecular electrostatic potential within a given molecular density isosurface surface, $V_{S,max}$, has also been classified based on the nature of the orbitals (s-, p- and d- orbitals) and it is associated with the corresponding deficiencies (σ_s-, σ_p- and σ_d-holes) [15].

There is a wide variety of non-covalent interactions due to most groups of the periodic table being associated with a certain type of NCI: starting from the archetypal hydrogen bond (HB) there are "alkali bonds" (group 1) [16,17], "alkaline earth bonds" (group 2) [18–20], "regium bonds" (groups 10 and 11) [21–23], "triel bonds" (group 13) [24], "tetrel bonds" (group 14) [25–28], "pnictogen (also called pnicogen) bonds" (group 15) [29–31], "chalcogen bonds" (group 16) [32–35], "halogen bonds" (group 17) [36,37], and "aerogen bonds" (group 18) [38]. Regium (or coinage-metal bonds) bonds involve mainly coinage metals, Cu, Ag and Au. These weak interactions are particularly interesting since they are associated with organometallic chemistry. However, it is necessary to clearly differentiate between clusters (e.g., Au_2 or Ag_{11}) [22,39] and molecules (e.g., AuX) [40,41]. Recent published works show an increased interest in regium bonds, some example of which are discussed here [42–45].

In this article, the dinuclear silver(I) pyrazolates (Scheme 1) have been considered following a previous paper on trinuclear silver(I) pyrazolates [46]. The effect of the ligands on the silver atom and the substituents on the pyrazole rings on the structure, energetic, electronic and magnetic properties have been analyzed.

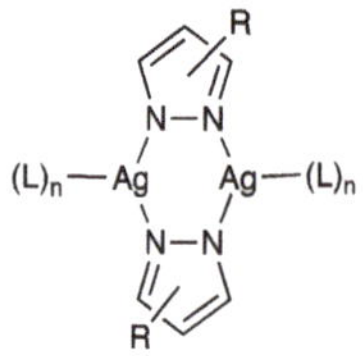

R = H, 4-NO₂,	$(PzAg)_2$	$(PzAg)_2(PH_3)_2$	$(4NO_2pzAg)_2(PH_3)$
3,5-diMe, 4-Cl,	$(PzAg)_2(N_2)_2$	$(PzAg)_2(CO)_2$	$(4NO_2pzAg)_2(PH_3)_2$
	$(PzAg)_2(OH_2)_2$	$(PzAg)_2(CNH)_2$	$(4NO_2pzAg)_2(PH_3)_4$
L = N₂, OH₂, NCH,	$(PzAg)_2(NCH)_2$	$(PzAg)_2(PH_3)$	$(DMepzAg)_2$
SH₂, NH₃, PH₃, CNH	$(PzAg)_2(SH_2)_2$	$(PzAg)_2(PH_3)_4$	$(4ClpzAg)_2$
	$(PzAg)_2(NH_3)_2$	$(4NO_2pzAg)_2$	$(DMepzAg)_2(PH_3)_2$
n = 0, 1, 2, 4			$(4ClpzAg)_2(PH_3)_2$

Scheme 1. General structure of the complexes under study and their corresponding names.

2. Materials and Methods

2.1. Cambridge Structural Database Search

The Cambridge Structural Database (CSD) [47] version 5.40 with updates from November 2018, May 2019, and August 2019 was searched for systems with cyclic (Pz-Ag)₂ structure. The geometrical characteristics of these systems have been analysed.

2.2. Ab Initio Calculations

The geometry of the systems have been fully optimized using the MP2 computational method [48–51] and the jul-cc-pVDZ basis set [52–54] for all the atoms except for the silver atoms, of which the aug-cc-pVDZ-PP effective core potential basis set [55] has been used. Frequency calculations have been carried out at the same computational level to verify that the obtained structures correspond to energetic minima. Dissociation energies have been obtained as the difference between the sum of the energies of the isolated monomers in their optimized geometry and the energy of the complex. These calculations have been carried out with the Gaussian-16 program [56].

The molecular electrostatic potential on the 0.001 au electron density isosurface (MESP) was calculated with the Gaussian16 program, analyzed with the Multiwfn program [57] and represented with the JMol program [58].

The topological analysis of the electron density was carried out by means of the quantum theory of atoms in molecules (QTAIM) method [59–62] using the AIMAll program [63]. This method identifies the points of space where the gradient of the electron density vanishes (critical points) and based on the number of positive curvatures characterizes them as nuclear attractors (3, −3), bond critical points (3, −1), ring critical points (3, +1) or cage critical points (3, +3). By connecting the bond critical points with the nuclear attractor, following the minimum gradient path, the molecular graph is obtained.

The natural bond orbital (NBO) method [64] was employed using the NBO-3 program to evaluate atomic charges, and to analyze charge-transfer interactions between occupied and unoccupied orbitals.

The TOPMOD program [65] has been used to analyze the areas of electron concentration in terms of the electron localization function (ELF) [66]. For the three-dimensional plots, a convenient ELF value of 0.7 was used [67].

Relativistic-corrected NMR chemical shielding values for the geometries optimized at MP2 level were obtained using the relativistic ZORA spin–orbit Hamiltonian [68,69], the BP86 functional [70–72], and the full electron QZ4P basis [73]. In addition, the nuclear-independent chemical shift (NICS) [74] at the centre and 1 Å above the centre of the ring formed by the two Ag atoms and the pyrazole moieties were calculated to study the aromaticity of all systems. In addition, a scan of the NICS values from 0 to 2 Å in steps of 0.25 Å for (PzAg)$_2$ was calculated [75,76]. These calculations have been performed with the ADF-2017 program [77].

3. Results and Discussion

In this paper we will report the study of nineteen compounds corresponding to the general formula depicted in Scheme 1 and named by a simple code that allows to easy identification: (R-pzAg)$_2$(L)$_n$. Firstly, the substituents (R) on the pyrazole ring are indicated: Pz, 4NO$_2$pz, DMepz and 4Clpz for the unsubstituted (H), 4-nitro, 3,5-dimethyl, and 4-chloro derivatives, respectively. Secondly, the ligands (L) interacting with the silver atoms are indicated. Thus, for instance, (4NO$_2$pzAg)$_2$(PH$_3$)$_4$ corresponds to the (4-nitropyrazole:Ag)$_2$ cyclic structure with four phosphines interacting with the silver atoms (two phosphines per each silver atom).

3.1. CSD Search

A search in the Cambridge Structural Database has been carried out and the resulting compounds with the structure represented in Scheme 1 were summarized and reported in Table 1 ordered by increasing Ag–Ag intramolecular distance. Two simplified views of these structures can be found in Table S1 of the Supplementary Materials. Structures with dinuclear silver(I) pyrazolates without ligands were not found within the CSD search.

As observed by the crystallographic data, the Ag–Ag distance ranges between 4.305 Å (ZIGSEQ) and 3.392 Å (FINWOR). The shortest distances correspond to complexes with only two ligands present concomitantly with the 3,5-bis-CF$_3$ substituent on the pyrazole. It is also clear that the larger the number of ligands, the longer the intramolecular Ag–Ag distance. Moreover, it is noteworthy that, in structures with four phosphines ligands, the Ag–Ag distance increases as the substitutes of the pyrazole are H < 4-Cl < 4-NO$_2$. These results will be compared against our theoretical calculations in the following sections.

Concerning the conformation of the six-membered rings in the crystal structures (similar to those of 1,4-cyclo-hexadiene, 9,10-dihydroanthracenes, phenothiazines, etc.), very little energy differences between those conformations in the crystals were found, in agreement what was proposed by Rasika Dias [78]. Regarding the tetrahedral configuration of Ag(I) atom, we have used Houser's τ_4 index [79] as recommended by Raptis [80]. This index is defined as:

$$\tau_4 = [360 - (\alpha + \beta)]/141 \tag{1}$$

where, α and β correspond to the two largest angles of the six angles around the tetrahedral silver atom. The values of τ_4 range from 1 (perfect tetrahedral geometry, i.e., $\tau_4 = [360 - (109.5 + 109.5)]/141 = 1$) to 0 (perfect square planar geometry). The τ_4 values obtained for three substituents, 0.9 (RATFEX and ZIGSIU) are very similar to those for 4 substituents (ZIGROZ and ZIGSEQ, 0.89 and 0.88 respectively) with both identical Ag atoms. The two lowest values (0.78) correspond to a substituent that differs from PPh$_3$ (see notes (a) and (b) of Table 1).

Table 1. Results of the Cambridge Structural Database (CSD) search. Abbreviations: nature of the pyrazolate anion, nature of the ligand (L), number of ligands; Ag(I)–Ag(I) distance in Å, silver–ligand distance in Å, conformation of the six-membered ring and tetrahedral configuration of the Ag (τ_4 index).

Refcode	Pyrazolate	Ligand (L)	N° of Ls	Ag–Ag	Ag–L	Conformation	τ_4 index
FINWOR	3,5-bis-CF_3	3,5-bis-CF_3-pyrazole-Ag	2	3.392	2.250	Nearly planar	—
PIRCUR	3,5-bis-CF_3	PPh_3	2	3.425	2.343	Boat	—
PIRCUR	3,5-bis-CF_3	PPh_3	2	3.425	2.414	Boat	—
FINWIL	3,5-bis-CF_3	PPh_3	2	3.479	2.366	Twisted boat	—
IPIGOD	3,5-bis-CF_3	2,4,6-collidine	2	3.499	2.277	Boat + flattened chair	—
IPIGOD	3,5-bis-CF_3	2,4,6-collidine	2	3.499	2.254	Boat + lattened chair	—
IPIGOD	3,5-bis-CF_3	2,4,6-collidine	2	3.502	2.269	Boat + flattened chair	—
IPIGOD01	3,5-bis-CF_3	2,4,6-collidine	2	3.562	2.253	Half boat	—
IPIGOD01	3,5-bis-CF_3	2,4,6-collidine	2	3.562	2.253	Half boat	—
SOJCIG	3,5-bis-CF_3	3,5-bis-CF_3-pyrazole-Ag	2	3.632	3.218	Chair	—
SOJCIG	3,5-bis-CF_3	3,5-bis-CF_3-pyrazole-Ag	2	3.632	2.258	Chair	—
RATFEX	Parent	PPh_3	3	3.706	2.370	Flattened chair	0.90
RATFEX	Parent	PPh_3	3	3.706	2.484	Flattened chair	
RATFEX	Parent	PPh_3	3	3.706	2.461	Flattened chair	
ZIGRUF	4-Cl	PPh_3	3	3.707	2.374	Flattened boat	0.83
ZIGRUF	4-Cl	PPh_3	3	3.707	2.475	Flattened boat	
ZIGRUF	4-Cl	PPh_3	3	3.707	2.485	Flattened boat	
ZIGSIU	4-NO_2	PPh_3	3	3.827	2.379	Boat	0.90
ZIGSIU	4-NO_2	PPh_3	3	3.827	2.495	Boat	
ZIGSIU	4-NO_2	PPh_3	3	3.827	2.469	Boat	
RATFAT	Parent	PPh_3	2	3.870	2.376	Planar	—
KIRXIV	Parent	(a)	2 (4)	3.900	2.532	Planar	0.78
KIRXIV	Parent	PPh_3	4	3.900	2.539	Planar	
ZIGSAM	4-Cl	(b): PR_3	4	4.205	2.460	Twisted	0.78
ZIGSAM	4-Cl	PR_3	4	4.205	2.534	Twisted	
ZIGROZ	4-Cl	PPh_3	4	4.209	2.484	Twisted	0.89
ZIGROZ	4-Cl	PPh_3	4	4.209	2.500	Twisted	
ZIGSEQ	4-NO_2	PPh_3	4	4.305	2.502	Twisted	0.88
ZIGSEQ	4-NO_2	PPh_3	4	4.305	2.487	Twisted	
max	—	—		4.305	3.218	—	—
min	—	—		3.392	2.250	—	—

(a) Two P ligands that corresponds to four P atoms. — 1,2-bis(diphenylphosphaneyl)benzene

(b) A P ligand in an adamantane type molecule: PTA = 1,3,5-triaza-7-phosphaadamantane

3.2. Electronic Properties of the Isolated (PzAg)₂ System

The isolated (PzAg)$_2$ system calculated at MP2 level presents a planar structure with D_{2h} symmetry, where the silver atoms are equidistant to the pyrazole rings with an intramolecular Ag–Ag distance of 2.801 Å (Table S2). This intramolecular distance is slightly larger than that of the Ag$_2$ cluster (2.514 Å) calculated at the same computational level. In fact, its molecular graph (Figure 1a) shows the presence of an Ag–Ag bond with the corresponding bond critical point. However, other electronic analysis carried out: ELF (Figure 1b) and NBO, do not indicate the presence of such bonds between the silver atoms due to the absence of the corresponding basin concomitantly with the Wiberg bond index of the Ag–Ag contact in (PzAg)$_2$—being 0.01, while that of Ag$_2$ is 0.92.

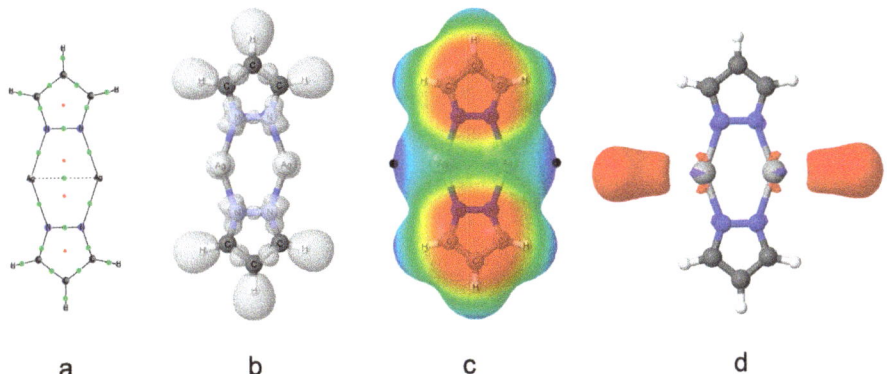

Figure 1. (a) Molecular graph, (b) electron localization function (ELF) at 0.7 isosurface value, (c) molecular electrostatic potential on the 0.001 au electron density isosurface (MESP) and (d) LUMO orbital of the (PzAg)$_2$ system. The green and red dots in the molecular graph indicate the location of the bond and ring critical points. The range of colors used in the MESP are red (MESP < −0.02 a.u.) and blue (MESP > 0.03 a.u.). The locations of the $V_{S,max}$ in the MESP are indicated with black dots.

The molecular electrostatic potential on the 0.001 au electron density isosurface (MESP) of the (PzAg)$_2$ system (Figure 1c) shows negative values above and below the pyrazole ring while the MESP in the extension of the CH and Ag–Ag bonds are positive. Moreover, the calculated maximum value on the MESP, $V_{S,max}$, in the vicinity of the silver atoms is +0.03 a.u. (marked with black dots) being the most positive value in the whole isosurface. Interestingly the LUMO orbital (Figure 1d) is solely associated to the Ag atoms and located in a region that coincides with the location of the $V_{S,max}$. Thus, the silver atoms of this molecule should act as a Lewis acid using both the electrostatic and orbital criteria.

3.3. The (PzAg)$_2$ Free Complex and the Effect of the Ag–Ag Distance on its Stability

The complex without ligands, (PzAg)$_2$, has in its minimum energy structure a distance between both silver atoms of 2.801 Å (Table S2) which is considerably shorter than any of the Ag–Ag distances found in the CSD. As we have already indicated in the CSD search, this distance is very sensitive to the environment (substituents of the pyrazole and ligands of the silver atoms) and can vary almost 1.0 Å from the shortest to the longest distance (3.392 and 4.305 Å, respectively). In order to explore the energetic penalty due to the elongation of the intramolecular Ag–Ag distance, the (PzAg)$_2$ system has been optimized while keeping fixed the Ag–Ag distance from 2.6 to 4.0 Å in steps of 0.2 Å (Table S2 and Figure 2). The energetic results indicate that lengthening the distance 1 Å, up to 3.8 Å, considerably decreases the stability of the system by 75 kJ·mol^{-1} which should be compensated by the interactions with the ligands.

3.4. Effect of the Ligands and Substituents on the Structure and Dissociation Energy (De)

After studying the free (PzAg)$_2$ system, the complexes with two ligands simultaneously interacting with the (PzAg)$_2$ system have been optimized, i.e., each silver atom interacts with a single ligand. The minimum structures obtained show that the interacting atom of the ligand is coplanar with the plane defined by the (PzAg)$_2$ system. The molecular graphs of two illustrative examples are shown in Figure 3 and the Cartesian coordinates of all of them were summarized in Table S3.

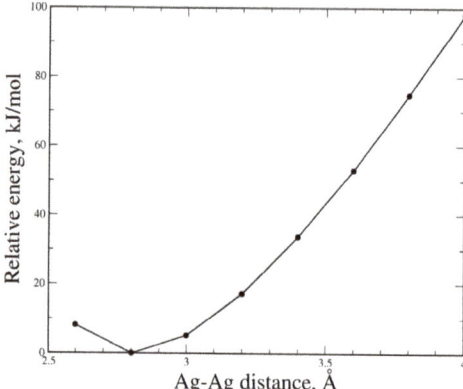

Figure 2. Relative energy vs. Ag–Ag distance in the (PzAg)$_2$ system.

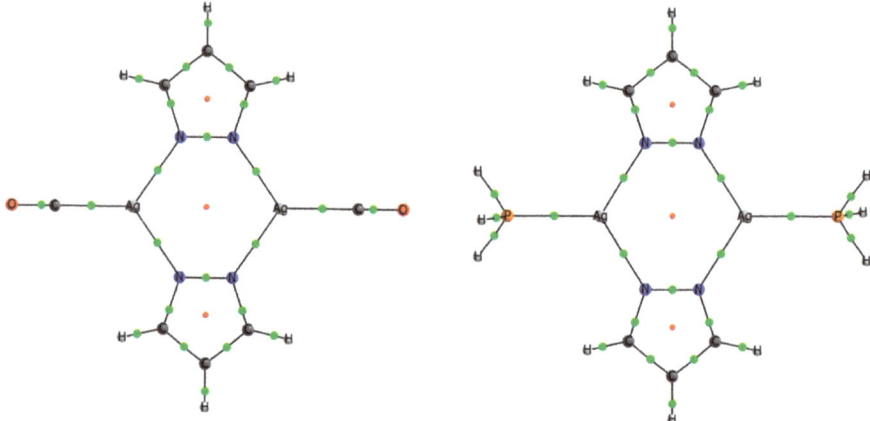

Figure 3. Molecular graph of the (PzAg)$_2$(CO)$_2$ and (PzAg)$_2$(PH$_3$)$_2$ complexes. The green and red dots indicate the location of the bond and ring critical points, respectively.

The complexation with ligands produces an elongation of the Ag–Ag distance up to 1.1 Å (Table 2) in agreement with the large variety of structures and the range of Ag–Ag distances found in the CSD search. For example, the crystal structure RATPAT (L = (PPh$_3$)$_2$) shows an Ag–Ag distance of 3.870 Å while in the calculated (PzAg)$_2$(PH$_3$)$_2$ complex this value is 3.689 Å. It is known that ligands coordinated by O, C or P atoms are strong, while those ligands coordinated by N atoms are weak. In the cases of CNH and HCN ligands are both -donors and -acceptors but the former is coordinated by the C atom while the latter is coordinated by the N atom. This results in larger dissociation energies for the (PzAg)$_2$(CNH)$_2$ complex in comparison with (PzAg)$_2$(NCH)$_2$. Similarly, this happens with CO and NH$_3$ ligands (both -donors), but while CO is a good -acceptor NH$_3$ is not, resulting in smaller values of De. In the case of the PH$_3$ ligand, it is both a -donor and -acceptor ligand, which is in agreement with the (PzAg)$_2$(PH$_3$)$_2$ complex, showing the second largest De. Chalcogen ligands, OH$_2$ (-donor) and SH$_2$ (-donor) present similar trends. In addition, it is also observed that the distance between both pyrazole rings decreases with the elongation of the Ag–Ag distance.

Table 2. Geometries and dissociation energy of (PzAg)$_2$ and (PzAg)$_2$L$_2$ systems.

Compound	Ag–Ag dist. (Å)	Pz-Pz dist. (Å) [a]	Ag–Z dist. (Å) [b]	De (kJ·mol^{-1})
(PzAg)$_2$	2.801	3.968	—	0.0
(PzAg)$_2$(N$_2$)$_2$	2.851	3.968	2.837	22.7
(PzAg)$_2$(OH$_2$)$_2$	2.908	3.994	2.630	54.5
(PzAg)$_2$(NCH)$_2$	3.078	3.933	2.425	56.3
(PzAg)$_2$(SH$_2$)$_2$	3.196	3.889	2.631	71.2
(PzAg)$_2$(NH$_3$)$_2$	3.255	3.905	2.358	91.7
(PzAg)$_2$(PH$_3$)$_2$	3.689	3.657	2.373	122.9
(PzAg)$_2$(CO)$_2$	3.845	3.519	1.998	90.0
(PzAg)$_2$(CNH)$_2$	3.880	3.534	2.018	136.6

[a] Measured as the distance between the smallest N–N distance in two pyrazole rings. [b] Z is the atom of L directly interacting with the silver atoms.

But, does this have any impact on the dissociation energy? Or, in other words, is there any relationship between the Ag–Ag intramolecular distance and the dissociation energy? The dissociation energy values corresponding to the (PzAg)$_2$L$_2$ complexes range between 23 kJ mol^{-1} for L = N$_2$ to 137 kJ mol^{-1} for L = CNH (Table 2). In general, despite observing a trend between the dissociation energy and the elongation of the Ag–Ag distance (Figure S1), no good fitting has been found (logarithmic fitting, R^2 = 0.73). The only outlier corresponds to the (PzAg)$_2$(CO)$_2$ complex, and when this point is neglected the fitting is more evident (R^2 = 0.90). A better exponential relationship has been found for the distance between the pyrazole rings vs. De (De = 5.55·e$^{-0.114Pz-Pz}$, R^2 = 0.92) which may suggest that the repulsion between pyrazole groups is partially responsible for the increase of the De energy but somehow compensated by the Ag–Ligand interaction.

The effect of the substituents (pyrazole ring) within the Ag–Ag distance and dissociation energies have been explored by considering four different pyrazole derivatives, R = H, 3,5-di(CH$_3$), 4-Cl and 4-NO$_2$. These isolated (R-pzAg)$_2$ systems have been fully optimized as well as in the presence of two phosphine molecules ligands with the results gathered in Table 3.

Table 3. Effect of the 3,5-dimethyl, 4-chloro and 4-nitro substituents (pyrazole ring) within the Ag–Ag distance and on the dissociation energy, De.

Compound	Ag–Ag dist. (Å)	De (kJ·mol^{-1})
(DMepzAg)$_2$	2.793	—
(PzAg)$_2$	2.801	—
(4ClpzAg)$_2$	2.819	—
(4NO$_2$pzAg)$_2$	2.841	—
(DMepzAg)$_2$(PH$_3$)$_2$	3.656	120.0
(PzAg)$_2$(PH$_3$)$_2$	3.689	122.9
(4ClpzAg)$_2$(PH$_3$)$_2$	3.730	135.1
(4NO$_2$pzAg)$_2$(PH$_3$)$_2$	3.782	150.8

In relation to the Ag–Ag distance in the isolated (R-pzAg)$_2$ systems, it increases in the following order: 3,5-di(CH$_3$) < H < 4-Cl < 4-NO$_2$ with a variation of 0.05 Å between both extremes. The same order was found within the (R-pzAg)$_2$(PH$_3$)$_2$ complexes but with a slightly larger range of 0.12 Å. This evolution was also found in the CSD search for complexes with three PPh$_3$ ligands: RATFEX (R = H) (3.706 Å) < ZIGRUF (R = 4-Cl) (3.707 Å) < ZIGSIU (R = 4-NO$_2$) (3.827 Å), and for complexes with four PPh$_3$ ligands: KIRVIX (R = H) (3.900 Å) < ZIGROG (R = 4-Cl) (4.209 Å) < ZIGSEQ (R = 4-NO$_2$) (4.305 Å).

The dissociation energies for the (R-pzAg)$_2$(PH$_3$)$_2$ complexes increase following the same trend of Ag–Ag distances (Table 3 and Figure S2). Furthermore, linear correlation with R^2 value of 0.97 was obtained between Ag–Ag distances and the De.

So far, only the systems with no ligands, (R-pzAg)$_2$, and complexes with two identical ligands, (R-pzAg)$_2$L$_2$, have been studied. In Table 4, the results corresponding to the complexes with 0, 1, 2 and 4 phosphine ligands bonded to (PzAg)$_2$ and (4NO$_2$pzAg)$_2$ systems are shown. In both series, the intramolecular Ag–Ag distance increases with the number of phosphine ligands bonded. When one PH$_3$ is bonded the increase in the Ag–Ag distance with respect to the isolated system is 0.286 and 0.311 Å for (PzAg)$_2$ and (4NO$_2$pzAg)$_2$ systems respectively. However, the change is two times larger when moving from 1 to 2 simultaneous ligands. Finally, the change is again moderate when 4 PH$_3$ ligands interacting with each Ag are considered. The CDS search also shows an increase of the Ag–Ag distance with the number of ligands. For example in the unsubstituted complexes (R=H) the Ag–Ag distance in RATFET (L = 3) is 3.706 Å and increases to 3.900 Å with four ligands (KIRXIV). This is also observed for substituted complexes with R=4-Cl: ZIGRUF (L = 3) 3.707 Å to ZIGROZ (L = 4) 4.209 Å and for R = 4NO$_2$: ZIGSIU (L = 3) 3.827 Å) to ZIGSEQ (L = 4) 4.305 Å. In addition, it is interesting to notice that in complexes with four phosphines, the (R-PzAg)$_2$ system adopts a chair conformation (Figure 4) vs. the planar one observed with one or two ligands. The experimental Ag–Ag distances are longer than the calculated ones; this could be due to the fact that the ligand found in the crystals are bulkier (for instance PPh$_3$ vs. PH$_3$). This can be also related with the ratio of σ-donor/π-acceptor capacity in phosphorus ligands. In principle P(CH$_3$)$_3$ is a better σ-donor than PH$_3$, while the latter is a better π-acceptor [81]. However, the σ-donor/π-acceptor ratio indicates that the P(CH$_3$)$_3$ is a stronger ligand than PH$_3$. The same can be expected for PPh$_3$ and therefore the Ag–Ag distance will be larger for complexes that are PPh$_3$-coordinated compared with those for PH$_3$ ones.

Table 4. Effect of the number of ligands.

Compound	Number of Ls	Ag–Ag dist. (Å)	De (kJ·mol^{-1})	τ$_4$ index
(PzAg)$_2$	0	2.801	0.0	—
(PzAg)$_2$(PH$_3$)	1	3.087	64.2	—
(PzAg)$_2$(PH$_3$)$_2$	2	3.689	122.9	—
(PzAg)$_2$(PH$_3$)$_4$	4	3.798	214.5	0.81 & 0.83
(4NO$_2$pzAg)$_2$	0	2.841	0.0	—
(4NO$_2$pzAg)$_2$(PH$_3$)	1	3.152	77.4	—
(4NO$_2$pzAg)$_2$(PH$_3$)$_2$	2	3.782	150.8	—
(4NO$_2$pzAg)$_2$(PH$_3$)$_4$	4	3.870	251.7	0.80 & 0.81

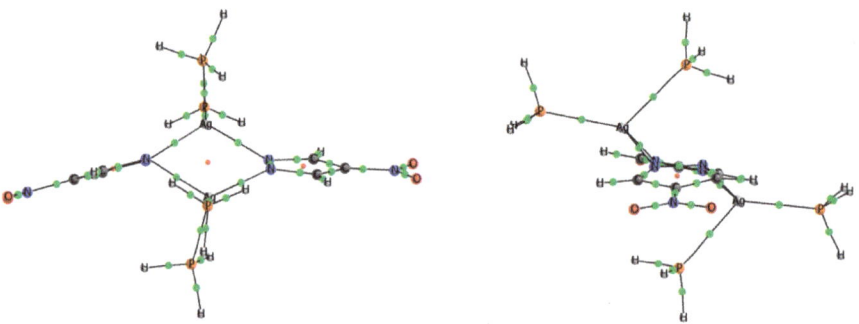

Figure 4. Two orthogonal views of the molecular graph of the (4NO$_2$pzAg)$_2$(PH$_3$)$_4$ complex.

Regarding the τ$_4$ index for (PzAg)$_2$(PH$_3$)$_4$ and (4NO$_2$pzAg)$_2$(PH$_3$)$_4$: the values for both Ag are 0.81 and 0.83 respectively, in between of those found for ZIGSAM (0.78) and ZIGROZ (0.89) crystal structures.

The values of De show an anticooperativity effect as the number of phosphines increases, thus the De's of the systems with two phosphines are smaller than twice the corresponding ones with one phosphine and the same happens when the results between the systems with two and four phosphines are compared. Looking at the values shown in Table 4, (PzAg)$_2$(PH$_3$) complex yields a De of 64.2 kJ·mol^{-1}, whereas (PzAg)$_2$(PH$_3$)$_2$ complex's is 122.9 kJ·mol^{-1}, 5.5 kJ·mol^{-1} smaller than twice the corresponding value for (PzAg)$_2$(PH$_3$). This is an indication of an anti-cooperativity effect, which is more evident when (PzAg)$_2$(PH$_3$)$_4$ complex is taken into account (~42 kJ·mol^{-1} less than four times the De of (PzAg)$_2$(PH$_3$)). This is also observed for (4NO$_2$pzAg)$_2$ and the corresponding complexes, but the differences, i.e., the anti-cooperativity, is rather smaller (4.0 kJ·mol^{-1}) for (4NO$_2$pzAg)$_2$(PH$_3$)$_2$ and slightly larger for (4NO$_2$pzAg)$_2$(PH$_3$)$_4$ (57.9 kJ·mol^{-1}).

3.5. Electron Density

The critical points of the electron density of all systems have been characterized using the quantum theory of atoms in molecules (QTAIM) method. As aforementioned, a bond critical point (BCP) is obtained for the (PzAg)$_2$ system in between both Ag atoms. A similar feature has been obtained for all the systems with Ag–Ag distances shorter than 3.3 Å. The electron density values, ρ_{BCP}, of this BCP (Table S4) range between 0.032 and 0.014 a.u., with positive values of the Laplacian, $\nabla^2\rho$ (between 0.098 and 0.041 a.u.) and negative values of the total energy density, H_{BCP}, except for the system with the largest Ag–Ag bond in this set ((PzAg)$_2$(NH$_3$)$_2$). Excellent exponential relationships are obtained between the ρ_{BCP} and $\nabla^2\rho_{BCP}$ and the interatomic distance in agreement with previous reports (Figure 5) [82,83].

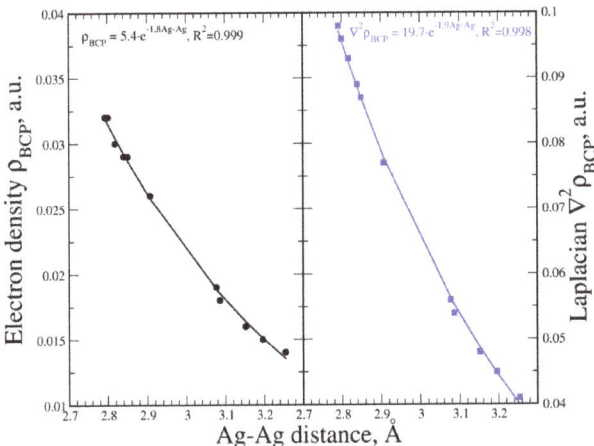

Figure 5. Electron density, ρ_{BCP}, and Laplacian, $\nabla^2\rho_{BCP}$, vs. the Ag–Ag distance (Å). The fitted exponential relationships are shown.

Concerning the Ag–L bonds, the corresponding bond critical points between the silver atom and the different ligands have been gathered in Table S5. In all the cases, the interactions exhibit positive values of the $\nabla^2\rho_{BCP}$ and negative values of H_{BCP}, which indicates a partial covalent character of the bond formed [84,85]. The only exception corresponds to the weakest complex, (PzAg)$_2$(N$_2$)$_2$, which shows a small positive H_{BCP} value. The 14 unique Ag–P contacts found in this set show similar relationships between ρ_{BCP} and $\nabla^2\rho_{BCP}$ vs. the Ag–P distance to those previously mentioned for Ag–Ag BCPs.

3.6. Magnetic Properties and Aromaticity

Among the different nuclei suitable for NMR spectra present in these systems (^{1}H, ^{13}C and ^{15}N), ^{109}Ag is the one with the largest range of chemical shifts. The calculated ^{109}Ag chemical shielding for all the systems studied in this article are listed in Table 5. It is worth noting that upon complexation ^{109}Ag can change its chemical shielding by more than 1200 ppm to lower field, from 3765 ppm ((PzAg)$_2$) to 2550.48 ppm ((PzAg)$_2$(PH$_3$)$_4$). Furthermore, a good relationship between the chemical shielding and the intramolecular Ag–Ag distance (only for complexes with two ligands) was found (Figure S3). Unfortunatelly, there are no experimental data on (PzAg)$_2$ compounds but a recent report on (PzAg)$_3$ derivatives show that the methodology used here provides δ^{109}Ag values within 10 ppm of the experimental ones [46].

Table 5. ^{109}Ag absolute chemical shielding (σ, ppm) and nuclear-independent chemical shift (NICS) values (ppm) of the six-membered and pyrazole rings.

System	σ^{109}Ag (ppm)	6-Membered Ring		Pz Ring	
		NICS(0)	NICS(1)	NICS(0)	NICS(1)
(PzAg)$_2$	3765.7	−7.39	−1.84	−13.06	−10.97
(PzAg)$_2$(N$_2$)$_2$	3648.3	−7.17	−1.70	−12.70	−10.95
(PzAg)$_2$(OH$_2$)$_2$	3752.8	−5.82	−1.48	−12.88	−11.22
(PzAg)$_2$(NCH)$_2$	3515.5	−4.76	−1.10	−12.67	−11.03
(PzAg)$_2$(SH$_2$)$_2$	3288.3	−3.55	−0.65	−12.75	−11.04
(PzAg)$_2$(NH$_3$)$_2$	3511.9	−2.40	−0.38	−12.74	−11.12
(PzAg)$_2$(PH$_3$)$_2$	2900.9	−0.14	0.12	−12.27	−11.11
(PzAg)$_2$(CO)$_2$	2706.9	−0.56	−0.33	−12.37	−11.13
(PzAg)$_2$(CNH)$_2$	2825.4	−0.16	−0.12	−12.29	−11.16
(DMepzAg)$_2$	3720.4	−7.98	−2.23	−10.50	−9.34
(4ClpzAg)$_2$	3815.9	−7.11	−1.75	−12.62	−9.98
(4NO$_2$pzAg)$_2$	3835.5	−6.95	−1.72	−11.98	−9.50
(DMepzAg)$_2$(PH$_3$)$_2$	2893.1	−0.73	−0.25	−9.32	−9.29
(4ClpzAg)$_2$(PH$_3$)$_2$	2954.4	−0.17	0.07	−11.74	−10.09
(4NO$_2$pzAg)$_2$(PH$_3$)$_2$	3010.2	0.12	0.19	−10.81	−9.41
(PzAg)$_2$(PH$_3$)$_4$	2561.6				
	2550.5				
(4NO$_2$pzAg)$_2$(PH$_3$)$_4$	2619.5				
	2627.7				

In order to explore the potential aromaticity of the six membered ring formed by the nitrogen atoms of the pyrazoles and the two silver atoms, the NICS(0) and NICS(1) have been calculated (Table 5). Despite NICS isotropic values being widely used and well established, there is still a controversy about the reliability of NICS values for assessment of the aromaticity of certain molecules [86,87]. Nevertheless, and following our previous experience, the isotropic values have in several cases been shown to present an accurate description of the aromatic behaviour in poly-aromatic systems [88–90].

Despite almost all the systems studied presenting negative NICS(0) values in the six-membered ring, only those with short Ag–Ag distances (2.8–3.0 Å) present very negative values close to the benzene molecule (−8 ppm) [91], and, also, those NICS(0) values decrease in absolute value as the Ag–Ag distance increases. NICS(1) are smaller, in absolute value, than NICS(0), but follow the same trend as the latter. Also, NICS(1) are very small compared with benzene ones (−10.2 ppm) [91] which suggest non-aromatic character. But, those values should be taken carefully, since the two silver atoms are very close and the proximity of the nuclei may affect the NICS(0) measure. To provide a further insight on this, Figure 6 clearly shows that there is a unique dependence between the NICS(0) and NICS(1), the distance between the location where the NICS is measured and the silver atom. The scan of the NICS values for (PzAg)$_2$ from 0 to 2.0 Å above the centre of the six-membered ring (Table S6) have also been plotted in Figure 6 showing a similar evolution to the NICS(0) and NICS(1) of the rest of the molecules. This indicates that, as aforementioned, this ring is not aromatic but the NICS values obtained are somehow affected by the proximity of the silver atom.

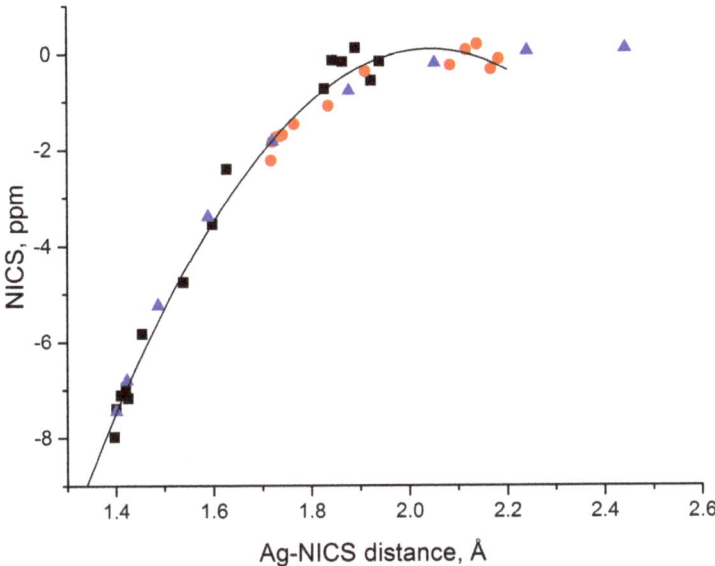

Figure 6. NICS(0), black squares, and NICS(1) values, red dots, of the six-membered ring versus the distance between the silver atom and the location where the NICS were measured (Ag–NICS distance). The blue triangles representes the NICS values for the (PzAg)$_2$ scan between 0.0 and 2.0 Å. The second order polynomial equation fitting the values of the NICS(0) and NICS(1) has an R^2 value of 0.989.

In contrast the NICS(0) and NICS(1) values of the pyrazole rings are negative and large in all cases and very close to the ones obtained by Kusakiewicz-Dawid (−13.5 and −11.4 ppm respectively) [92].

On the other hand, the substituents on the pyrazole ring have a greater effect on the NICS values than the different ligands on the (PzAg)$_2$L$_2$ complexes, most likely due to changes in the electron density on the ring current electrons. For instance the NICS(0) in the (Rpz-Ag)$_2$ systems ranges between −10.5 and −13.1 ppm while in the (PzAg)$_2$L$_2$ complexes it is between −12.3 and −13.1 ppm.

4. Conclusions

A theoretical study and CSD search of the different effects provoked by substituents and ligands upon complexation with dinuclear Ag(I) pyrazolates has been carried out and the structural, energetic, electron density and magnetic features analyzed.

The CSD search shows a great variability of the Ag–Ag distance in the crystal structures. These results have been rationalized based on the number of ligand interactions with the Ag atoms and the substituents of the pyrazole ring.

In the isolated (PzAg)$_2$ system, it was observed that for Ag–Ag the longer the distance, the lesser the stability of the unsubstituted complex with no ligands. This decrease in the stability is somehow compensated by the ligands upon complexation.

Furthermore, complexation with ligands through the Ag atoms increases the intramolecular distance Ag–Ag. In fact, considering the PH$_3$ ligand, the increase of the Ag–Ag distance was found moderate when going from no ligand to one ligand. However, when two simultaneous PH$_3$ are interacting, the Ag–Ag distance increases dramatically. Nevertheless, when four PH$_3$ are considered, the increase is again moderate.

In terms of the QTAIM analysis, it is noteworthy the presence of a BCP between both Ag atoms among all the systems with Ag-Ag distances shorter than 3.3 Å.

Finally, regarding the aromatic/non-aromatic properties, six-membered rings containing the Ag–Ag motif show negative NICS values but those reveal a non-aromatic character mainly affected by the proximity of the Ag nuclei. This was confirmed by the relationship found between the NICS values and the Ag–NICS distance. On the other hand, pyrazole rings maintain their aromatic behaviour with slight changes.

Supplementary Materials: The following are available online at http://www.mdpi.com/2073-4352/10/2/137/s1, Table S1: Two simplified views of the structures found in the CSD search, Table S2. Effect of the Ag(I)–Ag(I) distance (Å) on the relative energy (kJ·mol^{-1}) of the (PzAg)$_2$ system, Table S3: Cartesian coordinates (Å) of the optimized systems at MP2/jul-cc-pVDZ/jul-cc-pVDZ-PP level, Figure S1: Relative energy vs. Ag–Ag distance in the (PzAg)$_2$ system, Figure. S2: De (kJ·mol^{-1}) vs. Ag–Ag dist. (Å) in the (PzAg)$_2$L$_2$ complexes, Figure S3: De (kJ·mol^{-1}) vs. Ag–Ag dist. (Å) in the (R-pzAg)$_2$(PH$_3$)$_2$ complexes, Table S4: Distance (Å) and electron density properties (au) of the Ag–Ag BCPs, Table S5: Distance (Å) and electron density properties (au) of the Ag–L BCPs, Figure S4: ^{109}Ag chemical shielding vs. Ag–Ag distance in the (PzAg)$_2$L$_2$ complexes. Table S6. NICS values (ppm) from 0.0 to 2.0 Å of the center of the 6-membered ring in (PzAg)$_2$.

Author Contributions: Conceptualization and project design, I.A., J.E.; resources, I.A., C.T., G.S.-S, data curation, I.A, J.E, C.T., G.S.-S., writing—original draft preparation, I.A, J.E, C.T., G.S.-S.; writing—review and editing, I.A, J.E, C.T., G.S.-S.; funding acquisition, I.A. C.T.; All authors have read and agreed to the published version of the manuscript.

Funding: This work was carried out with financial support from the Spanish Ministerio de Ciencia, Innovación y Universidades (Projects PGC2018-094644-B-C2) and Dirección General de Investigación e Innovación de la Comunidad de Madrid (PS2018/EMT-4329 AIRTEC-CM), and Science Foundation of Ireland (SFI), grant number 18/SIRG/5517.

Acknowledgments: We are grateful to the Irish Centre for High-End Computing (ICHEC) and CTI(CSIC) for the provision of computational facilities. We are also grateful to Dr Lee O'Riordan for his continuous support and help.

Conflicts of Interest: The authors declare no conflict of interest.

References

1. Cavallo, G.; Metrangolo, P.; Pilati, T.; Resnati, G.; Terraneo, G. Naming Interactions from the Electrophilic Site. *Cryst. Growth Des.* **2014**, *14*, 2697–2702. [CrossRef]
2. Politzer, P.; Murray, J.S. *An Overview of σ-Hole Bonding, an Important and Widely-Occurring Noncovalent Interaction Practical Aspects of Computational Chemistry*; Leszczynski, J., Shukla, M.K., Eds.; Springer: Dordrecht, The Netherlands, 2010; pp. 149–163.
3. Politzer, P.; Murray, J.S.; Clark, T. Halogen bonding: An electrostatically-driven highly directional noncovalent interaction. *PCCP* **2010**, *12*, 7748–7757. [CrossRef] [PubMed]
4. Politzer, P.; Murray, J.S.; Clark, T. Halogen bonding and other [sigma]-hole interactions: A perspective. *PCCP* **2013**, *15*, 11178–11189. [CrossRef] [PubMed]
5. Bauzá, A.; Mooibroek, T.J.; Frontera, A. The Bright Future of Unconventional σ/π-Hole Interactions. *ChemPhysChem* **2015**, *16*, 2496–2517. [CrossRef] [PubMed]
6. Murray, J.; Lane, P.; Clark, T.; Riley, K.; Politzer, P. σ-Holes, π-holes and electrostatically-driven interactions. *J. Mol. Model.* **2012**, *18*, 541–548. [CrossRef] [PubMed]
7. Solimannejad, M.; Ramezani, V.; Trujillo, C.; Alkorta, I.; Sánchez-Sanz, G.; Elguero, J. Competition and Interplay between σ-Hole and π-Hole Interactions: A Computational Study of 1:1 and 1:2 Complexes of Nitryl Halides (O2NX) with Ammonia. *J. Phys. Chem. A* **2012**, *116*, 5199–5206. [CrossRef]
8. Grabowski, S.J. Hydrogen bonds, and σ-hole and π-hole bonds—Mechanisms protecting doublet and octet electron structures. *PCCP* **2017**, *19*, 29742–29759. [CrossRef]
9. Bauzá, A.; Mooibroek, T.J.; Frontera, A. Directionality of π-holes in nitro compounds. *Chem. Commun.* **2015**, *51*, 1491–1493. [CrossRef]
10. Zierkiewicz, W.; Michalczyk, M.; Wysokiński, R.; Scheiner, S. On the ability of pnicogen atoms to engage in both σ and π-hole complexes. Heterodimers of ZF2C6H5 (Z = P, As, Sb, Bi) and NH3. *J. Mol. Model.* **2019**, *25*, 152. [CrossRef]
11. Azofra, L.; Alkorta, I.; Scheiner, S. Noncovalent interactions in dimers and trimers of SO3 and CO. *Theor. Chem. Acc.* **2014**, *133*, 1–8. [CrossRef]

12. Blanco, F.; Alkorta, I.; Rozas, I.; Solimannejad, M.; Elguero, J. A theoretical study of the interactions of NF3 with neutral ambidentate electron donor and acceptor molecules. *PCCP* **2011**, *13*, 674–683. [CrossRef] [PubMed]
13. Alkorta, I.; Elguero, J.; Del Bene, J.E. Exploring the PX3:NCH and PX3:NH3 potential surfaces, with X=F, Cl, and Br. *Chem. Phys. Lett.* **2015**, *641*, 84–89. [CrossRef]
14. Bauzá, A.; Mooibroek, T.J.; Frontera, A. σ-Hole Opposite to a Lone Pair: Unconventional Pnicogen Bonding Interactions between ZF3 (Z=N, P, As, and Sb) Compounds and Several Donors. *ChemPhysChem* **2016**, *17*, 1608–1614. [CrossRef] [PubMed]
15. Stenlid, J.H.; Brinck, T. Extending the σ-Hole Concept to Metals: An Electrostatic Interpretation of the Effects of Nanostructure in Gold and Platinum Catalysis. *JACS* **2017**, *139*, 11012–11015. [CrossRef]
16. Shahi, A.; Arunan, E. Hydrogen bonding, halogen bonding and lithium bonding: An atoms in molecules and natural bond orbital perspective towards conservation of total bond order, inter- and intra-molecular bonding. *PCCP* **2014**, *16*, 22935–22952. [CrossRef]
17. Solimannejad, M.; Rabbani, M.; Ahmadi, A.; Esrafili, M.D. Cooperative and diminutive interplay between the sodium bonding with hydrogen and dihydrogen bondings in ternary complexes of NaC3N with HMgH and HCN (HNC). *Mol. Phys.* **2014**, *112*, 2017–2022. [CrossRef]
18. Yáñez, M.; Sanz, P.; Mó, O.; Alkorta, I.; Elguero, J. Beryllium Bonds, Do They Exist? *J. Chem. Theor. Comput.* **2009**, *5*, 2763–2771. [CrossRef]
19. Alkorta, I.; Legon, A.C. Non-Covalent Interactions Involving Alkaline-Earth Atoms and Lewis Bases B: An ab Initio Investigation of Beryllium and Magnesium Bonds, B·MR2 (M = Be or Mg, and R = H, F or CH3). *Inorganics* **2019**, *7*, 35. [CrossRef]
20. Montero-Campillo, M.M.; Mó, O.; Yáñez, M.; Alkorta, I.; Elguero, J. Chapter Three—The beryllium bond. In *Advances in Inorganic Chemistry*; van Eldik, R., Puchta, R., Eds.; Academic Press: Cambridge, MA, USA, 2019; Volume 73, pp. 73–121.
21. Frontera, A.; Bauzá, A. Regium–π bonds: An Unexplored Link between Noble Metal Nanoparticles and Aromatic Surfaces. *Chem. Eur. J.* **2018**, *24*, 7228–7234. [CrossRef]
22. Zierkiewicz, W.; Michalczyk, M.; Scheiner, S. Regium bonds between Mn clusters (M = Cu, Ag, Au and n = 2–6) and nucleophiles NH3 and HCN. *PCCP* **2018**, *20*, 22498–22509. [CrossRef]
23. Legon, A.C.; Walker, N.R. What's in a name? 'Coinage-metal' non-covalent bonds and their definition. *PCCP* **2018**, *20*, 19332–19338. [CrossRef] [PubMed]
24. Grabowski, S.J. Triel Bonds, π-Hole-π-Electrons Interactions in Complexes of Boron and Aluminium Trihalides and Trihydrides with Acetylene and Ethylene. *Molecules* **2015**, *20*, 11297–11316. [CrossRef] [PubMed]
25. Alkorta, I.; Rozas, I.; Elguero, J. Molecular Complexes between Silicon Derivatives and Electron-Rich Groups. *J. Phys. Chem. A* **2001**, *105*, 743–749. [CrossRef]
26. Alkorta, I.; Blanco, F.; Elguero, J.; Dobado, J.A.; Ferrer, S.M.; Vidal, I. Carbon·Carbon Weak Interactions. *J. Phys. Chem. A* **2009**, *113*, 8387–8393. [CrossRef] [PubMed]
27. Bauzá, A.; Mooibroek, T.J.; Frontera, A. Tetrel-Bonding Interaction: Rediscovered Supramolecular Force? *Angew. Chem. Int. Ed.* **2013**, *52*, 12317–12321. [CrossRef] [PubMed]
28. Grabowski, S.J. Tetrel bond-[sigma]-hole bond as a preliminary stage of the SN2 reaction. *Phys. Chem. Chem. Phys.* **2014**, *16*, 1824–1834. [CrossRef]
29. Del Bene, J.E.; Alkorta, I.; Elguero, J. The Pnicogen Bond in Review: Structures, Binding Energies, Bonding Properties, and Spin-Spin Coupling Constants of Complexes Stabilized by Pnicogen Bonds. In *Noncovalent Forces*; Scheiner, S., Ed.; Springer International Publishing: Cham, Switzerland, 2015; pp. 191–263. [CrossRef]
30. Zahn, S.; Frank, R.; Hey-Hawkins, E.; Kirchner, B. Pnicogen Bonds: A New Molecular Linker? *Chem. Eur. J.* **2011**, *17*, 6034–6038. [CrossRef]
31. Scheiner, S. The Pnicogen Bond: Its Relation to Hydrogen, Halogen, and Other Noncovalent Bonds. *Acc. Chem. Res.* **2013**, *46*, 280–288. [CrossRef]
32. Minyaev, R.M.; Minkin, V.I. Theoretical study of O- > X (S, Se, Te) coordination in organic compounds. *Can. J. Chem.* **1998**, *76*, 776–788. [CrossRef]
33. Wang, W.; Ji, B.; Zhang, Y. Chalcogen Bond: A Sister Noncovalent Bond to Halogen Bond. *J. Phys. Chem. A* **2009**, *113*, 8132–8135. [CrossRef]
34. Sanz, P.; Yáñez, M.; Mó, O. Competition between X···H···Y Intramolecular Hydrogen Bonds and X····Y (X = O, S, and Y = Se, Te) Chalcogen–Chalcogen Interactions. *J. Phys. Chem. A* **2002**, *106*, 4661–4668. [CrossRef]

35. Azofra, L.M.; Alkorta, I.; Scheiner, S. Strongly bound noncovalent (SO3)n:H2CO complexes (n = 1, 2). *Phys. Chem. Chem. Phys.* **2014**, *16*, 18974–18981. [CrossRef] [PubMed]
36. Legon, A.C. Prereactive Complexes of Dihalogens XY with Lewis Bases B in the Gas Phase: A Systematic Case for the Halogen Analogue B···XY of the Hydrogen Bond B···HX. *Angew. Chem. Int. Ed.* **1999**, *38*, 2686–2714. [CrossRef]
37. Cavallo, G.; Metrangolo, P.; Milani, R.; Pilati, T.; Priimagi, A.; Resnati, G.; Terraneo, G. The Halogen Bond. *Chem. Rev.* **2016**, *116*, 2478–2601. [CrossRef] [PubMed]
38. Bauzá, A.; Frontera, A. Aerogen Bonding Interaction: A New Supramolecular Force? *Angew. Chem. Int. Ed.* **2015**, *54*, 7340–7343. [CrossRef] [PubMed]
39. Halldin Stenlid, J.; Johansson, A.J.; Brinck, T. σ-Holes and σ-lumps direct the Lewis basic and acidic interactions of noble metal nanoparticles: Introducing regium bonds. *PCCP* **2018**, *20*, 2676–2692. [CrossRef]
40. Li, Q.; Li, H.; Li, R.; Jing, B.; Liu, Z.; Li, W.; Luan, F.; Cheng, J.; Gong, B.; Sun, J. Influence of Hybridization and Cooperativity on the Properties of Au-Bonding Interaction: Comparison with Hydrogen Bonds. *J. Phys. Chem. A* **2011**, *115*, 2853–2858. [CrossRef]
41. Sánchez-Sanz, G.; Alkorta, I.; Elguero, J.; Yáñez, M.; Mó, O. Strong interactions between copper halides and unsaturated systems: New metallocycles? Or the importance of deformation. *PCCP* **2012**, *14*, 11468–11477. [CrossRef]
42. Zhang, J.; Wang, Z.; Liu, S.; Cheng, J.; Li, W.; Li, Q. Synergistic and diminutive effects between triel bond and regium bond: Attractive interactions between π-hole and σ-hole. *Appl. Organomet. Chem.* **2019**, *33*, e4806. [CrossRef]
43. Sánchez-Sanz, G.; Trujillo, C.; Alkorta, I.; Elguero, J. Understanding Regium Bonds and their Competition with Hydrogen Bonds in Au2:HX Complexes. *ChemPhysChem* **2019**, *20*, 1572–1580. [CrossRef]
44. Wang, R.; Yang, S.; Li, Q. Coinage-Metal Bond between [1.1.1]Propellane and M2/MCl/MCH3 (M = Cu, Ag, and Au): Cooperativity and Substituents. *Molecules* **2019**, *24*, 2601. [CrossRef] [PubMed]
45. Zheng, B.; Liu, Y.; Wang, Z.; Zhou, F.; Liu, Y.; Ding, X.; Lu, T. Regium bonds formed by MX (M = Cu, Ag, Au; X = F, Cl, Br) with phosphine-oxide/phosphinous acid: Comparisons between oxygen-shared and phosphine-shared complexes. *Mol. Phys.* **2019**, *117*, 2443–2455. [CrossRef]
46. Alkorta, I.; Elguero, J.; Dias, H.V.R.; Parasar, D.; Martín-Pastor, M. An experimental and computational NMR study of organometallic nine-membered rings: Trinuclear silver(I) complexes of pyrazolate ligands. *Magn. Reson. Chem.* **2020**. [CrossRef] [PubMed]
47. Allen, F. The Cambridge Structural Database: A quarter of a million crystal structures and rising. *Acta Crystallogr. Sect. B Struct. Sci.* **2002**, *58*, 380–388. [CrossRef]
48. Pople, J.A.; Binkley, J.S.; Seeger, R. Theoretical models incorporating electron correlation. *Int. J. Quantum Chem.* **1976**, *10*, 1–19. [CrossRef]
49. Krishnan, R.; Pople, J.A. Approximate fourth-order perturbation theory of the electron correlation energy. *Int. J. Quantum Chem.* **1978**, *14*, 91–100. [CrossRef]
50. Bartlett, R.J.; Silver, D.M. Many-body perturbation theory applied to electron pair correlation energies. I. Closed-shell first-row diatomic hydrides. *J. Chem. Phys.* **1975**, *62*, 3258–3268. [CrossRef]
51. Bartlett, R.J.; Purvis, G.D. Many-body perturbation theory, coupled-pair many-electron theory, and the importance of quadruple excitations for the correlation problem. *Int. J. Quantum Chem.* **1978**, *14*, 561–581. [CrossRef]
52. Del Bene, J.E. Proton affinities of ammonia, water, and hydrogen fluoride and their anions: A quest for the basis-set limit using the Dunning augmented correlation-consistent basis sets. *J. Phys. Chem.* **1993**, *97*, 107–110. [CrossRef]
53. Dunning, T.H. Gaussian-Basis Sets for Use in Correlated Molecular Calculations. I. The Atoms Boron through Neon and Hydrogen. *J. Chem. Phys.* **1989**, *90*, 1007–1023. [CrossRef]
54. Woon, D.E.; Dunning, T.H. Gaussian basis sets for use in correlated molecular calculations. V. Core-valence basis sets for boron through neon. *J. Chem. Phys.* **1995**, *103*, 4572–4585. [CrossRef]
55. Peterson, A.K.; Puzzarini, C. Systematically convergent basis sets for transition metals. II. Pseudopotential-based correlation consistent basis sets for the group 11 (Cu, Ag, Au) and 12 (Zn, Cd, Hg) elements. *Theor. Chem. Acc.* **2005**, *114*, 283–296. [CrossRef]

56. Frisch, M.J.T.; Schlegel, H.B.; Scuseria, G.E.; Robb, M.A.; Cheeseman, J.R.; Scalmani, G.; Barone, V.; Petersson, G.A.; Nakatsuji, H.; Li, X.; et al. *Gaussian 16, Revision A.03*; Gaussian Inc.: Wallingford, CT, USA, 2016.
57. Lu, T.; Chen, F. Multiwfn: A multifunctional wavefunction analyzer. *J. Comput. Chem.* **2012**, *33*, 580–592. [CrossRef] [PubMed]
58. Jmol: An Open-Source Java Viewer for Chemical Structures in 3D. Available online: http://www.jmol.org/ (accessed on 24 February 2020).
59. Bader, R.F.W. *Atoms in Molecules: A Quantum Theory*; Clarendon Press: Oxford, UK, 1990.
60. Bader, R.F.W. A quantum theory of molecular structure and its applications. *Chem. Rev.* **1991**, *91*, 893–928. [CrossRef]
61. Popelier, P.L.A. *Atoms in Molecules An Introduction*; Prentice Hall: Harlow, UK, 2000.
62. Matta, C.F.; Boyd, R.J. *The Quantum Theory of Atoms in Molecules: From Solid State to DNA and Drug Design*; Wiley-VCH: Weinheim, Germany, 2007.
63. Keith, T.A. *AIMAll, 17.11.14 B*; Version 17.11.14 B; TK Gristmill Software: Overland Park, KS, USA, 2017.
64. Reed, A.E.; Curtiss, L.A.; Weinhold, F. Intermolecular Interactions from a Natural Bond Orbital, Donor-Acceptor Viewpoint. *Chem. Rev.* **1988**, *88*, 899–926. [CrossRef]
65. Noury, S.; Krokidis, X.; Fuster, F.; Silvi, B. *TopMod Package*; Universite Pierre et Marie Curie: Paris, France, 1997.
66. Silvi, B.; Savin, A. Classification of chemical bonds based on topological analysis of electron localization functions. *Nature* **1994**, *371*, 683–686. [CrossRef]
67. Savin, A.; Silvi, B.; Coionna, F. Topological analysis of the electron localization function applied to delocalized bonds. *Can. J. Chem.* **1996**, *74*, 1088–1096. [CrossRef]
68. Schreckenbach, G.; Ziegler, T. Calculation of NMR Shielding Tensors Using Gauge-Including Atomic Orbitals and Modern Density Functional Theory. *J. Phys. Chem.* **1995**, *99*, 606–611. [CrossRef]
69. Lenthe, E.V.; Baerends, E.J.; Snijders, J.G. Relativistic regular two-component Hamiltonians. *J. Chem. Phys.* **1993**, *99*, 4597–4610. [CrossRef]
70. Vosko, S.H.; Wilk, L.; Nusair, M. Accurate spin-dependent electron liquid correlation energies for local spin density calculations: A critical analysis. *Can. J. Phys.* **1980**, *58*, 1200–1211. [CrossRef]
71. Becke, A.D. Density-functional exchange-energy approximation with correct asymptotic behavior. *Phys. Rev. A* **1988**, *38*, 3098–3100. [CrossRef] [PubMed]
72. Perdew, J.P. Density-functional approximation for the correlation energy of the inhomogeneous electron gas. *Phys. Rev. B* **1986**, *33*, 8822–8824. [CrossRef] [PubMed]
73. Van Lenthe, E.; Baerends, E.J. Optimized Slater-type basis sets for the elements 1–118. *J. Comput. Chem.* **2003**, *24*, 1142–1156. [CrossRef] [PubMed]
74. Schleyer, P.V.; Maerker, C.; Dransfeld, A.; Jiao, H.J.; Hommes, N.J.R.V. Nucleus-independent chemical shifts: A simple and efficient aromaticity probe. *JACS* **1996**, *118*, 6317–6318. [CrossRef]
75. Stanger, A. Nucleus-Independent Chemical Shifts (NICS): Distance Dependence and Revised Criteria for Aromaticity and Antiaromaticity. *J. Org. Chem.* **2006**, *71*, 883–893. [CrossRef]
76. Gershoni-Poranne, R.; Stanger, A. Magnetic criteria of aromaticity. *Chem. Soc. Rev.* **2015**, *44*, 6597–6615. [CrossRef]
77. Te Velde, G.; Bickelhaupt, F.M.; Baerends, E.J.; Fonseca Guerra, C.; van Gisbergen, S.J.A.; Snijders, J.G.; Ziegler, T. Chemistry with ADF. *J. Comput. Chem.* **2001**, *22*, 931–967. [CrossRef]
78. Omary, M.A.; Rawashdeh-Omary, M.A.; Diyabalanage, H.V.K.; Dias, H.V.R. Blue Phosphors of Dinuclear and Mononuclear Copper(I) and Silver(I) Complexes of 3,5-Bis(trifluoromethyl)pyrazolate and the Related Bis(pyrazolyl)borate. *Inorg. Chem.* **2003**, *42*, 8612–8614. [CrossRef]
79. Yang, L.; Powell, D.R.; Houser, R.P. Structural variation in copper(i) complexes with pyridylmethylamide ligands: Structural analysis with a new four-coordinate geometry index, τ4. *Dalton Trans.* **2007**, 955–964. [CrossRef]
80. Kandel, S.; Stenger-Smith, J.; Chakraborty, I.; Raptis, R.G. Syntheses and X-ray crystal structures of a family of dinuclear silver(I)pyrazolates: Assessment of their antibacterial efficacy against P. aeruginosa with a soft tissue and skin infection model. *Polyhedron* **2018**, *154*, 390–397. [CrossRef]
81. Mitoraj, M.P.; Michalak, A. σ-Donor and π-Acceptor Properties of Phosphorus Ligands: An Insight from the Natural Orbitals for Chemical Valence. *Inorg. Chem.* **2010**, *49*, 578–582. [CrossRef] [PubMed]

82. Mata, I.; Alkorta, I.; Molins, E.; Espinosa, E. Universal Features of the Electron Density Distribution in Hydrogen-Bonding Regions: A Comprehensive Study Involving H⋯X (X=H, C, N, O, F, S, Cl, π) Interactions. *Chem. Eur. J.* **2010**, *16*, 2442–2452. [CrossRef] [PubMed]
83. Alkorta, I.; Solimannejad, M.; Provasi, P.; Elguero, J. Theoretical Study of Complexes and Fluoride Cation Transfer between N2F+ and Electron Donors. *J. Phys. Chem. A* **2007**, *111*, 7154–7161. [CrossRef] [PubMed]
84. Cremer, D.; Kraka, E. A description of the chemical bond in terms of local properties of electron density and energy. *Croat. Chem. Acta* **1984**, *57*, 1259–1281.
85. Rozas, I.; Alkorta, I.; Elguero, J. Behavior of ylides containing N, O, and C atoms as hydrogen bond acceptors. *J. Am. Chem. Soc.* **2000**, *122*, 11154–11161. [CrossRef]
86. Poater, J.; Solà, M.; Viglione, R.G.; Zanasi, R. Local Aromaticity of the Six-Membered Rings in Pyracylene. A Difficult Case for the NICS Indicator of Aromaticity. *J. Org. Chem.* **2004**, *69*, 7537–7542. [CrossRef]
87. Poater, J.; Bofill, J.M.; Alemany, P.; Solà, M. Role of Electron Density and Magnetic Couplings on the Nucleus-Independent Chemical Shift (NICS) Profiles of [2.2]Paracyclophane and Related Species. *J. Org. Chem.* **2006**, *71*, 1700–1702. [CrossRef]
88. Sánchez-Sanz, G. Aromatic behaviour of benzene and naphthalene upon pnicogen substitution. *Tetrahedron* **2015**, *71*, 826–839. [CrossRef]
89. Sánchez-Sanz, G.; Trujillo, C.; Rozas, I.; Alkorta, I. Influence of fluoro and cyano substituents in the aromatic and antiaromatic characteristics of cyclooctatetraene. *Phys. Chem. Chem. Phys.* **2015**, *17*, 14961–14971. [CrossRef]
90. Trujillo, C.; Sánchez-Sanz, G. A Study of π–π Stacking Interactions and Aromaticity in Polycyclic Aromatic Hydrocarbon/Nucleobase Complexes. *ChemPhysChem* **2016**, *17*, 395–405. [CrossRef]
91. Sánchez-Sanz, G.; Alkorta, I.; Trujillo, C.; Elguero, J. A theoretical NMR study of the structure of benzynes and some of their carbocyclic and heterocyclic analogs. *Tetrahedron* **2012**, *68*, 6548–6556. [CrossRef]
92. Kusakiewicz-Dawid, A.; Porada, M.; Dziuk, B.; Siodlak, D. Annular Tautomerism of 3(5)-Disubstituted-1H-pyrazoles with Ester and Amide Groups. *Molecules* **2019**, *24*, 2632. [CrossRef] [PubMed]

© 2020 by the authors. Licensee MDPI, Basel, Switzerland. This article is an open access article distributed under the terms and conditions of the Creative Commons Attribution (CC BY) license (http://creativecommons.org/licenses/by/4.0/).

Article

Intramolecular sp^2-sp^3 Disequalization of Chemically Identical Sulfonamide Nitrogen Atoms: Single Crystal X-ray Diffraction Characterization, Hirshfeld Surface Analysis and DFT Calculations of *N*-Substituted Hexahydro-1,3,5-Triazines

Alexey V. Kletskov [1], Diego M. Gil [2], Antonio Frontera [3,*], Vladimir P. Zaytsev [1], Natalia L. Merkulova [1], Ksenia R. Beltsova [1], Anna A. Sinelshchikova [4], Mikhail S. Grigoriev [4], Mariya V. Grudova [1] and Fedor I. Zubkov [1,*]

[1] Organic Chemistry Department, Faculty of Science, Peoples' Friendship University of Russia (RUDN University), 6 Miklukho-Maklaya St., Moscow 117198, Russia; avkletskov@gmail.com (A.V.K.); vzaitsev@sci.pfu.edu.ru (V.P.Z.); fraumerk@gmail.com (N.L.M.); k.beltsova17@mail.ru (K.R.B.); shokoi@mail.ru (M.V.G.)

[2] INBIOFAL (CONICET-UNT), Instituto de Química Orgánica-Cátedra de Química Orgánica I, Facultad de Bioquímica, Química y Farmacia, Universidad Nacional de Tucumán, Ayacucho 471, San Miguel de Tucumán, Tucumán T4000INI, Argentina; diegomauriciogil@gmail.com

[3] Department de Química, Universitat de les Illes Balears, Crta de Valldemossa km 7.5, 07122 Palma de Mallorca (Baleares), Spain

[4] Frumkin Institute of Physical Chemistry and Electrochemistry, Russian Academy of Sciences, Leninsky pr. 31, bld. 4, Moscow 119071, Russia; asinelshchikova@gmail.com (A.A.S.); mickgrig@mail.ru (M.S.G.)

* Correspondence: toni.frontera@uib.es (A.F.); fzubkov@sci.pfu.edu.ru (F.I.Z.)

Received: 2 April 2020; Accepted: 23 April 2020; Published: 4 May 2020

Abstract: In this manuscript, the synthesis and single crystal X-ray diffraction characterization of four N-substituted 1,3,5-triazinanes are reported along with a detailed analysis of the noncovalent interactions observed in the solid state architecture to these compounds, focusing on C–H···π and C–H···O H-bonding interactions. These noncovalent contacts have been characterized energetically by using DFT calculations and also by Hirshfeld surface analysis. In addition, the supramolecular assemblies have been characterized using the quantum theory of "atoms-in-molecules" (QTAIM) and molecular electrostatic potential (MEP) calculations. The XRD analysis revealed a never before observed feature of the crystalline structure of some molecules: symmetrically substituted 1,3,5-triazacyclohexanes possess two chemically identical sulfonamide nitrogen atoms in different sp^2 and sp^3-hybridizations.

Keywords: triazinane; 1,3,5-Triazacyclohexane; Hirshfeld surface analysis; DFT study; H-bonding; C–H···π interaction; hybridization of a nitrogen atom in sulfonamides

1. Introduction

N-substituted triazinanes are interesting molecules that are used as efficient aminomethylation reagents and as formal 1,4- and 1,2-dipolar adducts in annulation reactions [1–10]. Moreover, this type of molecules presents remarkable antimicrobial activity [11]. While the access to symmetric N-substituted triazinanes is simple, there was no convenient method for the synthesis of triazinanes bearing different substituents on nitrogen atoms. Recently, we have described a straightforward approach to N-alkyl-N',N''-substituted triazinanes that is based on a one-pot multi-component reaction of amines, paraformaldehyde and sulfonamides or thioureas [12].

In this manuscript, the synthesis, single crystal X-ray diffraction characterization, Hirshfeld surface analysis and density functional theory (DFT) calculations of four triazinanes (see Scheme 1) are reported. The combination in the same structure of butyl substituents (n-Bu or t-Bu) with two aromatic rings facilitates the formation of a variety of C–H···π interactions in combination with C–H···O/N bonds. These noncovalent interactions have been studied using Hirshfeld surface analysis and DFT calculations. Moreover, they have been rationalized using the quantum theory of atoms in molecules (QTAIM) and molecular electrostatic potential (MEP) surfaces.

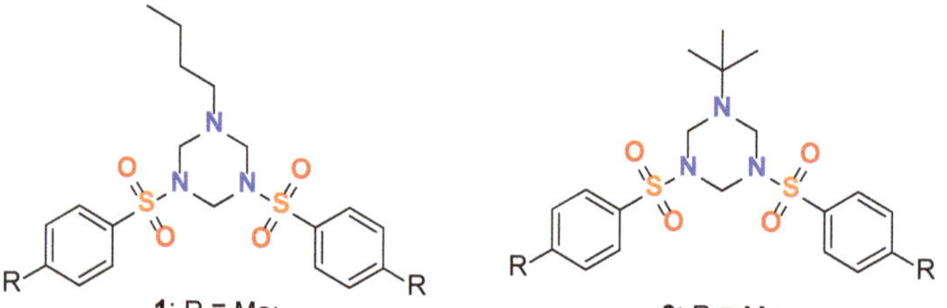

1: R = Me;
2: R = H

3: R = Me;
4: R = H

Scheme 1. Compounds **1–4** studied in this work.

2. Materials and Methods

2.1. Experimental Details

As it was mentioned above, the main objects of this work, N,N′-disulfamide substituted triazinanes **1–4**, were prepared according to the method described in our preliminary communication [12] using the three-component Mg(ClO$_4$)$_2$ catalyzed condensation of arylsulfonamides with paraformaldehyde and n- or tert-butyl amine (Scheme 2, see the Supplementary Materials for detail of the experimental procedures and spectral data). The tert-butyl- and n-butylamines were chosen as the amino-components providing the highest yield of the target triazinanes.

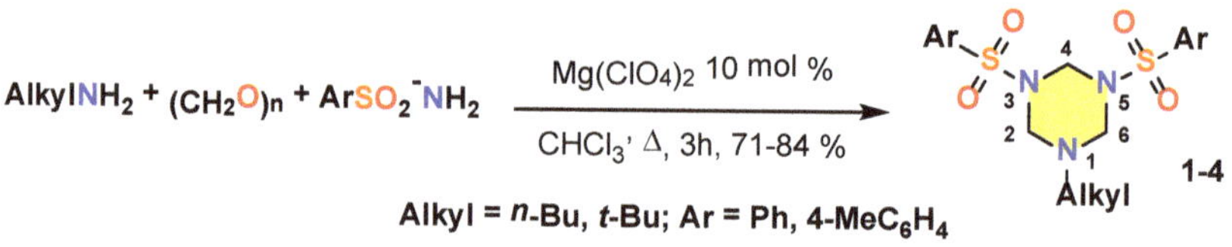

AlkylNH$_2$ + (CH$_2$O)$_n$ + ArSO$_2^-$NH$_2$ $\xrightarrow[\text{CHCl}_3, \Delta, 3h, 71-84\%]{\text{Mg(ClO}_4)_2 \text{ 10 mol \%}}$ **1-4**

Alkyl = n-Bu, t-Bu; Ar = Ph, 4-MeC$_6$H$_4$

Scheme 2. Synthesis of 1,3,5-triazacyclohexanes **1–4**.

All obtained triazinanes are well-crystallized solids that allowed the growth of crystals suitable for XRD analysis.

2.2. Crystallographic Details

Single crystal X-ray diffraction experiments were performed at the Center for Shared Use of Physical Methods of Investigation at the Frumkin Institute of Physical Chemistry and Electrochemistry. The single crystal X-ray diffraction data for 1,3,5-triazacyclohexanes (**1–4**) were collected on a Bruker Kappa Apex II automatic four-circle diffractometer (Bruker AXS, Madison, WI, USA) equipped with an area detector (Mo-Kα sealed-tube X-ray source, λ = 0.71073 Å, graphite monochromator) at 100 K for all compounds. The principal crystallographic data and structural refinements are summarized in Table 1. Atomic coordinates for compounds **1–4**, have been deposited with the CCDC (number 1992667–1992670). The supplementary crystallographic data are available in the ESI section. The comparison of the crystal structure parameters with the analogous compounds were performed using ConQuest search in Cambridge Structural Database (CSD, Version 5.40). The histograms of angles values were obtained from a graphical search of sulfonamides (C–S(=O)$_2$–NC$_2$) with 3D parameters for angles. More than 7000 hits were analyzed.

Table 1. Crystal data and structure refinement for 1–4.

Identification Code	1	2	3	4
CCDC number	1992667	1992668	1992669	1992670
Empirical formula	C$_{21}$H$_{29}$N$_3$O$_4$S$_2$	C$_{19}$H$_{25}$N$_3$O$_4$S$_2$	C$_{21}$H$_{29}$N$_3$O$_4$S$_2$	C$_{19}$H$_{25}$N$_3$O$_4$S$_2$
Formula weight	451.59	423.54	451.59	423.54
Temperature/K	100(2)	100(2)	100(2)	100(2)
Crystal system	monoclinic	monoclinic	monoclinic	orthorhombic
Space group	$P2_1/c$	$P2_1/n$	$P2_1/n$	$P2_12_12_1$
a/Å	13.2871(4)	8.4284(2)	5.955(4)	10.7298(3)
b/Å	10.3261(3)	25.9248(8)	15.378(12)	11.1010(3)
c/Å	15.9595(4)	9.5601(3)	23.915(19)	16.9303(5)
α/°	90	90	90	90
β/°	90.511(2)	106.639(1)	90.968(16)	90
γ/°	90	90	90	90
Volume/Å^3	2189.62(11)	2001.46(10)	2190(3)	2016.59(10)
Z	4	4	4	4
ρ$_{calc}$g/cm^3	1.370	1.406	1.370	1.395
μ/mm^{-1}	0.276	0.297	0.276	0.295
F(000)	960.0	896.0	960.0	896.0
Crystal size/mm^3	0.440 × 0.360 × 0.320	0.400 × 0.320 × 0.260	0.500 × 0.180 × 0.030	0.420 × 0.400 × 0.360
Radiation	MoKα (λ = 0.71073)	MoKα (λ = 0.71073)	MoKα (λ = 0.71073)	MoKα (λ = 0.71073)
2Θ range for data collection/°	7.126 to 59.998	7.392 to 59.994	8.476 to 55	8.16 to 69.998
Index ranges	−18 ≤ h ≤ 18, −14 ≤ k ≤ 14, −20 ≤ l ≤ 22	−6 ≤ h ≤ 11, −35 ≤ k ≤ 36, −13 ≤ l ≤ 13	−4 ≤ h ≤ 7, −19 ≤ k ≤ 19, −31 ≤ l ≤ 31	−17 ≤ h ≤ 16, −17 ≤ k ≤ 16, −19 ≤ l ≤ 27
Reflections collected	33914	28404	14104	35553
Independent reflections	6383 [R$_{int}$ = 0.0390, R$_{sigma}$ = 0.0303]	5835 [R$_{int}$ = 0.0351, R$_{sigma}$ = 0.0280]	4940 [R$_{int}$ = 0.1432, R$_{sigma}$ = 0.1883]	8855 [R$_{int}$ = 0.0303, R$_{sigma}$ = 0.0303]
Data/restraints/parameters	6383/0/273	5835/0/253	4940/6/274	8855/0/253
Goodness-of-fit on F^2	1.029	1.035	1.049	1.042
Final R indexes [I >= 2σ (I)]	R$_1$ = 0.0350, wR$_2$ = 0.0877	R$_1$ = 0.0334, wR$_2$ = 0.0824	R$_1$ = 0.1398, wR$_2$ = 0.3472	R$_1$ = 0.0283, wR$_2$ = 0.0698
Final R indexes [all data]	R$_1$ = 0.0457, wR$_2$ = 0.0939	R$_1$ = 0.0423, wR$_2$ = 0.0873	R$_1$ = 0.2253, wR$_2$ = 0.4075	R$_1$ = 0.0316, wR$_2$ = 0.0715
Largest diff. peak/hole/e Å^{-3}	0.38/−0.37	0.39/−0.37	0.94/−0.55	0.36/−0.28

2.3. Hirshfeld Surface Calculations

The Hirshfeld surface (HS) analysis [13–15] and their associated 2D fingerprint plots (full and decomposed) [16] were carried out employing the CrystalExplorer 17 program [17] in order to visualize and quantify various non-covalent interactions that stabilize the crystal packing. The HS was mapped over d_{norm} property. The d_{norm} property is a symmetric function of distances to the surface from nuclei inside and outside the Hirshfeld surface (d_i and d_e, respectively), relative to their respective van der Waals radii. The regions with red and blue color on the d_{norm} represent the shorter and longer inter contacts while the white color indicates the contacts around the van der Waals radii. 2D fingerprint plots provide relevant information of intermolecular contacts in the crystal. The d_{norm} surface was mapped with the color scale in the range −0.050 au (red) to 0.600 au (blue). 2D fingerprint plots (d_i vs. d_e) were displayed using the expanded 0.6–2.8 Å range.

2.4. Theoretical Methods

All DFT calculations were carried out using the Gaussian-16 program [18] at the PBE1PBE-D3/def2-TZVP level of theory and using the crystallographic coordinates. The formation energies of the assemblies were evaluated by calculating the difference between the total energy of the assembly and the sum of the monomers that constitute the assembly, which have been maintained frozen. That is $\Delta E_{AB} = E_{AB} - E_A - E_B$, where ΔE_{AB} is the interaction energy; E_{AB} is the energy of the dimer and E_A and E_B are the energy of the monomers. The BSSE has been used to correct the interaction energies by using the counterpoise =2 keyword in the Gaussian-19 program [18]. The molecular electrostatic potential was computed at the same level of theory and plotted onto the 0.001 a.u. isosurface. The Quantum Theory of Atoms-in-Molecules (QTAIM) [19] analysis was carried out at the same level of theory by means of the AIMAll program [20] to obtain the distribution of bond critical points (CPs) and bond paths [21].

3. Results

3.1. Structural Description

According to the single crystal X-ray diffraction data, molecules **1–4** comprise the 1,3,5-triazacyclohexane ring bearing three substituents at the nitrogen atoms (see Figure 1).

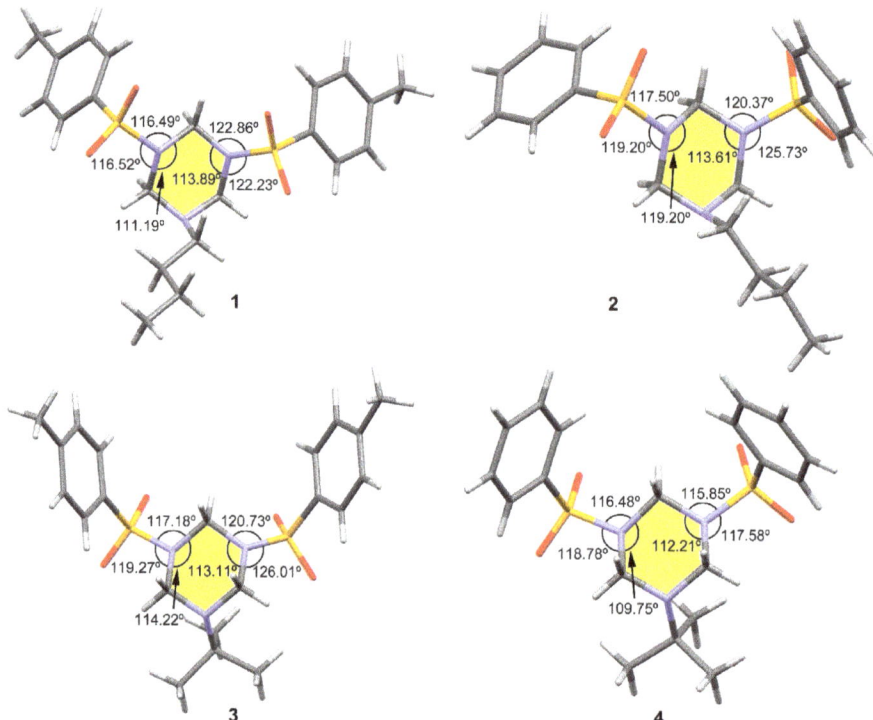

Figure 1. Single crystal X-ray diffraction structures of triazinanes **1–4**.

The asymmetric unit contains one molecule of each triazine. The general geometrical features of these systems are similar, which are the slightly distorted chair conformation of the six-membered ring, with the N-alkyl substituents (n-Bu or t-Bu) occupying the axial position and the two N-sulfamide fragments occupying the sterically favorable pseudo-equatorial orientation (see Table S2 for torsion angles of N-substituents). The atoms C2, C6, N3, and N5 of the triazinane cycle in all structures lie nearly in one plane (the deviation of one atom from the plane of the other three is less than 0.03 Å), while the deviation of N1 and C4 atoms from this plane range from −0.633(2) Å to −0.661(2) Å and from 0.643(14) Å to 0.712(2) Å, correspondingly, therefore the molecules have a classical chair-conformation of the central heterocycle (see Scheme 2 for atom numbering scheme). All CH_2-N bond lengths and CH_2-N-CH_2 bond angles are typical for 1,3,5-triazinanes and are listed in the corresponding tables in (Supplementary Materials Table S2–S15). The torsion angles CNCN in the triazacyclohexane ring are close to 60° (see Table S2).

The most intriguing and distinguishing feature of the triazinanes under discussion is the unprecedented geometry of the N3 and N5 nitrogen atoms of the sulfonamide fragments in the N-butyl substituted heterocycles **1** and **2**. As is generally known, the nitrogen atom in a sulfonamide group can adopt both sp^2 and sp^3 hybridization depending on substituents at the nitrogen atom [22–24]. However, according to the data of the CCDC, there are no known examples of 1,3,5-triazinanes or other saturated six-membered azaheterocycles simultaneously possessing two chemically identical sulfonamide nitrogen atoms in different hybridization. Analysis of the values of the sums of valence angles at nitrogen atoms in positions 3 and 5 allows to clearly identify atoms in sp^2 or in sp^3 hybridization (Figure 1 and Table 2). The N3-atoms in compounds **1–3** are sp^2-hybridizated and, as a result, they assume the flat trigonal configuration (the sum of the angles is close to 360°). N5-Atoms, chemically equivalent to N3-atoms, in the same molecules adopt the tetrahedral configuration (the sum of the angles lies in the diapason of 344–351°) and, therefore, are sp^3-hybridizated. This is most clearly seen in the examples of compounds **1** and **2**. The presence of the *tert*-butyl group at the N1 position in compounds **3** and **4**, probably due to its high steric volume, symmetrizes the molecules, leveling the difference between both sulfonamide nitrogen atoms in a crystal. This is also observed in the equalization of the S–N distances of the sulfonamide groups in compounds **3** and **4** compared to **1** and **2**, see Table 2. The largest difference between both S–N distances is observed in compound **2**, i.e., 0.016 Å. In fact, the short S2–N5 distance is an indication of a partial double bond character, in agreement with the sp^2-hybridization.

Table 2. Sums of angles at the sulfonamide nitrogen atoms N3 and N5 and S–N distances in 1,3,5-triazinanes **1–4**.

Compound	1	2	3	4
Alkyl	Bu	Bu	t-Bu	t-Bu
SO$_2$Ar	SO$_2$C$_6$H$_4$Me	SO$_2$Ph	SO$_2$C$_6$H$_4$Me	SO$_2$Ph
Sum of angles around N3 (°)	359.0	359.7	359.8	345.0
Sum of angles around N5 (°)	344.2	347.4	350.7	345.6
S1–N3 distance (Å)	1.642(1)	1.630(1)	1.632(9)	1.640(1)
S2–N5 distance (Å)	1.632(1)	1.614(1)	1.629(9)	1.638(1)

Compound **1** crystalizes in the monoclinic crystal system in the space group $P2_1/c$. Selected bond lengths, angles and torsional angles are shown in Tables S2–S5. The S2–C21 and S1–N3 bond lengths of 1.758(1) and 1.642(1) Å respectively are in agreement with a single bond character of these bonds [24]. The S–O bond distances of the sulfamide moiety are in the range 1.432(1)–1.436(1) Å, which is typical for N-sulfamides. The SO$_2$ group has typical angles if compare with CSD data for N-sulfamides (Figure S1 in Supplementary Materials): O–S–O angle is around 120° [O1–S1–O2 is 119.95(6)°], while N–S–O angles are between 105° and 115° [N3–S1–O2 is 106.07(6)°, N3–S1–O1 is 106.63(5)°].

The crystal structure of this compound exhibits interesting assemblies in the solid state (see Table S3 for H-bonds). For instance, Figure 2a shows a self-assembled dimer dominated by C–H···O interactions where the methyl group in para acts as a H-bond donor. The acidity of these protons is higher than usual for a methyl group due to the presence of the electron withdrawing sulfamide group. Quite remarkable is the ternary assembly shown in Figure 2b, where the aromatic π-cloud interacts simultaneously with the methyl group at one side and an aromatic C–H bond at the opposite side, thus forming a C–H···π···H–C assembly. It is worth noting that the C–H···π distances are very short (2.60 and 2.70 Å) thus, confirming their relevance in the solid state of this compound.

Compound **2** crystalizes in the monoclinic crystal system in the space group $P2_1/n$ and the main difference with compound **1** is the absence of methyl groups. In addition, the sulfonamide groups are attached to the 1,3,5-triazacyclohexane ring in different orientations as reflected by the C6–N5–S2–C21 and C2–N3–S1–C11 torsional angles of −105.0(1)° and −65.5(1)°, respectively. It also forms self-assembled dimers in the solid state, where both C–H···π and C–H···O (Table S3) interactions are established, as shown in Figure 3a. Moreover, it also forms infinite 1D supramolecular chains in the solid state promoted by C–H···O interactions involving the butyl chain and the sulfonamide group (see Figure 3b).

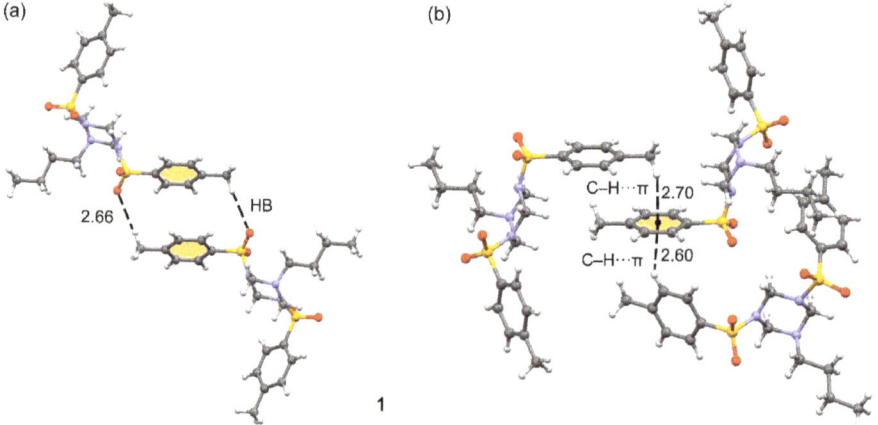

Figure 2. (a) Self assembled dimer in compound **1**. (b) C–H···π···H–C assembly in the solid state of structure **1**. Distances in Å. For the C–H···π interactions, the distances are measured from the H-atom to the ring centroid.

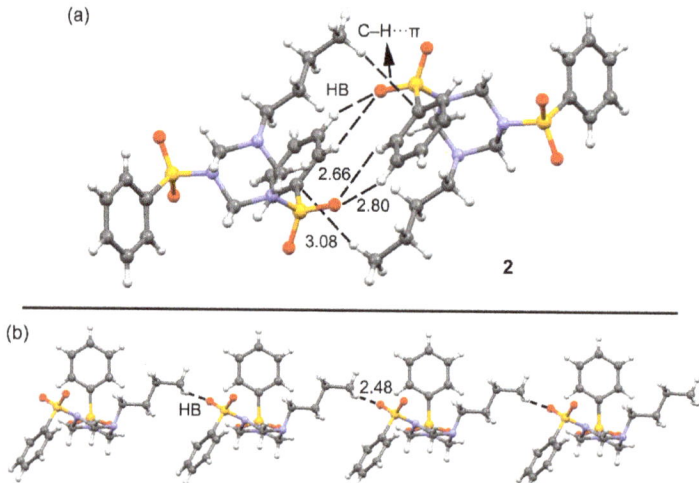

Figure 3. (a) Self-assembled dimer of compound **2**. (b) 1D supramolecular chain in **2**. Distances in Å. The C–H···π interaction distances are measured from the H-atom to the closest C-atom of the ring.

Compound **3** crystalizes in the monoclinic crystal system in the space group $P2_1/n$ and, similarly to compound **1**, also forms ternary assemblies where the same aromatic ring establishes C–H···π interactions at both sides of the ring, thus forming a C–H···π···H–C assembly (see Figure 4). This compound also forms 1D supramolecular chains in the solid state where the H-atoms of the triazinane ring interact with the O-atoms of the sulfonamide groups (Table S3), as shown in Figure 4b. The formation of this assembly is facilitated by the relative orientation of the p-methyl-benzene-sulfonamide groups.

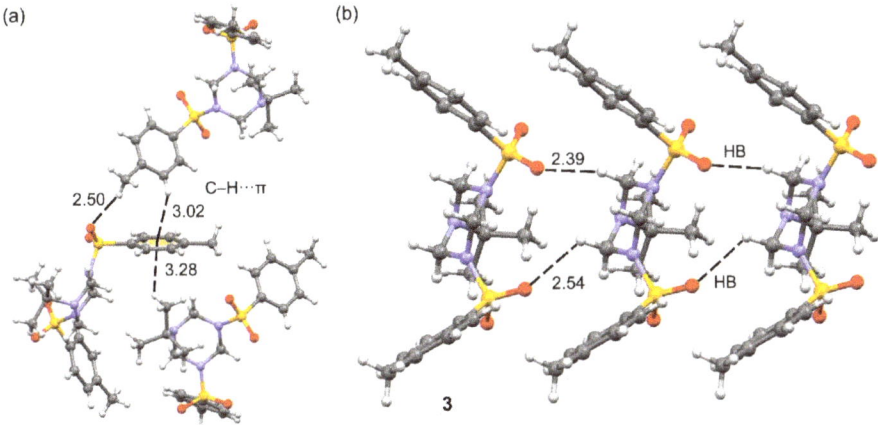

Figure 4. (a) C–H···π···H–C assembly in the solid state of structure **3**. Distances in Å. (b) 1D supramolecular chain in **3**. Distances in Å. For the C–H···π interactions, the distances are measured from the H-atom to the ring centroid.

Compound **4** crystallizes in the orthorhombic crystal system in the space group $P2_12_12_1$ and also exhibits several motifs in the solid state that are mainly dominated by C–H···O interactions (Table S3). As examples, two motifs are given in Figure 5, one corresponds to a discrete dimer where three H-bonds are formed and the other one to a 1D supramolecular polymer also governed by the formation of C–H···O bonds involving the aromatic H-atom located in para to the sulfonamide group (Figure 5b).

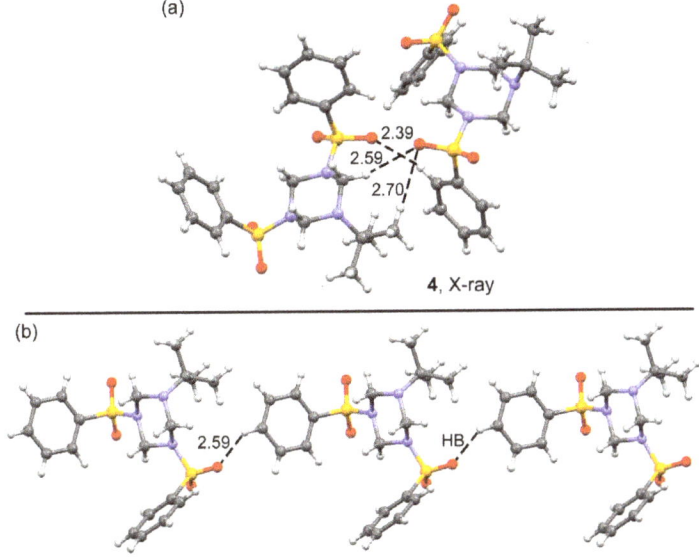

Figure 5. (a) Self-assembled dimer of compound **4**. (b) 1D supramolecular chain in **4**. Distances in Å.

3.2. Hirshfeld Surfaces

The Hirshfeld surface analysis is a very convenient tool for analyzing intermolecular interactions. The HS surfaces mapped over d_{norm} property are displayed in Figure 6 highlighting the main intermolecular interactions and scheme of labels. The patterns of intermolecular interactions are similar in all structures, which prompted us to evaluate the contributions of the weak non-covalent contacts in the supramolecular assembly, as well as the importance of C–H···π interactions in stabilization of the crystal packing. The 2D fingerprint plots (Figure 7) of the molecules illustrate significant differences between the intermolecular interaction patterns. The surfaces are shown as transparent to allow the visualization of the molecules. Contacts with distances equal to the sum of van der Waals (vdW) radii are represented as white regions and contacts with distances shorter than and longer than the vdW radii are shown as red and blue colors, respectively.

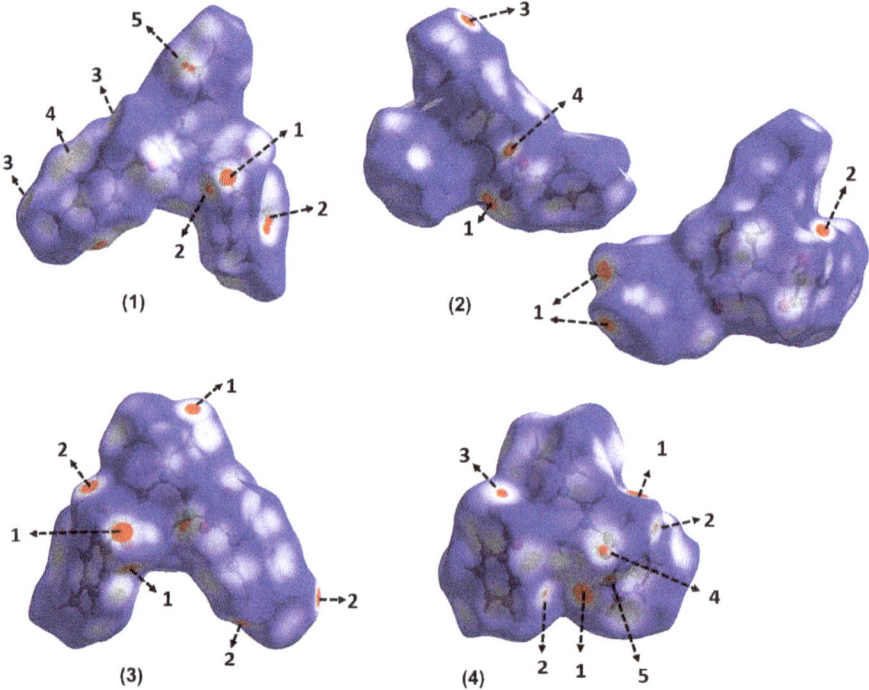

Figure 6. Hirshfeld surfaces mapped over d_{norm} function for compounds **1–4**. The labels are discussed in the text. For compound (**2**), the second molecule is rotated by 180° around the vertical axis of the plot.

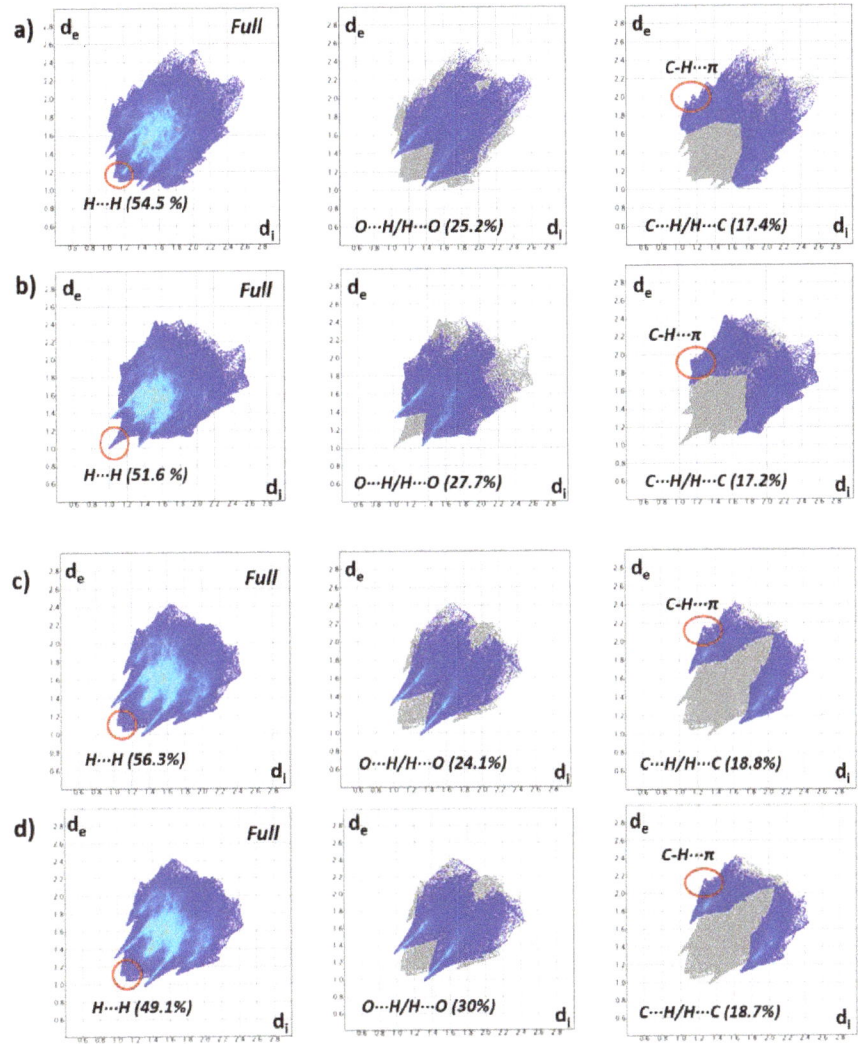

Figure 7. Full and decomposed 2D fingerprint plots for compounds: (a) **1**; (b) **2**; (c) **3** and (d) **4**.

The vdW forces (H···H contacts) have the largest contribution to the HS, and they are highlighted in the scattered middle points in the fingerprint plots with a minimum value of $(d_e + d_i) \sim 2.2$ Å (Figure 7), which is the sum of the vdW radii. All red areas that are visible on the surfaces mapped over d_{norm} function correspond to C–H···O contacts. For (**1**), the largest bright-red spot labeled 1 on the HS shows O···H/H···O contact associated with C17–H17C···O1 interaction, which constitutes the strongest among all interactions present in this compound. Two medium sized red spots labeled as 2 and 3 are associated with C15–H15A···O1 and C27–H27C···O4, respectively. These interactions are also visible as symmetrical sharp spikes centered at $(d_e + d_i) \cong 2.4$ Å in the fingerprint plots (Figure 7a) with 25.2% contribution to the Hirshfeld contact surface. The intermolecular C25–H25A···N1 contact is visible in the HS as a red spot labelled 4, which comprises 2.3% of the total HS area. The HS of (**2**) mapped over d_{norm} function (see Figure 6) shows four red spots, indicating the presence of C–H···O hydrogen bonds

[C13–H13A⋯O1 (1), C26–H16A⋯O4 (2), C10–H10B⋯O1 (3), C14–H14A⋯O2 (4)]. The decomposed fingerprint plots (Figure 7b) show that intermolecular O⋯H/H⋯O contacts contribute 27.7% to the total HS area. The O⋯H/H⋯O contacts appeared as sharp spikes with (de + di) ≅ 2.35 Å. In the HS of (3), the O⋯H/H⋯O contacts (Figure 6) are visible as six red spots attributed to C6–H6A⋯O3 (labeled 1), C10–H10B⋯O3 (labeled 1), C15–H15A⋯O4 (labeled 2) and C17–H17C⋯O4 (labeled 2). These interactions comprise the 24.1% of the HS area. A similar behavior was observed in the HS mapped over d_{norm} function for (4), which the six red spots observed (Figure 6) are attributed to C26–H26⋯O4 (1), C16–H16⋯O4 (2), C24,H24⋯O2 (3), C12–12⋯O3 (4) and C6–H6A⋯O3 (5). Intermolecular interactions O⋯H/H⋯O are observed around 2.3 Å which is slightly shorter than those of other compounds with 30% contribution to the Hirshfeld contact surface.

As was described previously, the structure of (1) is also stabilized by C–H⋯π interactions. The red area labeled 5 in the HS mapped over d_{norm} is attributed to C9–H9B⋯π. These C–H⋯π interactions are also evident from a pair of "wings" in the top left and bottom right region of the fingerprint plots for compounds **1–4** (Figure 7). The shape of the wings and the sum of d_e and d_i show the importance of this interaction. The decomposition of the fingerprint plots shows that the C⋯H/H⋯C contributions comprising 17.4%, 17.2%, 18.8% and 18.7% of the total HS for each molecule of **1, 2, 3** and **4**, respectively.

3.3. DFT Calculations

The DFT study is focused to analyze the supramolecular assemblies commented above in Figures 2–5, where combinations of C–H⋯π and C–H⋯O H-bonding networks are commonly formed in compounds **1–4**. First of all, the molecular electrostatic potential (MEP) surfaces of compounds **1** and **2** have been computed in order to analyze the electron rich and electron poor regions of the molecules. The surfaces are represented in Figure 8 and it can be observed that the most negative regions correspond to the O-atoms of the sulfonamide group. The N-atoms of the triazinane ring are not good H-bond acceptors, likely because either the lone pair is delocalized into the SO_2-group, in accordance with the S2–N5 and S1–N3 bond lengths of 1.632(1) and 1.642(1) Å, respectively indicating a double bond character of these bonds. The most positive region corresponds to the middle of the three axial H-atoms of the triazinane ring (+27 kcal/mol). The aromatic H-atoms and the CH_3 substituents also present positive MEP values (+23 and +20 kcal/mol, respectively). Finally, the MEP value over the aromatic rings is negative (−8 kcal/mol), thus adequate for establishing C–H⋯π interactions. The MEP analysis evidences that the most favored interactions from an electrostatic point of view are those involving the O-atoms as electron donors and either aromatic or triazinane protons as electron acceptors.

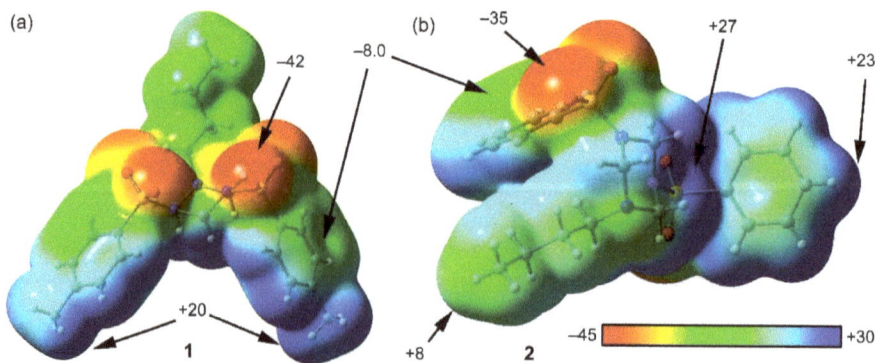

Figure 8. MEP surfaces for compounds **1** (a) and **2** (b). The energies at selected points of the surfaces are given in kcal/mol.

We have selected two supramolecular assemblies commented above in Figure 2 to analyze the energetic features of the H-bonds and C–H···π interactions in **1**. The QTAIM distribution of critical points and bond paths are also given in Figure 9. The existence of a bond CP and bond path connecting two atoms is a good indicator of interaction [21]. For the self-assembled dimer (Figure 9a), in addition to the symmetrically related H-bonds (characterized by a bond CP (critical point) and bond path interconnecting the H and O-atoms), the QTAIM analysis reveals the existence of a π···π stacking interaction that further stabilizes the formation of the dimer. The dimerization energy is moderately strong ($\Delta E_1 = -9.1$ kcal/mol) due to the contribution of both H-bonds and the π-stacking. We have also analyzed the other motif, where C–H···π interactions are established. The interaction energy is very strong ($\Delta E_2 = -20.4$ kcal/mol), because in addition to the C–H···π contacts (two bond CPs and bond paths connect two aromatic H-atoms to two carbon atoms of the adjacent ring) an intricate network of H-bonds is established where six C–H···O and one C–H···N contacts are formed, which are highlighted in Figure 9b by yellow circles.

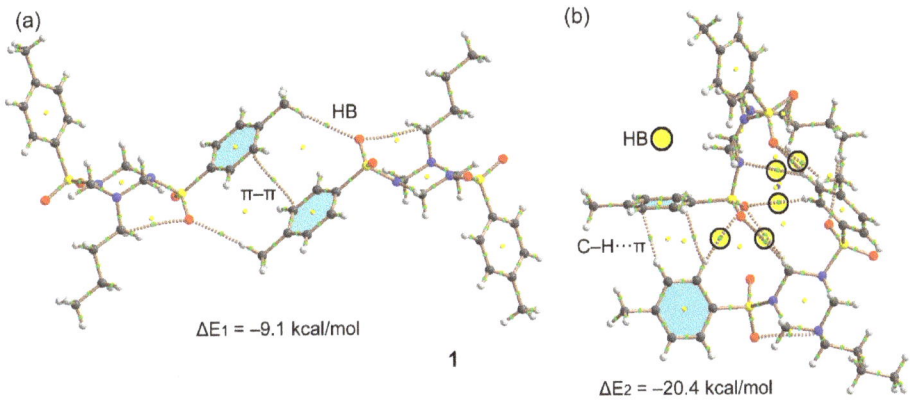

Figure 9. (**a**,**b**) Distribution of bond, ring and cage critical points (green, yellow and blue spheres, respectively) and bond paths in two dimers of complex **1**. The C–H···O bonds are highlighted in (**b**).

Figure 10a shows the self-assembled dimer of compound **2** where up to eight C–H···O contacts are established between either aromatic or aliphatic H-atoms and the O-atoms of sulfonamide (each one characterized by a bond CP and bond path, see yellow circles in Figure 10a). Moreover, two symmetrically distributed C–H···π interactions are also present and characterized by a bond CP and bond path connecting the aliphatic H-atom to one C-atom of the aromatic ring. As a consequence of this combination of interactions, the dimerization energy is very large $\Delta E_3 = -16.7$ kcal/mol, thus confirming the importance of this motif in the solid state of compound **2**. Figure 10b shows a dimer extracted from the infinite 1D chain represented in Figure 3b. In this case, the interaction energy is modest ($\Delta E_4 = -3.4$ kcal/mol) because only one H-bond is established. The distribution of bond CPs and bond path also reveals the existence of van der Waals interactions between the alkyl chain and the aromatic ring.

Figure 11 shows two dimers of compound **3**, one corresponds to the C–H···π assembly commented above in Figure 4, where in addition to the C–H···π interaction (characterized by a bond CP and bond path) the assembly is further characterized by a C–H···O bond involving the methyl group. This assembly presents a modest interaction energy of $\Delta E_4 = -5.9$ kcal/mol. In contrast, the dimer shown in Figure 11b, extracted from the infinite 1D assembly, exhibits a strong interaction energy ($\Delta E_5 = -17.3$ kcal/mol) due to the formation of four C–H···O contacts, which are characterized by a bond CP and bond path (see yellow circles in Figure 11b). The strong interaction energy agrees well with the MEP surface analysis commented above, since the H-bond donors belong to the triazinane ring that exhibit the most positive MEP values. Moreover, the H-bond acceptors are the O-atoms of the sulfonamide groups that present the most negative MEP values (see Figure 8).

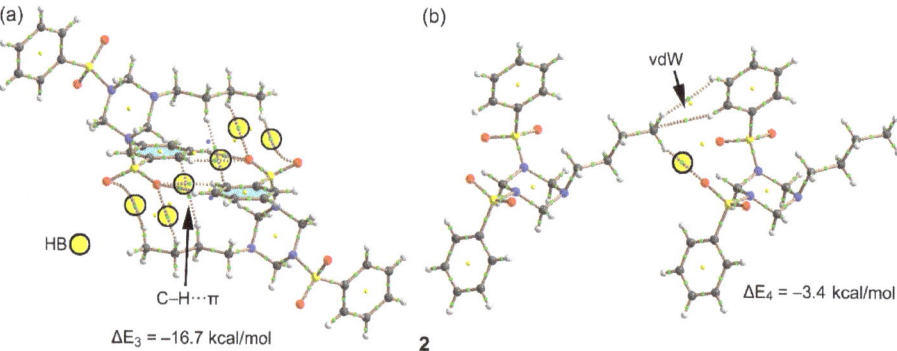

Figure 10. (a,b) Distribution of bond, ring and cage critical points (green, yellow and blue spheres, respectively) and bond paths in two dimers of complex **2**. The C–H···O bonds are highlighted.

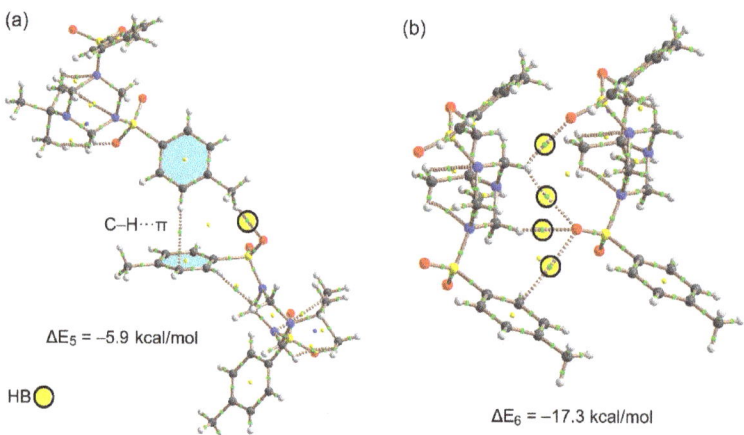

Figure 11. (a,b) Distribution of bond, ring and cage critical points (green, yellow and blue spheres, respectively) and bond paths in two dimers of complex **3**. The C–H···O bonds are highlighted.

Finally, Figure 12 shows the dimeric motifs analyzed in compound 4. The dimer of Figure 12a presents an intricate combination of C–H···O bonds in addition to two C–H···π interactions involving the t-butyl group. As a consequence of the formation of six concurrent H-bonds, the dimerization energy is very large, $\Delta E_7 = -20.2$ kcal/mol. Figure 12b shows the dimer extracted from the 1D supramolecular polymer (see Figure 5b), which presents a modest interaction energy due to the formation of a single H-bond along with van der Waals contacts between the aromatic and aliphatic C–H bonds.

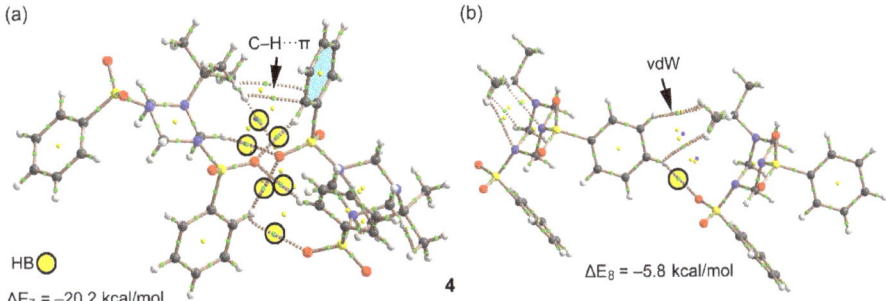

Figure 12. (a,b) Distribution of bond, ring and cage critical points (green, yellow and blue spheres, respectively) and bond paths in two dimers of complex **4**. The C–H···O bonds are highlighted.

4. Concluding Remarks

In summary, the synthesis and single crystal X-ray diffraction characterization of four N-substituted 1,3,5-triazinanes are reported along with a detailed analysis of the noncovalent interactions observed in the solid state. All complexes have in common the formation of several motifs characterized by a network of C–H···O interactions that exhibits very strong binding energies as a consequence of these cooperative H-bonds. Moreover, several structures also form interesting C–H···π···H–C ternary assemblies that have been described in detail. Besides, the MEP surfaces have been used to rationalize the noncovalent interactions and the QTAIM method to confirm the existence of the intricate combinations of H-bonds. Finally, the Hirshfeld surface analysis provides further evidence for the importance of C–H···π and C–H···O in the crystal packing of compounds **1–4**. We assume, that observed case of intramolecular sp^2-sp^3 disequalization makes the corresponding family of N-alkyl-N',N''-substituted triazinanes an interesting object for research in the domain of local molecular disorder in organic crystals [25].

Supplementary Materials: The following are available online at http://www.mdpi.com/2073-4352/10/5/369/s1, Figure S1: Histogram of O–S–O angle in N-sulfamides (a) N–S–O angle in N-sulfamides (b) according CSD analysis. Table S1: Data and structure refinement for **1–4**. Table S2: Selected distances (Å), angles (°) and torsion angles (°) for compounds **1–4**. Table S3: Hydrogen bonds for compounds **1–4**. Table S4: Bond Lengths for **1**. Table S5: Bond Angles for **1**. Table S6: Torsion Angles for **1**. Table S7: Bond Lengths for **2**. Table S8: Bond Angles for **2**. Table S9: Torsion Angles for **2**. Table S10: Bond Lengths for **3**. Table S11: Bond Angles for **3**. Table S12: Torsion Angles for **3**. Table S13: Bond Lengths for **4**. Table S14: Bond Angles for **4**. Table S15. Torsion Angles for **4**.

Author Contributions: Conceptualization, D.M.G., V.P.Z., N.L.M. and A.F.; formal analysis, D.M.G., K.R.B. and A.A.S. investigation, D.M.G., A.F., F.I.Z., M.V.G. and A.V.K.; Methodology, M.S.G.; writing—original draft preparation, D.M.G. and A.F.; writing—review and editing, D.M.G., A.F., F.I.Z. and A.V.K.; supervision, A.F. and F.I.Z.; funding acquisition, D.M.G., A.F. and F.I.Z. All authors have read and agreed to the published version of the manuscript.

Funding: This research was partly funded by the MICIU/AEI of Spain (project CTQ2017-85821-R FEDER funds). The publication has been prepared with the support of the "RUDN University Program 5-100" and with the support of the Russian Foundation for Basic Research (project No 19-03-00807 A).

Acknowledgments: We thank the "Centre de Tecnologies de la Informació" (CTI), Universitat de les Illes Balears for computational facilities.

Conflicts of Interest: The authors declare no conflict of interest.

References

1. Ha, H.-J.; Koo Lee, W. Synthetic applications of Lewis acid-induced N-methyleneamine equivalents. *Heterocycles* **2002**, *57*, 1525–1538. [CrossRef]
2. Ji, D.; Sun, J. [3+2]-Cycloaddition of azaoxyallyl cations with hexahydro-1,3,5-triazines: Access to 4-imidazolidinones. *Org. Lett.* **2018**, *20*, 2745–2748. [CrossRef]

3. Ji, D.; Wang, C.; Sun, J. Asymmetric [4+2]-cycloaddition of copper–allenylidenes with hexahydro-1,3,5-triazines: access to chiral tetrahydroquinazolines. *Org. Lett.* **2018**, *20*, 3710–3713. [CrossRef]
4. Chen, L.; Liu, K.; Sun, J. Catalyst-free synthesis of tetrahydropyrimidines via formal [3+3]-cycloaddition of imines with 1,3,5-hexahydro-1,3,5-triazines. *RSC Adv.* **2018**, *8*, 5532–5535. [CrossRef]
5. Zhu, C.; Xu, G.; Sun, J. Gold-catalyzed formal [4 + 1]/[4 + 3] cycloadditions of diazo esters with triazines. *Angew. Chem. Int. Ed.* **2016**, *55*, 11867–11871. [CrossRef]
6. Liu, S.; Yang, P.; Peng, S.; Zhu, C.; Cao, S.; Li, J.; Sun, J. Gold-catalyzed sequential annulations towards 3,4-fused bi/tri-cyclic furans involving a [3 + 2 + 2]-cycloaddition. *Chem. Commun.* **2017**, *53*, 1152–1155. [CrossRef]
7. Liu, P.; Xu, G.; Sun, J. Metal-free [2 + 1 + 2]-cycloaddition of tosylhydrazones with hexahydro-1,3,5-triazines to form imidazolidines. *Org. Lett.* **2017**, *19*, 1858–1861. [CrossRef]
8. Peng, S.; Ji, D.; Sun, J. Gold-catalyzed [2 + 2 + 2 + 2]-annulation of 1,3,5-hexahydro-1,3,5-triazines with alkoxyallenes. *Chem. Commun.* **2017**, *53*, 12770–12773. [CrossRef]
9. Zeng, Z.; Jin, H.; Song, X.; Wang, Q.; Rudolph, M.; Rominger, F.; Hashmi, A.S.K. Gold-catalyzed intermolecular cyclocarboamination of ynamides with 1,3,5-triazinanes: En route to tetrahydropyrimidines. *Chem. Commun.* **2017**, *53*, 4304–4307. [CrossRef]
10. Garve, L.K.B.; Jones, P.G.; Werz, D.B. Ring-opening 1-amino-3-aminomethylation of donor-acceptor cyclopropanes via 1,3-diazepanes. *Angew. Chem. Int. Ed.* **2017**, *56*, 9226–9230. [CrossRef]
11. Yin, B. Synergistic Antimicrobial Composition. WO Patent 2,011,049,761, 28 April 2011.
12. Kletskov, A.V.; Frontera, A.; Sinelshchikova, A.A.; Grigoriev, M.S.; Zaytsev, V.P.; Grudova, M.V.; Bunev, A.S.; Presnukhina, S.; Shetnev, A.; Zubkov, F.I. Straightforward three-component synthesis of N-alkyl-N',N"-substituted hexahydro-1,3,5-triazines. *Synlett* **2020**. [CrossRef]
13. Spackman, M.A.; Jayatilaka, D. Hirshfeld surface analysis. *CrystEngComm* **2009**, *11*, 19–32. [CrossRef]
14. McKinnon, J.J.; Spackman, M.A.; Mitchell, A.S. Novel tools for visualizing and exploring intermolecular interactions in molecular crystals. *Acta Cryst. B* **2004**, *60*, 627–668. [CrossRef]
15. McKinnon, J.J.; Jayatilaka, D.; Spackman, M.A. Towards quantitative analysis of intermolecular interactions with Hirshfeld surfaces. *Chem. Commun.* **2007**, 3814–3816. [CrossRef]
16. Spackman, M.A.; McKinnon, J.J. Fingerprinting intermolecular interactions in molecular crystals. *CrystEngComm* **2002**, *4*, 378–392. [CrossRef]
17. Turner, M.J.; McKinnon, J.J.; Wolf, S.K.; Grimwood, D.J.; Spackman, P.R.; Jayatilaka, D.; Spackman, M.A. *CrystalExplorer17*; University of Western Australia: Perth, Australia, 2017.
18. Frisch, M.J.; Trucks, G.W.; Schlegel, H.B.; Scuseria, G.E.; Robb, M.A.; Cheeseman, J.R.; Scalmani, G.; Barone, V.; Petersson, G.A.; Nakatsuji, H.; et al. *Gaussian 16, Revision A.01*; Gaussian, Inc.: Wallingford, CT, USA, 2016.
19. Bader, R.F.W. A quantum theory of molecular structure and its applications. *Chem. Rev.* **1991**, *91*, 893–928. [CrossRef]
20. Keith, T.A. *AIMAll (Version 13.05.06)*; TK Gristmill Software: Overland Park, KS, USA, 2013.
21. Bader, R.F.W. A Bond Path: A Universal Indicator of Bonded Interactions. *J. Phys. Chem. A* **1998**, *102*, 7314–7323. [CrossRef]
22. Breneman, C.M.; Weber, L.W. Charge and Energy Redistribution in Sulfonamides Undergoing Conformational Changes. Hybridization as a Controlling Influence over Conformer Stability. *Can. J. Chem.* **1996**, *74*, 1271–1282. [CrossRef]
23. Wilson, A.J.C.; Geist, V. International Tables for Crystallography. Volume C: Mathematical, Physical and Chemical Tables (See, Table 9.5.1.1.). *Cryst. Res. Technol.* **1993**, *28*, 110. [CrossRef]
24. Caine, B.A.; Bronzato, M.; Popelier, P.L.A. Experiment Stands Corrected: Accurate Prediction of the Aqueous pKa Values of Sulfonamide Drugs Using Equilibrium Bond Lengths. *Chem. Sci.* **2019**, *10*, 6368–6381. [CrossRef]
25. Habgood, M.; Grau-Crespo, R.; Price, S.L. Substitutional and Orientational Disorder in Organic Crystals: A Symmetry-Adapted Ensemble Model. *Phys. Chem. Chem. Phys.* **2011**, *13*, 9590. [CrossRef]

© 2020 by the authors. Licensee MDPI, Basel, Switzerland. This article is an open access article distributed under the terms and conditions of the Creative Commons Attribution (CC BY) license (http://creativecommons.org/licenses/by/4.0/).

Article

Does Chlorine in CH₃Cl Behave as a Genuine Halogen Bond Donor?

Pradeep R. Varadwaj [1,2,*], Arpita Varadwaj [1,2] and Helder M. Marques [3]

1. Department of Chemical System Engineering, School of Engineering, The University of Tokyo 7-3-1, Tokyo 113-8656, Japan; varadwaj.arpita@gmail.com
2. The National Institute of Advanced Industrial Science and Technology (AIST), Tsukuba 305-8560, Japan
3. Molecular Sciences Institute, School of Chemistry, University of the Witwatersrand, Johannesburg 2050, South Africa; helder.marques@wits.ac.za
* Correspondence: prv.aist@gmail.com or pradeep@t.okayama-u.ac.jp

Received: 25 January 2020; Accepted: 15 February 2020; Published: 26 February 2020

Abstract: The CH$_3$Cl molecule has been used in several studies as an example purportedly to demonstrate that while Cl is weakly negative, a positive potential can be induced on its axial surface by the electric field of a reasonably strong Lewis base (such as O=CH$_2$). The induced positive potential then has the ability to attract the negative site of the Lewis base, thus explaining the importance of polarization leading to the formation of the H$_3$C–Cl···O=CH$_2$ complex. By examining the nature of the chlorine's surface in CH$_3$Cl using the molecular electrostatic surface potential (MESP) approach, with MP2/aug-cc-pVTZ, we show that this view is not correct. The results of our calculations demonstrate that the local potential associated with the axial surface of the Cl atom is inherently positive. Therefore, it should be able to inherently act as a halogen bond donor. This is shown to be the case by examining several halogen-bonded complexes of CH$_3$Cl with a series of negative sites. In addition, it is also shown that the lateral portions of Cl in CH$_3$Cl features a belt of negative electrostatic potential that can participate in forming halogen-, chalcogen-, and hydrogen-bonded interactions. The results of the theoretical models used, viz. the quantum theory of atoms in molecules; the reduced density gradient noncovalent index; the natural bond orbital analysis; and the symmetry adapted perturbation theory show that Cl-centered intermolecular bonding interactions revealed in a series of 18 binary complexes do not involve a polarization-induced potential on the Cl atom.

Keywords: halogen bonding; hydrogen bonding sigma-hole interactions; theoretical studies; characterizations

---

1. Introduction

Clark, Murray, Politzer, and their colleagues have analyzed the surface reactivity of several molecular systems using the molecular electrostatic surface potential (MESP) model [1–20]. They utilized density functional theory (DFT) with a variety of functionals (B3LYP, B3PW91, M06-2X) and a standard double/triple-ζ quality Gaussian basis set to compute the electrostatic potential [1–10]. They concluded that DFT, together with an 0.001 a.u. isodensity envelope on which to compute the potential, is adequate to reveal the nature of the electrostatic potential on the surface of any atom in a molecule [7,8]. In 1992, some of these authors considered several systems in their study of noncovalent interactions, including molecules such as CH$_3$F and CH$_3$Cl. It was contended in that study that "the potentials of CH$_3$F and CF$_4$ are indicative of fluorine interacting only with electrophiles, as is found experimentally" [1]. In this, and in a later study [2], the authors pondered why a σ-hole is not found when X = F in CF$_4$, as well as in other instances, such as in CH$_3$Cl. (A σ-hole is an electron density deficient region on the outer surface of X along the extension of the R–X bond, where R is remaining part of the molecule [2,7,15,21].) They concluded that the higher electronegativity of fluorine gives it a disproportionately large share of the σ-bonding electrons, which helps to neutralize the σ-hole. This also applies to chlorine in CH_3Cl, which does not have a σ-hole and does not halogen bond [2].

A halogen bond is formed when there is a favorable attractive interaction between a positive site (viz. a positive σ-hole) on a halogen in one molecule and a negative site on another molecule [21–24]. Such a broad view is applicable to other interactions such as the hydrogen bond [25], chalcogen bond [26], pnictogen bond [27], or any other σ-hole interaction [6,15,21] since a positive site on the hydrogen, chalcogen, pnictogen, or halogen atom in the molecule attracts a negative site on the other to form such an interaction.

Contrary to their earlier assertions, Politzer and co-workers have more recently found that the F atom in CH_3F molecule does indeed have a σ-hole, but it is negative [7]; similarly, Cl in CH_3Cl was also found to have a negative σ-hole [18].

The contention that Cl atom in CH_3Cl does not have a σ-hole on its own [2,3], has appeared quite frequently [5,7,10,20,28]. This is sometimes done when proposing that the CH_3Cl molecule is a good model system to understand the effect of electrostatic polarization in noncovalent interactions. For example, to explain what causes the formation of a $H_3C-Cl···O=CH_2$ complex, it was argued that despite the potential on the outer axial surface of the Cl atom in H_3C-Cl being weakly negative in the isolated molecule, this can be transformed and become positive through the electrostatic polarizing effect of the negative site interacting with it [5,7,10,28,29].

Such a provocative view led to the suggestion that the MESP model is superior to other computational methods such as the second-order natural bonding orbital analysis (NBO) [30], the quantum theory of atoms in molecules in molecules (QTAIM) [31–33], and the density functional theory symmetry adapted perturbation theory energy decomposition analysis (DFT-SAPT-EDA) [34,35]. While the reliability of these latter methods has been questioned [36–39], such claims have been rebutted by others [40–46]. Some of these conflicting views have been briefly highlighted in one of our recent reviews [21].

In contrast with the arguments given by Politzer and co-workers [1,2], some of us have shown that each fluorine in CF_4 conceives a positive σ-hole along each of the four C–F bond extensions [47]. CF_4 can not only form a 1:1 cluster with Lewis bases such as H_2O, NH_3, $H_2C=O$, HF, and HCN but also 1:2, 1:3, and 1:4 clusters with the last three (randomly chosen) Lewis bases. There are many known fluorinated compounds in which F conceives a positive or a negative σ-hole that has the ability to engage in a σ-hole centered noncovalent interaction [48–55]. This also applies not only to Cl in H_3C-Cl [56,57], but also to O in a variety of molecules as reported recently [58,59], despite claims on several occasions that O does not conceive a σ-hole and does not participate in chalcogen bonding [60–64].

In this study we use the ab initio Møller–Plesset second-order perturbation theory (MP2) method in combination with the Dunning's correlated consistent aug-cc-pVTZ basis set and the MESP model to investigate the detailed nature of various local potential maxima and minima on the electrostatic surface of a CH_3Cl molecule. The critical point (cp) topology of the Laplacian of the charge density is calculated within the QTAIM framework to see whether this model is capable of providing insights into the reactivity of the molecule, and whether these are comparable with the predictions of the MESP model. We consider 10 Lewis bases to examine whether these are capable of sustaining an attractive intermolecular interaction with the axial and/or lateral sites of the Cl atom in CH_3Cl. We consider whether the various intermolecular interactions revealed (viz. halogen bonds, chalcogen bonds, hydrogen bonds, and pnictogen bonds) can be unambiguously regarded as σ-hole interactions, as has been claimed [28]. We also explore whether the various arguments advanced [2,3] to support the idea that the positive potential on the Cl atom in CH_3Cl can be induced by the electric field of the Lewis base during the course of an intermolecular interaction is tenable.

We utilize the NBO, QTAIM, DFT-SAPT-EDA, and RDG (reduced density gradient) noncovalent index [65] theoretical tools to explore and discuss the reliability of and the agreement between the results of these approaches in elucidating intermolecular interactions in the 18 complexes of H_3C-Cl

molecules studied. Based on our results, we argue that combining an inappropriate theoretical method with an arbitrarily chosen isodensity envelope can be misleading insofar as the sign of the potential on the axial portion of the Cl atom is concerned, and when such a result is used for the interpretation of the origin of an intermolecular interaction, misleading conclusions can be reached.

2. Computational Details

Using the Gaussian 09 code [66], 10 monomers and 18 binary complexes were fully energy-minimized with MP2 [67] and the aug-cc-pVTZ basis set. A Hessian second derivative calculation was performed for each of them to identify the nature of the structure; positive eigenvalues were found.

To evaluate the effect of the isodensity envelope on the nature of the electrostatic potential, four different isodensity values, viz. 0.0005, 0.0010, 0.0015, and 0.0020 a.u., were chosen on which to compute the electrostatic potential. The local maxima and minima of potential ($V_{s,max}$ and $V_{s,min}$, respectively) on the electrostatic surface of the CH$_3$Cl monomer were identified and characterized. The MP2 energy-minimized geometry of the monomer was used. The positive ($V_{s,max} > 0$ or $V_{s,min} > 0$) and negative signs ($V_{s,max} < 0$ or $V_{s,min} < 0$) of the potential on an atom X in a molecule generally represent the electrophilic and nucleophilic regions on any molecule, respectively [47,49–52,58–60]. Regions described by $V_{s,max} > 0$ (or $V_{s,max} < 0$) on the outer axial portion of the atom X represent a positive (or a negative) σ-hole (as on X in X$_2$ and CX$_4$, where X = F, Cl, Br, I [21,39,68,69] or on F in H–F and H$_3$C–F [51]) and those described by $V_{s,max} = 0$ on the outer axial portion of the atom X represent to a neutral σ-hole [2,70].

A selected number of charge density-based descriptors of bonding interaction were evaluated using QTAIM [31–33], including the charge density (ρ_b), the Laplacian of the charge density ($\nabla^2 \rho_b$), and the total energy density (H_b) at the bond critical points (bcps). The model assumes that an open system is bounded by a surface **S**(r_s) of local zero-flux in the gradient vector field of the charge density $\rho(r)$ (Equation (1), where **n**(r) is a unit vector normal to the surface at r).

$$\nabla \rho(r).n(r) = 0 \quad \forall r \varepsilon S(r_s) \tag{1}$$

The analysis of the delocalization indices (DIs) between atom pairs was also performed within the interacting quantum atoms (IQA) model of QTAIM [71,72]. DI is a measure of bond order since it represents the extent of the delocalization of electron pairs between two atomic basins in any closed-shell system [73]. Since noncovalent interactions are a result of very minimal charge density localization between the lump and hole, the DI values are typically small (< 0.05 for weakly bound interactions) [73,74].

The RDG [65] based isosurface plots were evaluated using the MP2 equilibrium geometries of the 18 complexes. This method uses the sign of the second eigenvalue λ_2 of the Hessian second derivative charge density matrix to recognize the nature of the chemical interaction. At the same time, it uses the value of charge density ρ to measure the strength of the interaction. As such, the signature sign(λ_2) × ρ < 0 represents a closed-shell interaction (attraction). Similarly, sign(λ_2) × $\rho \approx 0$ represents a van der Waals (attraction) and sign(λ_2) × ρ > 0 a steric interaction (repulsion). The AIMAll [75], Multiwfn [76], and VMD [77] suite of programs, together with some in-house codes, were used for the analysis of the topological properties of the charge density, the RDG isosurfaces, and the electrostatic surface potentials.

The binding energy ΔE for each complex A···B was calculated using the supermolecular procedure proposed by Pople [78], described by Equation (2). The terms $E_T(A)$ and $E_T(B)$ in Equation (2) are, respectively, the electronic total energies of the two isolated monomers A and B in the complex A···B that has an electronic total energy of $E_T(A \cdots B)$. The ΔE was corrected for the basis set superposition error energy, E(BSSE), using the counterpoise procedure proposed by Boys and Bernardi [79]. Equation (3) was used for the calculation of the BSSE corrected energy, ΔE(BSSE).

$$\Delta E(A\cdots B) = E_T(A\cdots B) - E_T(A) - E_T(B) \qquad (2)$$

$$\Delta E(BSSE) = \Delta E(A\cdots B) + E(BSSE), \qquad (3)$$

The zeroth-order DFT SAPT-EDA analysis [34,35] was performed using the Psi4 code [80] and the MP2 geometries of the monomers in the complexes. The aug-pVDZ-JKFIT [81] DF basis was used for SCF calculations, whereas the aug-cc-pVDZ-RI DF basis was used for the evaluation of the SAPT0 electrostatics, induction and dispersion components. The frozen core as well as asyncronous I/O was invoked while forming the DF integrals and CPHF coefficients. Equation (4) represents the SAPT0 interaction energy, E(SAPT0), which is the sum of the component energies arising from electrostatics (E_{eles}), repulsion (E_{exch}), induction (E_{ind}), and dispersion (E_{disp}).

$$E(SAPT0) = E_{eles} + E_{exch} + E_{ind} + E_{disp}, \qquad (4)$$

3. Results and Discussion

3.1. The Reactive Surface Profile of the CH_3Cl Monomer

Figure 1a shows the 2D contour plot of the Laplacian of the charge density ($\nabla^2\rho$) for the CH_3Cl molecule, obtained using a Cl-C-H plane. The positive contours (green solid lines) indicate areas of charge depletion, and the negative contours (red dashed lines) indicate areas of charge concentration. As such, the charge depletion is significant near C along the outer extension of the Cl–C covalent bond, thus showing a prominent "hole". In QTAIM representation, one might call this "hole" a region of valence shell charge depletion (VSCD). The same feature is less noticeable on Cl along the outer extension of the C–Cl bond. One might conclude that there is no "hole" on the Cl atom. We therefore carried out the critical point (cp) analysis of $\nabla^2\rho$ to provide some insight into the exact nature of charge density concentration and depletion around the Cl atom, since the minimum and maximum of $\nabla^2\rho$ represent the open- and closed-shell structure, respectively, of any specific region [59,82].

Although many cps of $\nabla^2\rho$ were identified, only a selected number are illustrated in Figure 1b. The tiny blue spheres represent the (3, −3) cps and are equivalent to the (3, +3) critical points of $-\nabla^2\rho$. The tiny pink spheres represent the (3, +3) critical point of $\nabla^2\rho$ and are equivalent to the (3, −3) critical point of $-\nabla^2\rho$. The (3, −3) critical point of $\nabla^2\rho$ is a local maximum of $\nabla^2\rho$; it is a point of locally maximal "charge depletion" when $\nabla^2\rho > 0$, and is a point of locally minimal "charge concentration" when $\nabla^2\rho < 0$. Similarly, the (3,+3) cp is a local minimum of $\nabla^2\rho$ and is a point of locally maximal "charge concentration" when $\nabla^2\rho < 0$, and of locally minimal "charge depletion" when $\nabla^2\rho > 0$. The $\nabla^2\rho$ at the (3,−3) cps on the extension of the C–Cl and Cl–C bond are both positive ($\nabla^2\rho$ = +0.1423 a.u. on Cl and +0.2221 a.u. on C); therefore the outer axial regions on the Cl and C atoms in the H_3C–Cl molecule are well characterized as regions of VSCD. These are therefore "holes", which may interact with the lumps localized on Lewis base molecules to form complexes. A similar conclusion might be arrived at when (3, −3) cps of $\nabla^2\rho$ are analyzed along the C–H bond extensions since (3, −3) critical point of $\nabla^2\rho$ are all positive ($\nabla^2\rho$ = +0.1114 a.u. on H along the C–H bond extension).

By contrast, the lateral portions of the Cl atom in the H_3C–Cl molecule are characterized by three (3, +3) cps of $\nabla^2\rho$. These are all negative ($\nabla^2\rho$ = −0.8297 a.u. each). They are associated with the lone-pairs on Cl; these "lumps" may have the ability to attract "holes" on an interacting molecule.

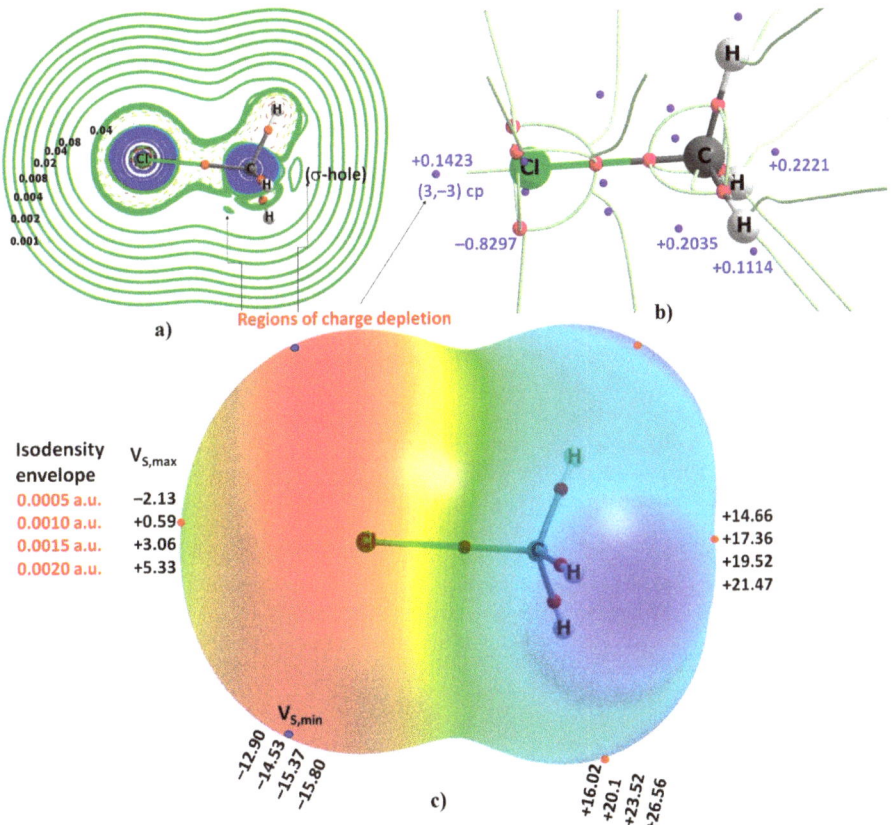

Figure 1. (a) The Laplacian of the charge density plot of CH$_3$Cl. (b) Selected critical points of the Laplacian of the charge density (values in a.u.). (c) The 0.001 a.u. isodensity envelope mapped potential on the surface of the CH$_3$Cl molecule. Values of potential extrema ($V_{S,min}$ and $V_{S,min}$ in kcal mol^{-1}) obtained via mapping with various isodensity envelopes are also shown.

The insight gained from an evaluation of the cps of $\nabla^2\rho$ is virtually no different from what might be inferred from the results of the MESP model. Figure 1c depicts the 0.001 a.u. (electrons bohr^{-3}) isodensity envelope mapped potential on the electrostatic surface of the H$_3$C–Cl molecule. It shows the axial outer portion of the Cl atom has a positive potential $V_{S,max}$ of +0.59 kcal mol^{-1}. This potential is associated with what might be called a positive, albeit weak, "σ-hole". The σ-hole region is surrounded by a belt of negative potential. The local minima associated with the lateral portions of the atom are characterized by a $V_{S,min}$ of −14.53 kcal mol^{-1}.

Passing from the 0.0010 a.u. through the 0.0015 a.u. to the 0.0020 a.u. isodensity envelope did not change the nature (sign) of the potential on the outer surface of the Cl atom in H$_3$C–Cl noted above, although the negative sites on Cl became more negative and the positive site becomes more positive. This is expected given that on moving closer to the nucleus of the atom one generally comes up with a relatively tiny electron density deficient surface.

The 0.0005 a.u. isodensity envelope mapped potentials are also included in Figure 1c. Passing from the 0.0005 a.u to the 0.0010 a.u. isodentiy surface has indeed had a notable effect on both the sign and magnitude of potential on Cl along the C–Cl bond extension. For instance, the $V_{S,max}$ was computed to be −2.13 kcal mol^{-1} on the 0.0005 a.u. isodensity envelope, which is completely different from that of +0.59 kcal mol^{-1} computed on the 0.001 a.u. isodensity envelope. This result unequivocally shows that the choice of the isodensity surface is arbitrary, which can lead to change in the sign of the potential. There is a somewhat less negative potential ($V_{S,min}$ = −12.90 kcal mol^{-1}) on the lateral portions of the same atom.

The negative potential on the axial portion of the Cl atom may be misleading given the 0.0005 a.u. isodensity envelope does not totally encompass the van der Waals surface of the molecule. This is consistent with the views of Bader et al. [83] and others [44,84], who have advocated the use of two contour values (0.0010 and 0.0020 a.u.) that should be large enough to encompass > 96% of a molecule's electronic charge density.

Based on the concern of a reviewer, and to confirm the reliability of [MP2/aug-cc-pVTZ] results above, we examined the nature of the local most potentials on the Cl atom using the aug-cc-pV(T + d)Z basis set. We used 10 different computational models, including the CCSD and nine popular density functionals. The results summarized in Table 1 demonstrate that the axial and lateral portions of the Cl atom on the C–Cl bond extensions are always positive and negative, respectively. Except for the PBE1 (PBE1PBE) functional, all other DFT and DFT-D3 functionals slightly underestimated the magnitude of $V_{S,max}$ on Cl compared to that obtained with CCSD. In addition, both the H and C atoms along the C–H and Cl–C bond extensions are positive, indicating that these can be sites for hydrogen bond and chalcogen bond formation when placed in close proximity to negative sites on another molecule.

Table 1. The 0.001 a.u. isodensity envelope mapped electrostatic potential on the outer surface of various bonded atoms in CH$_3$Cl, computed using various computational approaches in conjunction with the aug-cc-pV(T + d)Z basis set.

Method/Basis Set	$V_{S,max}$	$V_{S,min}$	$V_{S,max}$	$V_{S,max}$
	C–Cl	C–Cl	Cl–C	C–H
[CCSD/aug-cc-pV(T + d)Z]	0.86	−14.78	17.52	20.17
[MP2/aug-cc-pV(T + d)Z]	0.71	−14.58	17.44	20.19
[PBE0/aug-cc-pV(T + d)Z]	0.72	−14.74	17.03	20.37
[PBE1/aug-cc-pV(T + d)Z]	1.00	−14.16	16.74	19.70
[M062X/aug-cc-pV(T + d)Z]	0.50	−14.83	17.25	20.57
[wB97XD/aug-cc-pV(T + d)Z]	0.52	−15.12	17.62	20.88
[B97D3/aug-cc-pV(T + d)Z]	0.61	−14.56	17.68	19.74
[B3PW91/aug-cc-pV(T + d)Z]	0.69	−14.83	17.18	20.26
[B3LYP/aug-cc-pV(T + d)Z]	0.47	−14.71	17.83	19.94
[B3LYP-D3/aug-cc-pV(T + d)Z]	0.49	−14.71	17.91	19.96

3.2. Geometries of Intermolecular Complexes of H$_3$C–Cl with 10 Lewis Bases

Figure 2 shows the optimized geometries of 18 binary complexes formed between H$_3$C–Cl and nine Lewis bases. In many of these complexes, both the axial and lateral portions of the Cl atom in H$_3$C–Cl are involved in the attractive engagement with negative and positive sites, respectively, on the bases. The behavior of the Cl atom in H$_3$C–Cl towards the acids and bases in the interacting monomers is clearly similar in all cases. This is consistent with the reactivity profile predicted by cps of $\nabla^2\rho$ (viz. a "lump" attracts a "hole" and vice-versa), and that predicted using the MESP model. Since the potential on the Cl atom is essentially positive, there is certainly no transformation (induction) from a negative potential to a positive potential when the axial portion of the Cl atom is in close proximity to the negative site of the Lewis base. This result clearly negates the suggestion that a positive potential is induced on the Cl atom by the electric field of the interacting partner to promote a mutual Coulomb-type attractive engagement between them [2,3]. It should be pointed out that the previous studies used a lower-level of theory and a double-ζ Gaussian basis set to compute the electrostatic potential on the surface of the H$_3$C–Cl molecule [2,3]. This combination predicted an incorrect (negative) potential associated with the σ-hole of the Cl atom in H$_3$C–Cl and led the authors to offer a different interpretation of the nature of the surface reactivity of the molecule, thus exaggerating the importance of the idea of "electrostatic polarization" in complex formation.

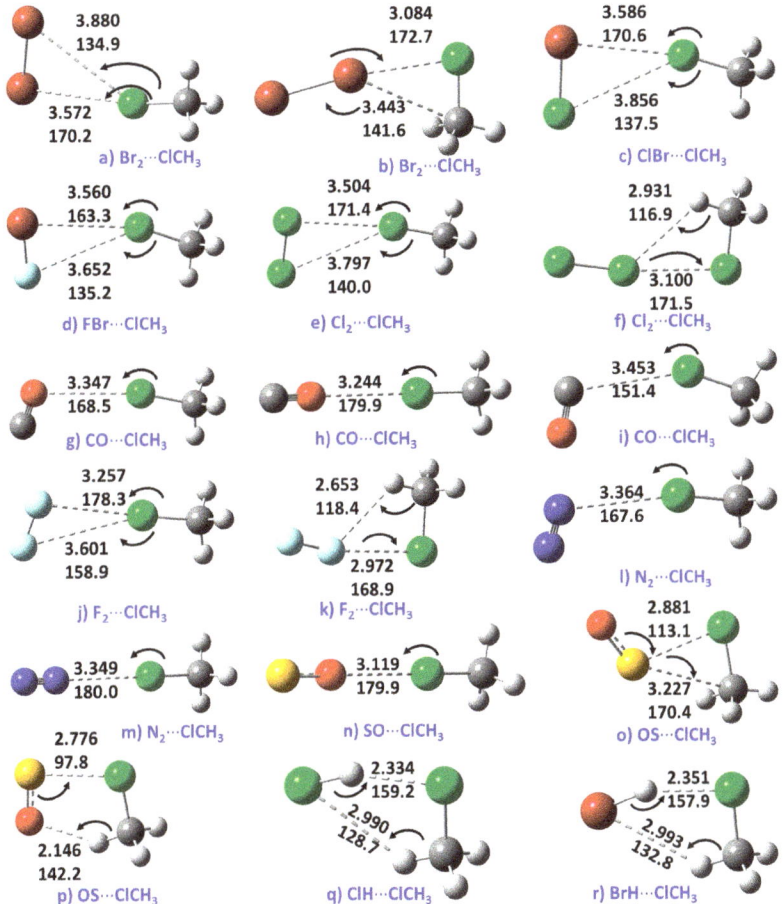

Figure 2. MP2/aug-cc-pVTZ energy-minimized geometries of the 18 binary complexes of CH_3Cl examined in this study. The intermolecular distance in Å (upper entry) and the angle of approach in degree (lower entry) between the monomers in each complex are given.

From the intermolecular geometries shown in Figure 2 between the monomers of the complexes, it is apparent that the Cl atom in CH_3Cl forms directional interactions with Br_2 (a); ClBr (c); FBr (d); Cl_2 (e); CO (g–i); F_2 (j);N_2 (l–m); and SO (n). The directionality of each contact is realized based on the angles of approach of the electrophile on the Cl atom, viz., 170.2, 170.6, 163.3, 171.4, 168.5, 179.9, 151.4, 178.3, 167.6, 180.0, and 179.9° for these complexes, respectively. These angles vary between 150 and 180°, and are typical of Type II contacts [21].

The intermolecular contact distances in all the complexes of Figure 2 are less than the sum of the van der Waals radii of the respective bonded atomic basins, (r_{vdW} (H) = 1.20 Å; r_{vdW} (F) = 1.46 Å; r_{vdW} (N) = 1.66 Å; r_{vdW} (O) = 1.50 Å; r_{vdW} (S) = 1.89 Å; r_{vdW} (Cl) = 1.82 Å; and r_{vdW} (Br) = 1.86 Å) [85]. This is consistent with the geometry-based criterion recommended for hydrogen bonding [25], halogen bonding [24], and chalcogen bonding [26]. For instance, the IUPAC recommendation advises that in "a chalcogen-bonded complex R–Ch···A, the interatomic distance between the chalcogen donor atom Ch and the nucleophilic site in the acceptor A tends to be less than the sum of the van der Waals radii and more than the sum of covalent radii" [26].

For the complexes $H_3C-Cl···Br_2$, $H_3C-Cl···Cl-Cl$ and $H_3C-Cl···F-F$ shown in Figure 2b,f,k, respectively, the lateral portion of the Cl atom in H_3C-Cl acts as a lump for making an attractive engagement with the "hole" on the partner molecules. The attraction is arguably due to the outer axial surfaces of the halogen atoms in the Br–Br, Cl–Cl, and F–F molecules, characterized by positive electrostatic potentials [86], interacting with the lateral negative site on the Cl atom in H_3C-Cl in the aforementioned complexes, resulting in the formation of the Cl···Br-Br, Cl···Cl-Cl, and Cl···F-F halogen bond interactions, respectively. The intermolecular distances associated with these interactions are 3.084, 3.100, and 2.972 Å, respectively, while the intermolecular angles are 172.7, 17.5, and 168.9°, respectively. The angular feature indicates not only the presence of Type II contacts, but also clarifies why the intermolecular distances are smaller than the sum of the van der Waals radii of the interacting atomic basins. For example, the intermolecular distances 3.084 (Cl···Br), 3.100 (Cl···Cl) and 2.972 Å (F···Cl) in Figure 2b,e,k are less than the sum of the van der Waals radii of 3.68, 3.64, and 3.28 Å, respectively. Clearly, the feasibility of positive potentials on the Br, Cl, and F atoms in Br_2, Cl_2, and F_2 causing the formation of these three complexes is certainly not developed by induction caused by the electric field of the lumps of the Cl atom in H_3C-Cl. The potentials on the bimolecular halogen atoms are inherently positive (as observed on Cl in H_3C-Cl), thus helping with the development of the intermolecular interaction with the lumps of the Cl atom in H_3C-Cl.

The intermolecular bonding features shown in the complexes (a), (c–e), (g–j), and (l–n) can also be regarded as halogen bonding. However, the only difference between these and the above set of three complexes (b, e, k) is that the Cl atom in H_3C-Cl acts as an electrophile in the former complexes but as a nucleophile in the latter. The results provide evidence of the amphoteric nature of the charge density profile on the surface of the Cl atom in H_3C-Cl, in excellent agreement with the nature of the surface reactivity predicted by the cp topology of $\nabla^2\rho$.

The complexes $H_3CCl···SO$ and $H_3CCl···SO$ shown in (o) and (p) are not the consequence of halogen bonding. They both feature a Cl···S intermolecular contact. For this, the lump on the lateral portion of the Cl atom in H_3C-Cl interacts with the S atom in SO. The intermolecular distances associated with the Cl···S contacts are very short, with r(Cl···S) of 2.881 Å and 2.776 Å for complexes in (o) and (p), respectively. These are significantly smaller than the sum of the van der Waals radii of the Cl and S atoms, 3.71 Å (r_{vdW} (S) = 1.89 Å; r_{vdW} (Cl) = 1.82 Å). Moreover, an examination of the intermolecular angular geometry suggests that Type I bonding topologies promote the formation of these contacts. Type I contacts are generally characterized by a contact angle that varies between 90° and 150°, and the participating atoms that form the contact are generally either both positive or both negative [21]. Previous studies have demonstrated that Type I contacts are dispersion driven [21,23,87]. This view has been advanced because the σ-hole model fails to provide true insight into the origin of this interaction; in this case, the Coulombic model description of noncovalent interactions [10,17,18,28] does not work very well.

The chalcogen bonded contacts identified in the $H_3CCl···SO$ complexes provide unequivocal evidence that the newly identified Type I contact can be formed not only between sites of opposite polarity, but also feature the fact that the Coulomb description (viz. positive site attracts a negative one!) can be utilized for its effective realization.

For the $H_3CCl···HCl$ and $H_3CCl···HBr$ complexes shown in (q) and (r), respectively, the "hole" on the hydrogen atom in HX (X = Cl, Br), which is described by the (3, –3) cp of $\nabla^2\rho$ (Figure 1b), interacts with the lump of the Cl in H_3CCl. The intermolecular distance associated with the resulting X···H (X = Cl, Br) contact (Cl···H = 2.334 Å and Br···H = 2.351 Å) is shorter than the sum of van der Waals radii of the X and H atoms (r_{vdW} (Cl + H) = 3.02 Å and and r_{vdW} (Br + H) = 3.06 Å). The approach angle of the electrophile identifies the interaction to be of Type II (∠Cl···H–Cl = 159.2° and ∠Cl···H–Br = 157.9°

in the respective complexes). These signify the presence of hydrogen bonding in the complexes of H$_3$CCl$\cdots$HCl and H$_3$CCl$\cdots$HBr.

3.3. QTAIM Description of Intermolecular Bonding Interactions in the Complexes of H$_3$C–Cl

The QTAIM molecular graphs of the 18 binary complexes of CH$_3$Cl studied are shown in Figure 3. They confirm the presence of primary interactions between the monomers in the complexes, as discussed above; there are well-defined bond paths and (3, −1) bond critical points between the bonded atomic basins in each complex. This is in good agreement with the recommendation of IUPAC [24–26]. The molecular graphs also indicate the possibility of secondary interactions in two cases: SO$\cdots$H in (p) and Br$\cdots$H in (r).

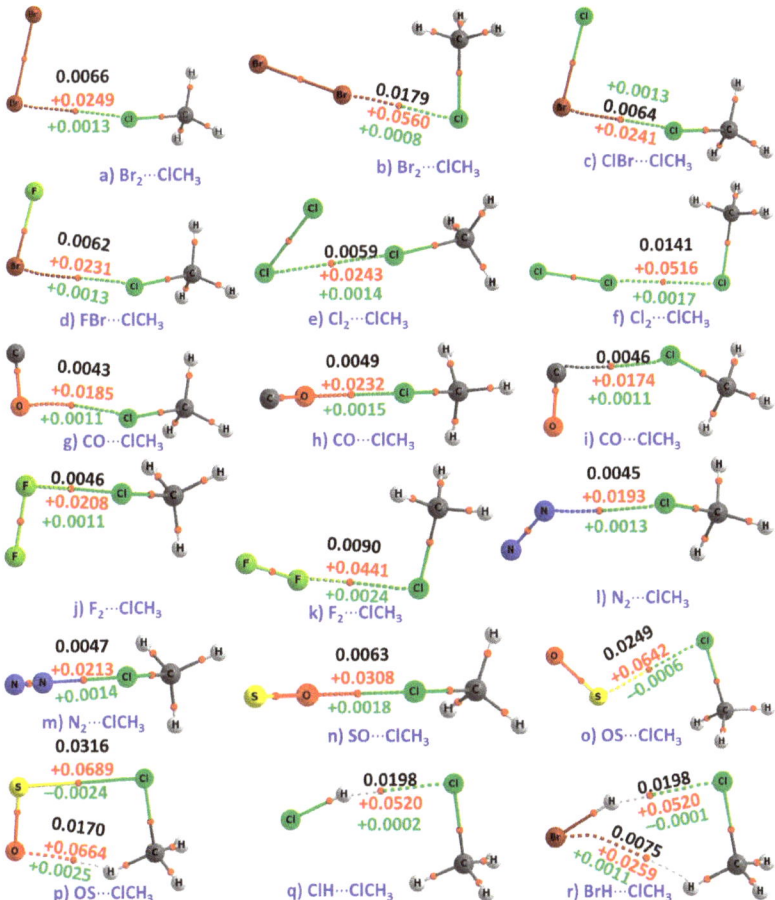

Figure 3. MP2/aug-cc-pVTZ calculated quantum theory of atoms in molecules in molecules (QTAIM) molecular graphs of the 18 binary complexes of CH$_3$Cl studied. The bond paths (solid and dotted lines) and the bond critical points (bcps) (tiny red spheres) are shown between the atomic basins. The charge density (ρ_b/a.u.), the Laplacian of the charge density ($\nabla^2\rho_b$/a.u.), and the total energy density (H_b/a.u.) at the bcps of the intermolecular interactions are shown in black, red, and green fonts, respectively.

The ρ_b values (ρ_b < 0.0198 a.u.) at the X···Cl (X = Cl, Br, F), Cl···X (X = Cl, Br, F), O···S, O···H, and Br···H bcps, are small for all the complexes (see Figure 3 for exact values). At all bcps $\nabla^2\rho_b$ > 0. These signatures indicate the closed-shell nature of the intermolecular interactions [42,43,47–52,57–59,88–91].

The total energy density, H_b, is another topological descriptor of bonding interactions; it is the sum of the "gradient" kinetic energy density and potential energy density (i.e., $H_b = G_b + V_b$) [88–91]. H_b > 0 indicates that G_b > V_b while H_b < 0 implies V_b > G_b. These are considered to be signatures of stabilizing and destabilizing interactions, respectively [88–91]. The H_b values are were found to be negative at the S···Cl and H···Br bcps of the complexes shown in (o), (p), and (r), respectively. This means that these interactions include partial shared (covalent) character. On the other hand, the H_b values were positive at the X···Cl (X = Cl, Br, F), Cl···X (X = Cl, Br, F), O···H, and H···Cl bcps of the remaining complexes of Figure 2, which is indicative of closed-shell ionic interactions.

It was recently argued [28] that many classical and non-classical interactions, variously referred to as proper and improper, blue-shifted and red-shifted, dihydrogen and anti-hydrogen, resonance-assisted and polarization-assisted, and so on, are straightforward σ-hole interactions. What then can be said about the Cl···S interactions identified in the H_3CCl···SO complexes (Figure 3o,p)? It would be misleading to refer to them as σ-hole interactions. The results of the MESP model suggests that the lateral portion of the S atom in SO is described by four extrema of potential. Two of them are positive, each with the $V_{S,max}$ of +34.2 kcal mol^{-1}. The other two are negative, each with a $V_{S,min}$ of −8.9 kcal mol^{-1}. There is no extremum of positive potential identified on the S atom along the outer extension of the O–S bond. The site on S that is interacting with the negative lateral site on Cl in H_3CCl is positive, thus forming the OS···ClCH$_3$ complexes. Since $V_{S,max}$ on S is a result of the depopulation of a π-type orbital, its engagement with the Cl atom in H_3CCl does not lead to the formation of a σ-hole interaction. As indicated above, the interaction cannot be regarded as a Type II interaction (∠Cl···S–O is 113.1° in (o) and in 97.8° in (p)).

To provide further insight into the orbital origin of the Cl···S interaction, we carried out an analysis of the second-order perturbative estimates of "donor-acceptor" (bond-antibond) interaction energies using the NBO approach [30]. Our results suggest that the Cl···S interaction in (o) is described by the combined effects of $n(3)Cl \rightarrow \sigma^*(S–O)$ and $n(3)Cl \rightarrow \pi^*(S–O)$ charge transfer delocalizations, where n refers to the lone-pair bonding orbital, and σ* and π* are the anti-bonding σ- and π-type orbitals, respectively. These charge transfer delocalizations are accompanied by second order perturbative lowering energy $E^{(2)}$ of 2.4 and 8.7 kcal mol^{-1}, respectively. Similarly, the $E^{(2)}$ for the charge transfer delocalizations responsible for the formation of the Cl···S interaction in (p) were found to be 0.4 and 23.9 kcal mol^{-1}, respectively. These results demonstrate that the origin of the Cl···S interactions in (o) and (p) cannot be understood by the oversimplified Coulombic arguments of the MESP model.

3.4. RDG Isosurface Topologies of the Complexes of $H_3C–Cl$

The results of the RDG isosurface analysis, summarized in Figure 4, show that the intermolecular bonding region in each complex is characterized by one (or two) RDG isosurface domain(s). These domains are colored either in bluish-green, green, light brown, or dark red. The coloring scheme is based on the combined effect of the extent of the electron density delocalization between the atomic basins and the sign of the second eigenvalue λ_2 of the Hessian second derivative charge density matrix. The signature sign $(\lambda_2) \times \rho < 0$ represents an attractive interaction; sign $(\lambda_2) \times \rho \approx 0$ represents a van der Waals interaction; and sign $(\lambda_2) \times \rho > 0$ represents a repulsive interactions. The spread of the isosurface (volume) is tuned by the extent of the charge density delocalization around the critical bonding region.

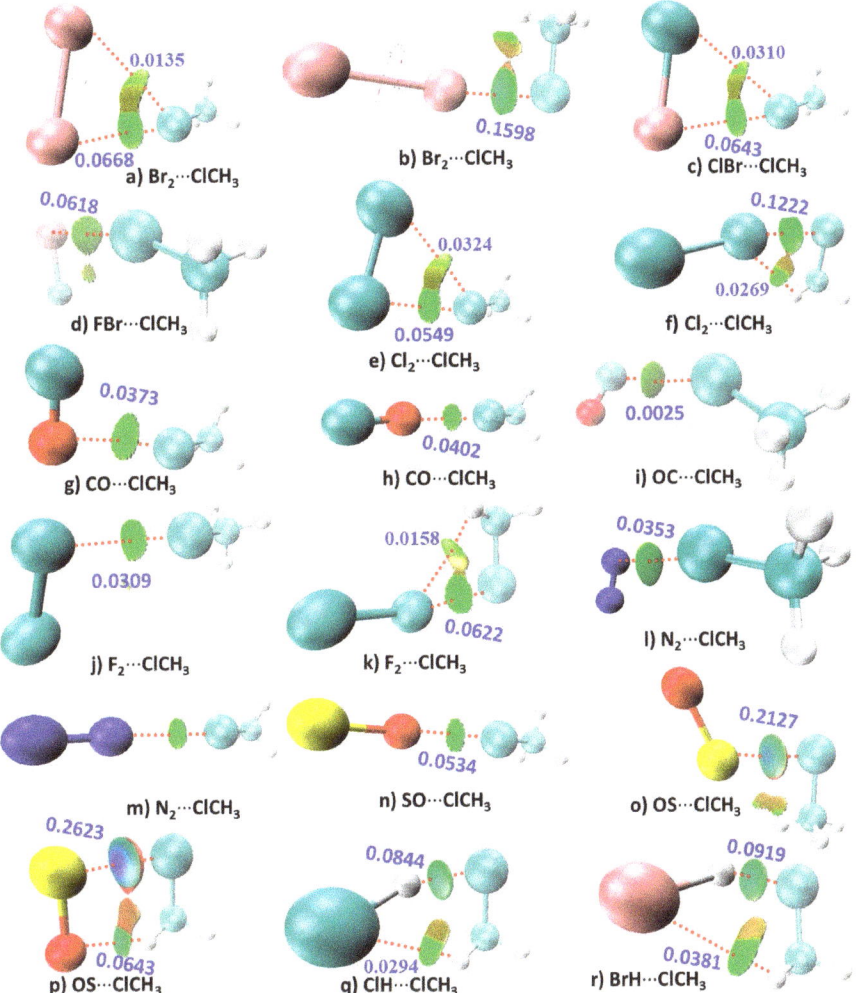

Figure 4. The MP2/aug-cc-pVTZ computed reduced density gradient (RDG) isosurface topologies for the 18 binary complexes of CH$_3$Cl studied. The delocalization indices (DIs) corresponding to selected atom–atom pairs are shown for each complex. The blue, green, and brownish isosurfaces represent strong, medium-strength, and weakly bound attractive interactions, respectively, whereas that in red represents repulsive interactions.

RDG predicts the presence of both primary and secondary contacts between the monomers in 14 complexes, except for (g)–(j) and (l)–(n). From the values of the angles of interaction shown in Figure 2, it is clear that the secondary interactions identified in most of the complexes follow the Type I topology of bonding and hence are dispersion driven.

The primary interactions in 14 of the 18 complexes are characterized by green isosurfaces. The interactions in the other four complexes are characterized by bluish RDG domains, including (o)–(r). The isosurface representing these interactions between the S and Cl atoms in OS···ClCH$_3$ (in (o) and (p)), and that between the H and Cl atoms in ClH···ClCH$_3$, (q), as well as that between the H and Br atoms in BrH···ClCH$_3$, (r), is bluish-green. It indicates that the strength of the intermolecular interaction in these four complexes is stronger than those in the remaining 14 complexes. This is consistent with the ρ_b, $\nabla\rho_b$, H_b, and DI values predicted for these interactions (see Figures 3 and 4 for values), suggesting that the stability of the intermolecular interaction in these four complexes is in the order S···Cl (p) >> S···Cl (o) > H···Cl (r) > H···Cl (q). Similarly, the preferential stability of the hydrogen bonds in the complexes (p), (q), and (r) is in the order H···O (p) > H···Br (r) > H···Cl (q). The positive $V_{S,max}$ on the donor atoms of the monomers responsible for these interactions predicted by the MESP model fail to provide such an insight, suggesting that the extrema of potential may not be reliable as a measure of bond stability.

QTAIM based bond path features shown in Figure 3 are in reasonable agreement with the RDG isosurface topologies for most of the complexes. The only discrepancy between them is in the complexes of ClCH$_3$ with Br$_2$ (k), ClBr (c), Cl$_2$ ((e) and (f)), F$_2$ (k), and ClH (q). This is apparently because the RDG method predicts the possibility of secondary interactions between interacting monomers in these complexes, but QTAIM does not recognize these as interactions since the bond path topologies between the bonded atomic basins are missing. The mismatch is not very surprising given that QTAIM sometimes underestimates weakly bound interactions in molecular complexes [58,59]. Even so, the QTAIM based delocalization results summarized in Figure 4 are in good agreement with the RDG's isosurface topologies of secondary interactions since the former recognizes all the interactions inferred by the latter.

3.5. Energy Stability

Table 2 summarizes the MP2 calculated binding energies for the 18 complexes of ClCH$_3$ examined in this study. As indicated above, the Cl-bonded complexes of ClCH$_3$ with X$_2$ (X = F, Cl, Br) are weaker than the X bonded complexes of ClCH$_3$. For example, the ΔE of the complex in (b) is –3.07 kcal mol^{-1} larger than that of complex (a) and of the complex (f) is –1.96 kcal mol^{-1} larger than that of (e). Similarly, the ΔE of complex (k) is –0.97 kcal mol^{-1} larger than that of (j). These results suggest that the weaker σ-hole on the Cl atom in ClCH$_3$ forms weaker complexes compared to those formed by the relatively stronger σ-holes on X in X$_2$. While this conclusion is rather qualitative, it has to be appreciated that the energy due to the secondary interactions does play a role to determine the overall strength of each of these complexes.

Table 2. Comparison of the MP2/aug-cc-pVTZ computed binding energies with the density functional theory symmetry adapted perturbation theory (DFT-SAPT) interaction energies for the 18 binary complexes of CH$_3$Cl [a].

Figure 2	Complex	ΔE	ΔE(BSSE)	E_{eles}	E_{exch}	E_{ind}	E_{disp}	E(SAPT0)
(a)	H$_3$CCl···Br$_2$	−1.83	−1.07	−0.87	2.77	−0.30	−2.37	−0.78
(b)	H$_3$CCl···Br$_2$	−4.90	−3.43	−6.49	10.88	−2.86	−4.66	−3.13
(c)	H$_3$CCl···BrCl	−1.58	−1.02	−0.73	2.33	−1.10	−2.10	−0.76
(d)	H$_3$CCl···BrF	−1.21	−0.71	−0.70	1.96	−0.24	−1.56	−0.53
(e)	Cl$_2$···ClCH$_3$	−1.27	−0.96	−0.59	1.97	−0.20	−1.94	−0.76
(f)	H$_3$CCl···Cl$_2$	−3.23	−2.68	−4.13	6.91	−1.74	−3.50	−2.46
(g)	OC···ClCH$_3$	−0.78	−0.61	−0.47	0.97	−0.08	−0.99	−0.57
(h)	OC···ClCH$_3$	−0.73	−0.5	−0.10	0.82	−0.09	−0.94	−0.30
(i)	OC···ClCH$_3$	−0.72	−0.56	−0.62	1.08	−0.15	−1.00	−0.70
(j)	F$_2$···ClCH$_3$	−0.55	−0.37	−0.21	0.68	−0.04	−0.77	−0.34
(k)	H$_3$CCl···F$_2$	−1.52	−1.15	−1.47	2.82	−0.58	−1.66	−0.89
(l)	N$_2$···ClCH$_3$	−0.68	−0.45	−0.29	0.90	−0.09	−0.97	−0.45
(m)	N$_2$···ClCH$_3$	−0.80	−0.6	−0.37	1.07	−0.09	−1.09	−0.49
(n) [b]	SO···ClCH$_3$	−4.52	−3.74	—	—	—	—	—
(o)	OS···ClCH$_3$	−0.74	−0.48	0.08	1.36	−0.24	−1.45	−0.25
(p) [b]	OS···ClCH$_3$	−6.79	−5.71	—	—	—	—	—
(q)	H$_3$CCl···HCl	−4.30	−3.65	−5.29	7.72	−2.70	−3.27	−3.55
(r)	H$_3$CBr···HBr	−4.59	−3.34	−5.37	8.86	−2.83	−3.80	−3.13

[a] Values in kcal mol^{-1}. [b] DFT-SAPT calculations could not be performed for these two complexes because of the convergence issues associated with the Psi4 code.

The OS···ClCH$_3$ complex (p), on the other hand, is found to be most stable in the series, with the ΔE of −6.79 kcal mol^{-1}. The complexes H$_3$CCl···Br$_2$ (b), H$_3$CCl···Cl$_2$ (f), SO···ClCH$_3$ (n), H$_3$CCl···HCl (q), and H$_3$CCl···HBr (r) are of intermediate strength, with the ΔE of −3.23, −4.52, −4.30, and −4.59 kcal mol^{-1}, respectively.

The BSSE has a significant effect on the binding energies of all the complexes. It is as large as 1.47, 1.08, and 1.25 kcal mol^{-1} for complexes (b), (p), and (r), respectively. Nevertheless, the BSSE corrected MP2 binding energies, ΔE(BSSE), are found to be comparable with the corresponding DFT-SAPT interaction energies (E(SAPT0)) for the 18 complexes. The marginal discrepancy between them can be attributed to the level of correlation effect accounted for by the DFT-SAPT formalism, together with the basis set utilized. The largest difference of 0.3 kcal mol^{-1} between E(SAPT0) and ΔE(BSSE) is found for the complexes of Br$_2$ with ClCH$_3$ ((a) and (b)). There is no obvious relationship between E(SAPT0) (or ΔE(BSSE)) and the extrema of the electrostatic potential responsible for the formation of the 18 complexes examined.

The interaction energies for nine of the 18 complexes were found to be smaller than −1.0 kcal mol^{-1}. Does this mean the complexes are unbound? Should one actually consider the link between the monomers in these complexes as an attraction? Since the interaction energy is negative, the answer to the first question is certainly "no", since a negative interaction energy provides a clear and unequivocal signature for any bound state. The answer to the second question is "yes". The obvious reason for this is that van der Waals complexes usually have a weak binding energy of less than −1.0 kcal mol^{-1} [92–98]. The importance of such weakly bound interactions have been much appreciated in many fields including polymer science, biology, and crystal engineering [92–98]. For instance, van der Waals interactions are always weaker than any other chemical interaction and are the determinant of structure of proteins or even the overall shape of polymer structures [92,95,96,98] and the significance of such weakly bound interactions cannot be overlooked thus assuming that only strong interactions are significant for materials design and weak interactions do not play an important role in the field of noncovalent interactions.

The results of the DFT-SAPT based decomposed energy components summarized in Table 2 suggest that dispersive attraction (E_{disp}) does not tend to exceed the electrostatic and polarization components (E_{eles} and E_{ind}, respectively) for 12 of the 18 complexes. These include complexes (a)–(f), (i), (k), (q), and (r). The formation of these complexes is not strictly electrostatically driven, but the contributions due to dispersion and repulsion also play a significant role in determining their overall interaction energies and hence promoting their overall stability.

By contrast, the dispersive attraction tends to exceed the electrostatic and polarization components in the other six complexes, viz. (a), (c), (d), (e), (g), (h), (j), (l), (m), and (o). This might prompt the suggestion that the weak attraction that does exist in these complexes is less the result of a specific interatomic interaction, and more a general, non-specific, fairly isotropic, attraction that would occur between any pair of molecules. However, one should not forget that the overall interaction energy in these six complexes is the sum of four specific interaction types, and that these interactions collectively work to determine and explain the directionality of the intermolecular interactions identified, as has been pointed out before [99]. There should be no ambiguity in the origin of the attractive forces that lead to the formation of the 18 complexes examined in this study.

4. Conclusions

This study has shown that the analysis of the critical points of the Laplacian of the charge density could be informative in revealing the actual nature of the surface reactivity of the chlorine atom in CH$_3$Cl. This is in line with the nature of the local extrema of electrostatic potential identified on the surface of the Cl atom in CH$_3$Cl using the MESP model. In particular, it is shown that the combination of a suitable isodensity envelope with an appropriate theoretical method is important to correctly identify the electrophilic nature the Cl atom in CH$_3$Cl.

The electronic charge density distributions around the lateral and axial sites of Cl in CH_3Cl is not isotropic, indicating the amphiphilic nature of the Cl atom. The negative lateral sites on the Cl are shown to display sufficient ability to attract positive sites on the interacting atoms to form halogen bonds, or chalcogen bonds, or hydrogen bonds.

The attractive interaction of the positive "hole" on the Cl atom in CH_3Cl with various "lumps" in the interacting bases has led to the conclusion that the positive electrostatic potential on the Cl is certainly not induced by the electric field of the interacting species as others have suggested [2,3,10,28]. Rather, it is an inherent property of this atom in the molecule.

The bond path and critical point topologies of QTAIM associated with the primary bonding interactions in the 18 complexes are shown to be consistent with an RDG isosurface analysis. Although these topologies did not appear between the weakly bound atoms in some complexes, the results of QTAIM's delocalization analysis were shown to be concordant with those of RDG.

The supermolecular and SAPT interaction energies were shown to be in agreement. The dispersion interaction was also shown to be one the most important driving forces responsible for the formation of the 18 complexes investigated.

As shown for the complexes between CH_3Cl and SO, all types of intermolecular contacts cannot be regarded as σ-hole interactions.

Author Contributions: Conceptualization and Project Design, P.R.V. and A.V.; Investigation, P.R.V. and A.V.; Supervision, P.R.V.; Writing—Original Draft, P.R.V. and A.V.; Writing—Review and Editing, P.R.V., A.V., and H.M.M. All authors have read and agreed to the published version of the manuscript.

Funding: This research received no external funding.

Acknowledgments: This work was entirely conducted using the various facilities provided by the University of Tokyo. P.R.V. is currently affiliated with AIST and thanks Koichi Yamashita for support. H.M.M. thanks the National Research Foundation, Pretoria, South Africa, and the University of the Witwatersrand for funding.

Conflicts of Interest: The authors declare no conflicts of interest.

References

1. Brinck, T.; Murray, J.S.; Politzer, P. Surface electrostatic potentials of halogenated methanes as indicators of directional intermolecular interactions. *Int. J. Quantum Chem.* **1992**, *44*, 57–64. [CrossRef]
2. Clark, T.; Hennemann, M.; Murray, J.S.; Politzer, P. Halogen bonding: The σ-hole. *J. Mol. Model.* **2007**, *13*, 291–296. [CrossRef] [PubMed]
3. Politzer, P.; Lane, P.; Conch, M.C.; Ma, Y.; Murray, J.S. An overview of halogen bonding. *J. Mol. Model.* **2007**, *13*, 305–311. [CrossRef]
4. Murray, J.S.; Politzer, P. The electrostatic potential: An overview. *WIREs Comput. Mol. Sci.* **2011**, *1*, 153–163. [CrossRef]
5. Politzer, P.; Murray, J.S.; Clark, T. Halogen bonding and other σ-hole interactions: A perspective. *Phys. Chem. Chem. Phys.* **2013**, *15*, 11178–11189. [CrossRef]
6. Politzer, P.; Murray, J.S.; Concha, M.C. σ-hole bonding between like atoms; a fallacy of atomic charges. *J. Mol. Model.* **2008**, *14*, 659–665. [CrossRef]
7. Politzer, P.; Murray, J.S.; Clark, T.; Resnati, G. The s-hole revisited. *Phys. Chem. Chem. Phys.* **2017**, *19*, 32166–32178. [CrossRef]
8. Murray, J.S.; Resnati, G.; Politzer, P. Close contacts and noncovalent interactions in crystals. *Faraday Discuss.* **2017**, *203*, 113–130. [CrossRef]
9. Politzer, P.; Murray, J.S.; Clark, T. Mathematical modeling and physical reality in noncovalent interactions. *J. Mol. Model.* **2015**, *21*, 52. [CrossRef]
10. Clark, T.; Politzer, P.; Murray, J.S. Correct electrostatic treatment of noncovalent interactions: The importance of polarization. *WIREs Comput. Mol. Sci.* **2015**, *5*, 169–177. [CrossRef]
11. Politzer, P.; Huheey, J.E.; Murray, J.S.; Grodzicki, M. Electronegativity and the concept of charge capacity. *J. Mol. Str. THEOCHEM* **1992**, *259*, 99–120. [CrossRef]
12. Clark, T.; Murray, J.S.; Politzer, P. The σ-Hole Coulombic Interpretation of Trihalide Anion Formation. *ChemPhysChem* **2018**, *19*, 3044–3049. [CrossRef]

13. Politzer, P.; Riley, K.E.; Bulat, F.A.; Murray, J.S. Perspectives on Halogen Bonding and Other Sigma-Hole Interactions: Lex Parsimoniae (Occam's Razor). *Comput. Theor. Chem.* **2012**, *998*, 2–8. [CrossRef]
14. Politzer, P.; Murray, J.S.; Clark, T. σ-Hole Bonding: A Physical Interpretation. In *Halogen Bonding I: Impact on Materials Chemistry and Life Sciences*; Metrangolo, P., Resnati, G., Eds.; Springer International Publishing: Cham, Switzerland, 2015; pp. 19–42.
15. Clark, T. σ-Holes. *WIREs Comput. Mol. Sci.* **2013**, *3*, 13–20. [CrossRef]
16. Politzer, P.; Murray, J.S. The Hellmann-Feynman theorem: A perspective. *J. Mol. Model.* **2018**, *24*, 266. [CrossRef]
17. Politzer, P.; Murray, J.S.; Lane, P.; Concha, M.C. Electrostatically driven complexes of SiF4 with amines. *Int. J. Quant. Chem.* **2009**, *109*, 3773–3780. [CrossRef]
18. Politzer, P.; Murray, J.S.; Clark, T. Halogen bonding: An electrostatically-driven highly directional noncovalent interaction. *Phys. Chem. Chem. Phys.* **2010**, *12*, 7748–7757. [CrossRef]
19. Murray, J.S.; Macaveiu, L.; Politzer, P. Factors affecting the strengths of σ-hole electrostatic potentials. *J. Comput. Sci.* **2014**, *5*, 590–596. [CrossRef]
20. Murray, J.S.; Riley, R.E.; Politzer, P.; Clark, T. Directional weak intermolecular interactions: Sigma-hole bonding. *Aust. J. Chem.* **2010**, *63*, 1598–1607. [CrossRef]
21. Varadwaj, P.R.; Varadwaj, A.; Marques, H.M. Halogen Bonding: A Halogen-Centered Noncovalent Interaction Yet to Be Understood. *Inorganics* **2019**, *7*, 40. [CrossRef]
22. Metrangolo, P.; Resnati, G. Halogen Versus Hydrogen. *Science* **2008**, *321*, 918–919. [CrossRef]
23. Cavallo, G.; Metrangolo, P.; Milani, R.; Pilati, T.; Priimagi, A.; Resnati, G.; Terraneo, G. The halogen bond. *Chem. Rev.* **2016**, *116*, 2478–2601. [CrossRef]
24. Desiraju, G.R.; Shing Ho, P.; Kloo, L.; Legon, A.C.; Marquardt, R.; Metrangolo, P.; Politzer, P.; Resnati, G.; Rissanen, K. Definition of the halogen bond (IUPAC Recommendations 2013). *Pure Appl. Chem.* **2013**, *85*, 1711–1713. [CrossRef]
25. Arunan, E.; Desiraju, G.R.; Klein, R.A.; Sadlej, J.; Scheiner, S.; Alkorta, I.; Clary, D.C.; Crabtree, R.H.; Dannenberg, J.J.; Hobza, P.; et al. Definition of the hydrogen bond (IUPAC Recommendations 2011). *Pure Appl. Chem.* **2011**, *83*, 1637–1641. [CrossRef]
26. Aakeroy, C.B.; Bryce, D.L.; Desiraju, R.G.; Frontera, A.; Legon, A.C.; Nicotra, F.; Rissanen, K.; Scheiner, S.; Terraneo, G.; Metrangolo, P.; et al. Definition of the chalcogen bond (IUPAC Recommendations 2019). *Pure Appl. Chem.* **2019**, *91*, 1889–1892. [CrossRef]
27. Legon, A.C. Tetrel, pnictogen and chalcogen bonds identified in the gas phase before they had names: A systematic look at non-covalent interactions. *Phys. Chem. Chem. Phys.* **2017**, *19*, 14884–14896. [CrossRef]
28. Politzer, P.; Murray, J.S. An Overview of Strengths and Directionalities of Noncovalent Interactions: σ-Holes and π-Holes. *Crystals* **2019**, *9*, 165. [CrossRef]
29. Clark, T.; Murray, J.S.; Politzer, P. Role of Polarization in Halogen Bonds. *Austr. J. Chem.* **2014**, *67*, 451–456. [CrossRef]
30. Weinhold, F.; Landis, C.R. *Discovering Chemistry with Natural Bond Orbitals*; John Wiley & Sons, Inc.: Hoboken, NJ, USA, 2012.
31. Bader, R.F.W.; Nguyen-Dang, T.T.; Tai, Y. A topological theory of molecular structure. *Rep. Prog. Phys.* **1981**, *44*, 893–947. [CrossRef]
32. Bader, R.F.W. A quantum theory of molecular structure and its applications. *Chem. Rev.* **1991**, *91*, 893–928. [CrossRef]
33. Bader, R.F.W. Atoms in Molecules. In *Encyclopedia of Computational Chemistry*; Schleyer, P.V., Ed.; John Wiley and Sons: Chichester, UK, 1998; Volume 1, pp. 64–86.
34. Jeziorski, B.; Moszynski, R.; Szalewicz, K. Perturbation Theory Approach to Intermolecular Potential Energy Surfaces of van der Waals Complexes. *Chem. Rev.* **1994**, *94*, 1887–1930. [CrossRef]
35. Parker, T.M.; Burns, L.A.; Parrish, R.M.; Ryno, A.G.; Sherrill, C.D. Levels of symmetry adapted perturbation theory (SAPT). I. Efficiency and performance for interaction energies. *J. Chem. Phys.* **2014**, *140*, 094106. [CrossRef]
36. Wick, C.R.; Clark, T. On bond-critical points in QTAIM and weak interactions. *J. Mol. Model.* **2018**, *24*, 142. [CrossRef]
37. Clark, T.; Murray, J.S.; Politzer, P. A perspective on quantum mechanics and chemical concepts in describing noncovalent interactions. *Phys. Chem. Chem. Phys.* **2018**, *20*, 30076–30082. [CrossRef]

38. Politzer, P.; Murray, J.S. A look at bonds and bonding. *Struct. Chem.* **2019**, *30*, 1153–1157. [CrossRef]
39. Clark, T.; Heßelmann, A. The coulombic σ-hole model describes bonding in CX3I···Y− complexes completely. *Phys. Chem. Chem. Phys.* **2018**, *20*, 22849–22855. [CrossRef]
40. Andrés, J.; Ayers, P.W.; Boto, R.A.; Carbó-Dorca, R.; Chermette, H.; Cioslowski, J.; Contreras-García, J.; Cooper, D.L.; Frenking, G.; Gatti, C.; et al. Nine questions on energy decomposition analysis. *J. Comp. Chem.* **2019**, *40*, 2248–2283. [CrossRef]
41. Thirman, J.; Engelage, E.; Huber, S.M.; Head-Gordon, M. Characterizing the interplay of Pauli repulsion, electrostatics, dispersion and charge transfer in halogen bonding with energy decomposition analysis. *Phys. Chem. Chem. Phys.* **2018**, *20*, 905–915. [CrossRef]
42. Varadwaj, A.; Marques, H.M.; Varadwaj, P.R. Nature of halogen-centered intermolecular interactions in crystal growth and design: Fluorine-centered interactions in dimers in crystalline hexafluoropropylene as a prototype. *J. Comp. Chem.* **2019**, *40*, 1836–1860. [CrossRef]
43. Varadwaj, P.R.; Varadwaj, A.; Marques, H.M.; Yamashita, K. Can Combined Electrostatic and Polarization Effects Alone Explain the F···F Negative-Negative Bonding in Simple Fluoro-Substituted Benzene Derivatives? A First-Principles Perspective. *Computation* **2018**, *6*, 51. [CrossRef]
44. Varadwaj, A.; Marques, H.M.; Varadwaj, P.R. Is the Fluorine in Molecules Dispersive? Is Molecular Electrostatic Potential a Valid Property to Explore Fluorine-Centered Non-Covalent Interactions? *Molecules* **2019**, *24*, 379. [CrossRef]
45. Sánchez–Sanz, G.; Alkorta, I.; Elguero, J. Theoretical Study of Intramolecular Interactions in Peri-Substituted Naphthalenes: Chalcogen and Hydrogen Bonds. *Molecules* **2017**, *22*, 227. [CrossRef]
46. Wolters, L.P.; Schyman, P.; Pavan, M.J.; Jorgensen, W.L.; Bickelhaupt, F.M.; Kozuch, S. The many faces of halogen bonding: A review of theoretical models and methods. *WIREs: Comput. Mol. Sci.* **2014**, *4*, 523–540. [CrossRef]
47. Varadwaj, A.; Varadwaj, P.R.; Jin, B.-Y. Fluorines in tetrafluoromethane as halogen bond donors: Revisiting address the nature of the fluorine's σ-hole. *Int. J. Quantum Chem.* **2015**, *115*, 453–470. [CrossRef]
48. Varadwaj, P.R.; Varadwaj, A.; Jin, B.-Y. Unusual bonding modes of perfluorobenzene in its polymeric (dimeric, trimeric and tetrameric) forms: Entirely negative fluorine interacting cooperatively with entirely negative fluorine. *Phys. Chem. Chem. Phys.* **2015**, *17*, 31624–31645. [CrossRef]
49. Varadwaj, A.; Varadwaj, P.R.; Jin, B.-Y. Can an entirely negative fluorine in a molecule, viz. perfluorobenzene, interact attractively with the entirely negative site (s) on another molecule (s)? Like liking like! *RSC Adv.* **2016**, *6*, 19098–19110. [CrossRef]
50. Varadwaj, A.; Varadwaj, P.R.; Marques, H.M.; Yamashita, K. Comment on "Extended Halogen Bonding between Fully Fluorinated Aromatic Molecules: Kawai et al., ACS Nano, 2015, 9, 2574–2583". *arXiv* **2018**, arXiv:1802.09995.
51. Varadwaj, A.; Varadwaj, P.R.; Marques, H.M.; Yamashita, K. Revealing Factors Influencing the Fluorine-Centered Non-Covalent Interactions in Some Fluorine-substituted Molecular Complexes: Insights from First-Principles Studies. *ChemPhysChem* **2018**, *19*, 1486–1499. [CrossRef]
52. Varadwaj, A.; Varadwaj, P.R.; Yamashita, K. Do surfaces of positive electrostatic potential on different halogen derivatives in molecules attract? like attracting like! *J. Comput. Chem.* **2018**, *39*, 343–350. [CrossRef]
53. Bauzá, A.; Frontera, A. Theoretical study on σ- and π-hole carbon···carbon bonding interactions: Implications in CFC chemistry. *Phys. Chem. Chem. Phys.* **2016**, *18*, 32155–32159. [CrossRef]
54. Bauzá, A.; Frontera, A. Electrostatically enhanced F···F interactions through hydrogen bonding, halogen bonding and metal coordination: An ab initio study. *Phys. Chem. Chem. Phys.* **2016**, *18*, 20381–20388. [CrossRef]
55. Eskandari, K.; Lesani, M. Does fluorine participate in halogen bonding? *Chem. Eur. J.* **2015**, *21*, 4739–4746. [CrossRef]
56. Kolář, M.H.; Hobza, P. Computer Modeling of Halogen Bonds and Other σ-Hole Interactions. *Chem. Rev.* **2016**, *116*, 5155–5187. [CrossRef]
57. Varadwaj, P.R.; Varadwaj, A.; Jin, B.-Y. Halogen bonding interaction of chloromethane withseveral nitrogen donating molecules: Addressing thenature of the chlorine surface σ-hole. *Phys. Chem. Chem. Phys.* **2014**, *16*, 19573–19589. [CrossRef]
58. Varadwaj, P.R. Does Oxygen Feature Chalcogen Bonding? *Molecules* **2019**, *24*, 3166. [CrossRef]

59. Varadwaj, P.R.; Varadwaj, A.; Marques, H.M.; MacDougall, P.J. The chalcogen bond: Can it be formed by oxygen? *Phys. Chem. Chem. Phys.* **2019**, *21*, 19969–19986. [CrossRef]
60. Beno, B.R.; Yeung, K.-S.; Bartberger, M.D.; Pennington, L.D.; Meanwell, N.A. A Survey of the Role of Noncovalent Sulfur Interactions in Drug Design. *J. Med. Chem.* **2015**, *58*, 4383–4438. [CrossRef]
61. Bauzá, A.; Mooibroek, T.J.; Frontera, A. The Bright Future of Unconventional σ/π-Hole Interactions. *ChemPhysChem* **2015**, *16*, 2496–2517. [CrossRef]
62. Mukherjee, A.J.; Zade, S.S.; Singh, H.B.; Sunoj, R.B. Organoselenium Chemistry: Role of Intramolecular Interactions. *Chem. Rev.* **2010**, *110*, 4357–4416. [CrossRef]
63. Huang, H.; Yang, L.; Facchetti, A.; Marks, T.J. Organic and Polymeric Semiconductors Enhanced by Noncovalent Conformational Locks. *Chem. Rev.* **2017**, *117*, 10291–10318. [CrossRef]
64. Benz, S.; Mareda, J.; Besnard, C.; Sakai, N.; Matile, S. Catalysis with chalcogen bonds: Neutral benzodiselenazole scaffolds with high-precision selenium donors of variable strength. *Chem. Sci.* **2017**, *8*, 8164–8169. [CrossRef] [PubMed]
65. Johnson, E.R.; Keinan, S.; Mori-Sánchez, P.; Contreras-García, J.; Cohen, A.J.; Yang, W. Revealing Noncovalent Interactions. *J. Am. Chem. Soc.* **2010**, *132*, 6498–6506. [CrossRef] [PubMed]
66. Frisch, M.J.; Trucks, G.W.; Schlegel, H.B.; Scuseria, G.E.; Robb, M.A.; Cheeseman, J.R.; Scalmani, G.; Barone, V.; Mennucci, B.; Petersson, G.A.; et al. *Gaussian 09*; Gaussian, Inc.: Wallinford, UK, 2009.
67. Head-Gordon, M.; Head-Gordon, T. Analytic MP2 frequencies without fifth-order storage. Theory and application to bifurcated hydrogen bonds in the water hexamer. *Chem. Phys. Lett.* **1994**, *220*, 122–128. [CrossRef]
68. Oliveira, V.; Kraka, E.; Cremer, D. Quantitative Assessment of Halogen Bonding Utilizing Vibrational Spectroscopy. *Inorg. Chem.* **2017**, *56*, 488–502. [CrossRef]
69. Brinck, T.; Borrfors, A.N. Electrostatics and polarization determine the strength of the halogen bond: A red card for charge transfer. *J. Mol. Model.* **2019**, *25*, 125. [CrossRef] [PubMed]
70. Murray, J.S.; Lane, P.; Politzer, P. Expansion of the σ-hole concept. *J. Mol. Model.* **2009**, *15*, 723–729. [CrossRef] [PubMed]
71. Fradera, X.; Austen, M.A.; Bader, R.F.W. The Lewis Model and Beyond. *J. Phys Chem. A* **1999**, *103*, 304–314. [CrossRef]
72. Poater, J.; Duran, M.; Solà, M.; Silvi, B. Theoretical Evaluation of Electron Delocalization in Aromatic Molecules by Means of Atoms in Molecules (AIM) and Electron Localization Function (ELF) Topological Approaches. *Chem. Rev.* **2005**, *105*, 3911–3947. [CrossRef]
73. Outeiral, C.; Vincent, M.A.; Martín Pendás, Á.; Popelier, P.L.A. Revitalizing the concept of bond order through delocalization measures in real space. *Chem. Sci.* **2018**, *9*, 5517–5529. [CrossRef]
74. Hugas, D.; Guillaumes, L.; Duran, M.; Simon, S. Delocalization indices for non-covalent interaction: Hydrogen and diHydrogen bond. *Comput. Theor. Chem.* **2012**, *998*, 113–119. [CrossRef]
75. Keith, T.A. *AIMAll (version 17.01.25)*; TK Gristmill Software: Overland Park, KS, USA, 2016.
76. Lu, T.; Chen, F. A multifunctional wavefunction analyzer. *J. Comput. Chem.* **2012**, *33*, 580–592. [CrossRef] [PubMed]
77. Humphrey, W.; Dalke, A.; Schulten, K. VMD—Visual Molecular Dynamics. *J. Molec. Graphics* **1996**, *14*, 33–38. [CrossRef]
78. Pople, J.A. The Lennard-Jones lecture. Intermolecular binding. *Faraday Discuss.* **1982**, *73*, 7–17.
79. Boys, S.F.; Bernardi, F. The calculation of small molecular interactions by the differences of separate total energies. Some procedures with reduced errors. *Mol. Phys.* **1970**, *19*, 553–566. [CrossRef]
80. Turney, J.M.; Simmonett, A.C.; Parrish, R.M.; Hohenstein, E.G.; Evangelista, F.; Fermann, J.T.; Mintz, B.J.; Burns, L.A.; Wilke, J.J.; Abrams, M.L.; et al. Psi4: An open-source ab initio electronic structure program. *WIREs Comput. Mol. Sci.* **2012**, *2*, 556–565. [CrossRef]
81. Auxiliary Basis Sets. Available online: http://www.psicode.org/psi4manual/master/basissets_byfamily.html (accessed on 11 February 2020).
82. Popelier, P.L.A. On the full topology of the Laplacian of the electron density. *Coord. Chem. Rev.* **2000**, *197*, 169–189. [CrossRef]
83. Bader, R.F.W.; Carroll, M.T.; Cheeseman, J.R.; Chang, C. Properties of atoms in molecules: Atomic volumes. *J. Am. Chem. Soc.* **1987**, *109*, 7968–7979. [CrossRef]

84. Bauzá, A.; Frontera, A. On the Importance of Halogen–Halogen Interactions in the Solid State of Fullerene Halides: A Combined Theoretical and Crystallographic Study. *Crystals* **2017**, *7*, 191. [CrossRef]
85. Alvarez, S. A cartography of the van der Waals territories. *Dalton Trans.* **2013**, *42*, 8617–8636. [CrossRef]
86. Varadwaj, A.; Varadwaj, P.R.; Marques, H.M.; Yamashita, K. A DFT assessment of some physical properties of iodine-centered halogen bonding and other non-covalent interactions in some experimentally reported crystal geometries. *Phys. Chem. Chem. Phys.* **2018**, *20*, 15316–15329. [CrossRef]
87. Mukherjee, A.; Tothadi, S.; Desiraju, G.R. Halogen Bonds in Crystal Engineering: Like Hydrogen Bonds yet Different. *Acc. Chem. Res.* **2014**, *47*, 2514–2524. [CrossRef] [PubMed]
88. Cremer, D.; Kraka, E. A Description of the Chemical Bond in Terms of Local Properties of Electron Density and Energy. *Croat. Chem. Acta* **1984**, *57*, 1259–1281.
89. Cremer, D.; Kraka, E. Chemical Bonds without Bonding Electron Density—Does the Difference Electron-Density Analysis Suffice for a Description of the Chemical Bond? *Angew Chem. Int. Ed.* **1984**, *23*, 627–628. [CrossRef]
90. Kraka, E.; Cremer, D. *Chemical Implications of Local Features of the Electron Density Distribution*; Springer: Heidelburg, Germany, 1990; Volume 2, pp. 457–542.
91. Zou, W.; Zhang, X.; Dai, H.; Yan, H.; Cremer, D.; Kraka, E. Description of an unusual hydrogen bond between carborane and a phenyl group. *J. Organomet. Chem.* **2018**, *865*, 114–127. [CrossRef]
92. Sarkhel, S.; Desiraju, G.R. N–H ... O, O–H ... O, and C–H ... O hydrogen bonds in protein–ligand complexes: Strong and weak interactions in molecular recognition. *Proteins Struct. Funct. Bioinform.* **2004**, *54*, 247–259. [CrossRef]
93. Schneider, H.-J. Crystallographic searches for weak interactions–the limitations of data mining. *Acta Cryst.* **2018**, *B74*, 322–324. [CrossRef]
94. Steiner, T.; R. Desiraju, G. Distinction between the weak hydrogen bond and the van der Waals interaction. *Chem. Commun.* **1998**, 891–892. [CrossRef]
95. Ege, S. *Organic Chemistry: Structure and Reactivity*, 4th ed.; Houghton Mifflin College: Boston, MA, USA, 2003.
96. Jeffrey, G.A. *An Introduction to Hydrogen Bonding*; Oxford University Press: Oxford, UK, 1997.
97. Emsley, J. Very strong hydrogen bonding. *Chem. Soc. Rev.* **1980**, *9*, 91–124. [CrossRef]
98. Desiraju, G.R.; Steiner, T. *The Weak Hydrogen Bond in Structural Chemistry and Biology (International Union of Crystallography Monographs on Crystallography, 9)*; Oxford University Press: Oxford, UK; New York, NY, USA, 1999.
99. Stone, A.J. Are Halogen Bonded Structures Electrostatically Driven? *J. Am. Chem. Soc.* **2013**, *135*, 7005–7009. [CrossRef]

© 2020 by the authors. Licensee MDPI, Basel, Switzerland. This article is an open access article distributed under the terms and conditions of the Creative Commons Attribution (CC BY) license (http://creativecommons.org/licenses/by/4.0/).

Article

Unexpected Sandwiched-Layer Structure of the Cocrystal Formed by Hexamethylbenzene with 1,3-Diiodotetrafluorobenzene: A Combined Theoretical and Crystallographic Study

Yu Zhang, Jian-Ge Wang and Weizhou Wang *

College of Chemistry and Chemical Engineering, Luoyang Normal University, Luoyang 471934, China; yzhpaper@yahoo.com (Y.Z.); wang_jiange2@126.com (J.-G.W.)
* Correspondence: wzw@lynu.edu.cn; Tel.: +86-379-686-18320

Received: 15 April 2020; Accepted: 5 May 2020; Published: 7 May 2020

Abstract: The cocrystal formed by hexamethylbenzene (HMB) with 1,3-diiodotetrafluorobenzene (1,3-DITFB) was first synthesized and found to have an unexpected sandwiched-layer structure with alternating HMB layers and 1,3-DITFB layers. To better understand the formation of this special structure, all the noncovalent interactions between these molecules in the gas phase and the cocrystal structure have been investigated in detail by using the dispersion-corrected density functional theory calculations. In the cocrystal structure, the theoretically predicted π···π stacking interactions between HMB and the 1,3-DITFB molecules in the gas phase can be clearly seen, whereas there are no π···π stacking interactions between HMB molecules or between 1,3-DITFB molecules. The attractive interactions between HMB molecules in the corrugated HMB layers originate mainly in the dispersion forces. The 1,3-DITFB molecules form a 2D sheet structure via relatively weak C–I···F halogen bonds. The theoretically predicted much stronger C–I···π halogen bonds between HMB and 1,3-DITFB molecules in the gas phase are not found in the cocrystal structure. We concluded that it is the special geometry of 1,3-DITFB that leads to the formation of the sandwiched-layer structure of the cocrystal.

Keywords: molecular cocrystal; sandwiched-layer structure; C–I···F halogen bonds; π···π stacking interactions; PBE0-D3(BJ) calculations

1. Introduction

Noncovalent interactions play key roles in crystal growth and design. In 1989, Desiraju defined the term "crystal engineering" as "the understanding of intermolecular interactions in the context of crystal packing and the utilization of such understanding in the design of new solids with desired physical and chemical properties" [1]. There are many kinds of noncovalent interactions, such as the hydrogen bond, π···π stacking interaction, dispersive interaction, σ-hole interaction, π-hole interaction, etc. [2–6]. In the preceding paper of this Special Issue, Alkorta, Elguero, and Frontera presented an excellent review of the noncovalent interactions formed by the electron-deficient elements of groups 1, 2, and 10–18 in the periodic table [7]. Here, we want to stress that the term noncovalent interaction is more general than the term noncovalent bond, and some of the noncovalent interactions such as the dispersive interactions are not noncovalent bonds [4]. What is a noncovalent bond? According to Bader's theory of atoms in molecules, in a noncovalent bond there usually is a bond path connecting the two interacting atoms and a (3, −1) bond critical point between the two interacting atoms [8]. Evidently, crystal packing is the result of the synergistic contributions of different types of strong or weak noncovalent interactions. Hence, in the field of crystal engineering, it is always significant and important to study the cooperativity and competition of these noncovalent interactions.

The π···π stacking interaction is one of the most common noncovalent interactions in crystal engineering [9]. The benzene dimer is always considered as a model for the study of the π···π stacking interaction [10]. Comparing with the π···π stacking interaction in the benzene dimer, the π···π stacking interaction in the complex between hexamethylbenzene (HMB) and hexafluorobenzene (HFB) is much stronger due to the strong attractive quadrupole–quadrupole electrostatic interaction between the two monomers. The quadrupole–quadrupole interaction is repulsive in the face-to-face structure of the benzene dimer with an interaction energy of +13.68 kcal/mol, whereas the quadrupole–quadrupole interaction is attractive in the face-to-face structure of the complex between HMB and HFB with an interaction energy of −8.53 kcal/mol [11]. The crystal structure of the complex between HMB and HFB has been reported in 1972, in which the partner molecules are stacked alternately to form infinite columns [12]. Naturally, HMB can also form π-stacked complexes with other electron-deficient perfluoro aromatic compounds. Perfluoroiodobenzenes are such compounds that we are very interested in because they are always employed as the halogen atom donors for the halogen bonds [13]. Figure 1 shows the molecular electrostatic potentials on the 0.001 a.u. electron density isodensity surfaces of HMB and 1,3-diiodotetrafluorobenzene (1,3-DITFB) along with some selected surface minima and surface maxima. The computational details of the molecular electrostatic potentials are given in the following section. As shown in Figure 1, the most negative electrostatic potentials of −22.47 kcal/mol on the surface of HMB are located 1.76 Å above or below the center of mass of HMB; the most positive electrostatic potentials of +30.76 kcal/mol on the surface of 1,3-DITFB are located on the extensions of the C–I bonds. For the 1,3-DITFB, besides the two electropositive σ-holes on the extensions of the C-I bonds, there are also electropositive regions (π-holes) that are perpendicular to the molecular plane. As a result, 1,3-DITFB can form the C–I···π halogen bond with HMB on the one hand, and on the other hand it can also form the strong π···π stacking interaction with HMB [13–15]. Certainly, the π···π stacking interactions can also be formed between two HMB molecules or between two 1,3-DITFB molecules. What is the order of strengths of all these noncovalent interactions? Will one of them, some of them, or all of them contribute to the formation of the cocrystal between HMB and 1,3-DITFB? Will there be other noncovalent interactions that we could not predict in the cocrystal structure? In this study, we solve these issues by employing a combined theoretical and crystallographic method.

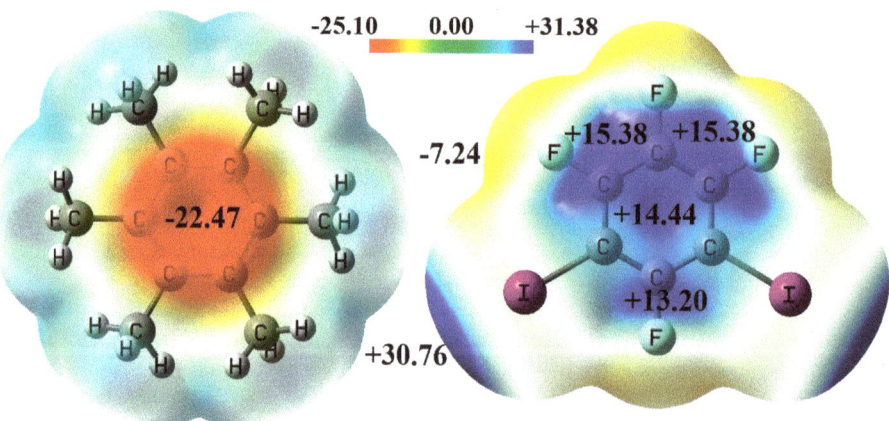

Figure 1. The molecular electrostatic potentials on the 0.001 a.u. electron density isodensity surfaces of hexamethylbenzene (HMB) (**left**) and 1,3-diiodotetrafluorobenzene (1,3-DITFB) (**right**). The numbers are in kcal/mol. Some selected surface minima and surface maxima are also shown.

This paper is organized as follows: First, we present and discuss the calculated results for the abovementioned noncovalent interactions in the gas phase; then, after describing the structure of the cocrystal between HMB and 1,3-DITFB, we calculate and analyze the noncovalent interactions in the crystal structure in detail. Finally, we give explanations for the formation of the special cocrystal structure.

2. Materials and Methods

2.1. Quantum Chemical Calculation

The geometries of the monomers and complexes in the gas phase were fully optimized at the PBE0-D3(BJ)/def2-TZVPP level of theory [16–19]. According to the corresponding frequency calculations, all the structures of these monomers and complexes are true minima on their respective potential energy surfaces. The interaction energies were calculated at the same theory level. For the complexes in the crystal structure, their geometries were directly extracted from the crystal structure, and only single-point interaction energies were calculated at the PBE0-D3(BJ)/def2-TZVPP theory level. All the interaction energies were calculated with the supermolecule method and corrected for the basis set superposition error using the conventional counterpoise method [20]. The molecular electrostatic potentials on the 0.001 a.u. electron density isodensity surfaces of HMB and 1,3-DITFB were also calculated at the PBE0-D3(BJ)/def2-TZVPP level of theory. The "ultrafine" integration grids were used for the PBE0-D3(BJ)/def2-TZVPP calculations to eliminate possible integration grid errors. All the calculations were performed with the GAUSSIAN 09 program package [21].

For the calculations of strong noncovalent interactions, many computational methods can give comparable results with experiments. The main challenges for the electronic structure calculations lie in the accurate descriptions of weak noncovalent interactions. In previous studies, we employed the PBE0-D3(BJ)/def2-TZVPP method to calculate the interaction energies of the different configurations of the complex between benzene and hexahalobenzene, the complex between benzene and naphthalene, and the complex between fullerene C_{60} and benzene [22–24]. It was found that the results from the PBE0-D3(BJ)/def2-TZVPP calculations are in excellent agreement with the results from the "gold standard" coupled-cluster calculations. Considering that the noncovalent interactions studied in this work are very similar to those in previously studied complexes, the results from the PBE0-D3(BJ)/def2-TZVPP calculations should be reliable throughout this paper.

2.2. Crystal Preparation

The chemical reagents HMB and 1,3-DITFB were purchased from J&K Scientific Ltd. in China and used as received. The solvent for the crystallization in this study was trichloromethane and also used without further purification. The HMB (0.0162 g, 0.10 mmol) and 1,3-DITFB (0.0402 g, 0.10 mmol) were dissolved in 10 mL trichloromethane, and the mixture was refluxed gently with stirring for half an hour. Then, the solution was filtered, and the filtrate was naturally volatilized at room temperature. After about three days, colorless block crystals that are suitable for the X-ray diffraction analyses were obtained.

2.3. Measurement

Single-crystal X-ray diffraction data were collected on a Rigaku AFC10 diffractometer (Rigaku Corporation, Tokyo, Japan) equipped with a Rigaku SuperNova X-ray generator (graphite-monochromatic Mo-Kα radiation, λ = 0.71073 Å). The structure of the cocrystal was solved and refined by a combination of direct methods and difference Fourier syntheses, employing the SHELX-2014 and Olex2.0 programs [25,26]. The hydrogen atoms of the methyl groups in HMB were placed in calculated positions and refined with the riding model approximation. Anisotropic thermal parameters were assigned to the nonhydrogen atoms. Crystallographic data have been deposited at the Cambridge Crystallographic Data Centre (deposition number CCDC 1996547). Copies of the data can be obtained free of charge via http://www.ccdc.cam.ac.uk/conts/retrieving.html.

3. Results and Discussion

3.1. Noncovalent Interactions in the Gas Phase

The study of the noncovalent interactions in the gas phase is significant and can provide useful information for the crystal growth and design, although in some cases the noncovalent interactions in the gas phase maybe be very different with the noncovalent interactions in the crystalline state. Figure 2 illustrates the PBE0-D3(BJ)/def2-TZVPP optimized structures and the corresponding interaction energies for the stacked complex between HMB and 1,3-DITFB, a stacked HMB dimer, a halogen-bonded complex between HMB and 1,3-DITFB, and a stacked 1,3-DITFB dimer. In fact, we also fully optimized the planar structures of the HMB dimer and the 1,3-DITFB dimer, but both of them were transformed into the stacked ones in Figure 2. This indicates that the planar structures of the HMB dimer and the 1,3-DITFB dimer are not stable in the gas phase.

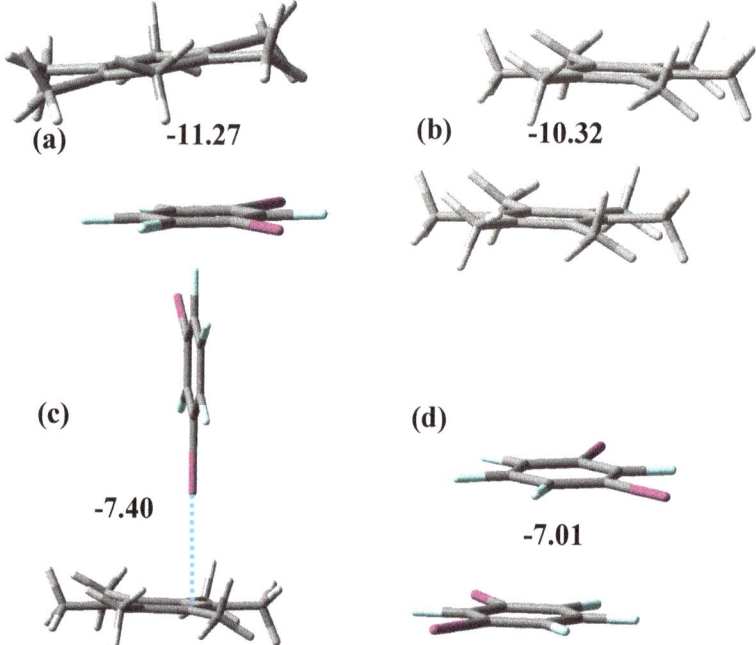

Figure 2. The interaction energies (black numbers, in kcal/mol) for the stacked complex between HMB and 1,3-DITFB (**a**), a stacked HMB dimer (**b**), a halogen-bonded complex between HMB and 1,3-DITFB (**c**), and a stacked 1,3-DITFB dimer (**d**).

It can be clearly seen from Figure 2 that the π···π stacking interaction between HMB and 1,3-DITFB is the strongest one among all the noncovalent interactions. The π···π stacking interaction energies for the complexes $C_6H_6···C_6X_6$ (X = F, Cl, Br, and I) are in the range of −9.70 to −5.50 kcal/mol [22]. Thus, the π···π stacking interaction between HMB and 1,3-DITFB is much stronger than the π···π stacking interactions in the complexes $C_6H_6···C_6X_6$ (X = F, Cl, Br, and I). This is understandable because the van der Waals surface area of HMB is larger than that of benzene, and the minimum value of the electrostatic potential of HMB is much more negative than that of benzene. The quadrupole–quadrupole electrostatic interactions in the HMB dimer and 1,3-DITFB dimer are repulsive, and this will weaken the π···π stacking interactions in the two dimers. The π···π stacking interaction energies for the HMB dimer and 1,3-DITFB dimer are −10.32 and −7.01 kcal/mol, respectively. As a contrast, the π···π stacking interaction energy for the complex between benzene and HFB is about −6.00 kcal/mol, and the π···π stacking interaction energy for the parallel-displaced configuration of the benzene dimer is about −2.70 kcal/mol [22,27]. The π-stacked HMB dimer and 1,3- DITFB dimer can also exist in the crystal structures. The CCDC database (version 5.41) was used in a search for the structures containing HMB or 1,3-DITFB [28]. It was found that there are 8 structures containing the π-stacked HMB dimer and 27 structures containing the π-stacked 1,3-DITFB dimer.

Another focus in Figure 2 is the halogen-bonded complex between HMB and 1,3-DITFB with the interaction energy of −7.40 kcal/mol. The binding energy of the conventional C–I···N halogen bond is below 7.00 kcal/mol [29]. Here, the strength of the C–I···π halogen bond is obviously close to or even stronger than the strength of the conventional strong C–I···N halogen bond. As shown in Figure 2c, the C–I bond does not point to the centroid of HMB but points to the site which is close to the carbon atom. Tsuzuki and coworkers calculated the C–I···π interaction energies for three orientations of the complex between benzene and pentafluoroiodobenzene, and they found that the difference of the interaction energies is not very marked [30]. Bosch and coworkers performed a statistical analysis of the C–I···π halogen bonds in the crystal structures by using the Cambridge Structural Database, and their results showed that the number of the structures in which the C–I bond points to the centroid of the benzene ring is very small [31]. In other words, the C–I···π halogen bond predicted in the gas phase may also exist in the crystal structure of the complex between HMB and 1,3-DITFB.

3.2. Noncovalent Interactions in the Crystal Structure

HMB and 1,3-DITFB form a 1:1 cocrystal. The cocrystal has an unexpected sandwiched-layer structure with alternating HMB layers and 1,3-DITFB layers (Figure 3). The HMB layer is corrugated, and the 1,3-DITFB layer is a 2D sheet. Crystal data for the cocrystal (M = 564.12 g/mol) are as follows: orthorhombic, space group *Cmcm* (no. 63), a = 16.3241(6) Å, b = 8.7254(5) Å, c = 13.6411(8) Å, β = 90°, V = 1942.96(18) Å^3, Z = 4, T = 290 K, μ(CuKα) = 3.270 mm^{-1}, D_{calc} = 1.928 g/cm^3, 11066 reflections measured (7.786° ≤ 2Θ ≤ 56.726°), 1219 unique (R_{int} = 0.0324, R_{sigma} = 0.0151), which were used in all calculations. The final R_1 was 0.0883 ($I > 2\sigma(I)$) and wR_2 was 0.2298 (all data).

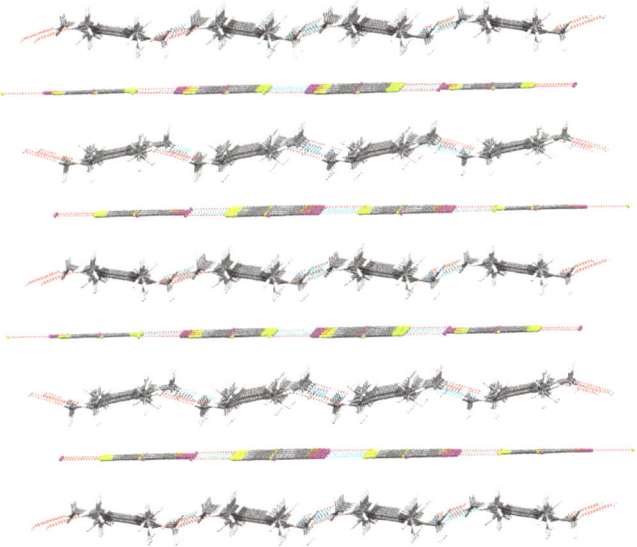

Figure 3. The side view of the sandwiched-layer structure of the cocrystal.

As expected from the gas-phase calculation, the π···π stacking interactions between HMB and 1,3-DITFB are found in the crystal structure. The interaction energy for the stacked two-body complex in the crystal structure is −11.16 kcal/mol, which is almost the same as the corresponding value of −11.27 kcal/mol) in the gas phase. In the crystal structure, the HMB and 1,3-DITFB molecules are stacked alternately in infinite columns. It is interesting to study the cooperativity of these $\pi\cdots\pi$ stacking interactions. Figure 4 shows the total interaction energies for the stacked two-body, three-body, and four-body complexes. Here, we use the three-body [$\Delta^3 E(123)$] and four-body [$\Delta^4 E(1234)$] interaction terms to assess the cooperativity of these $\pi\cdots\pi$ stacking interactions, such as the study of the benzene trimer and the benzene tetramer [32]. The three-body and four-body interaction terms can be defined as follows:

$$\Delta^3 E(123) = E(123) - \sum_i E(i) - \sum_{ij} \Delta^2 E(ij)$$

$$\Delta^4 E(1234) = E(1234) - \sum_i E(i) - \sum_{ij} \Delta^2 E(ij) - \sum_{ijk} \Delta^3 E(ijk)$$

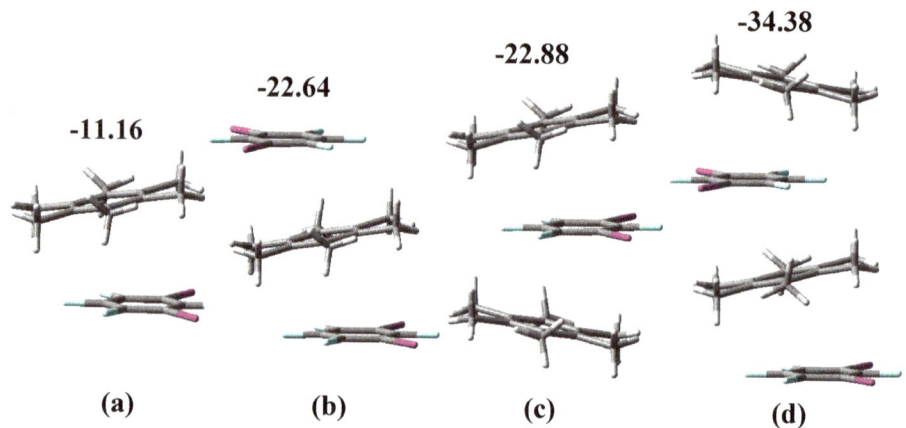

Figure 4. The interaction energies (black numbers, in kcal/mol) for the stacked two-body complex (**a**), a three-body complex (**b**), a three-body complex (**c**), and a four-body complex (**d**) with alternating HMB and 1,3-DITFB molecules.

The three-body interaction terms for the two three-body complexes are −0.32 and −0.56 kcal/mol, respectively. The four-body interaction term for the four-body complex is −0.90 kcal/mol. The three-body and four-body interaction terms are all negative and obviously have stabilizing contributions to the total interactions. Considering that the total interaction energy is very large, it is still reasonable to estimate the total interaction energy of a large complex simply from the sum of the two-body interaction energies.

Figures 5 and 6 show the noncovalent interactions in the HMB layer and 1,3-DITFB layer. Let us add here that these noncovalent interactions do not exist in the gas phase. The HMB molecules form the corrugated layers via dispersion forces. In the corrugated HMB layer, two methyl groups of HMB along the crystallographic a axis are disordered, and the other four methyl groups form four H$\cdots$H contacts with other HMB molecules. The disorder of the two methyl groups of the one HMB molecule indicates that the H$\cdots$H contacts make negligible contribution to the stability of the cocrystal from another perspective. The 1,3-DITFB molecules form the 2D sheets via the weak C–I$\cdots$F halogen bonds. One 1,3-DITFB molecule can form four C–I$\cdots$F halogen bonds with four neighboring 1,3-DITFB molecules. It is the special structure of 1,3-DITFB that leads to the formation of the 2D sheet and furthers the formation of the sandwiched-layer structure of the cocrystal. A similar structure can be found in the cocrystal formed between HMB and 1,2,4,5-tetracyanobenzene [33]. This cocrystal also has a layer structure. However, the 1,2,4,5-tetracyanobenzene layer is not a 2D sheet but a corrugated layer.

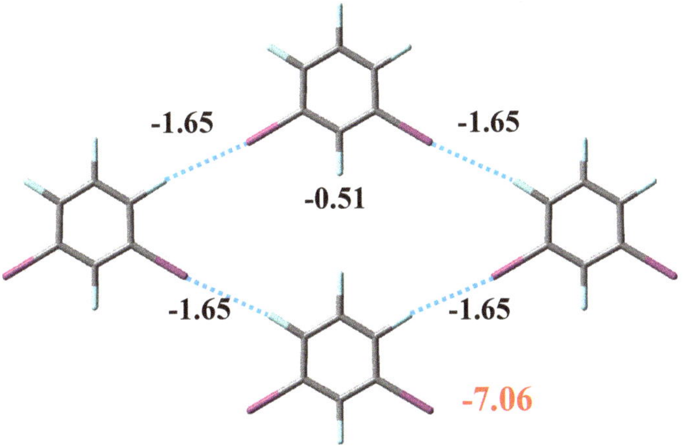

Figure 5. The four 1,3-DITFB molecules involved in a C–I···F halogen-bonded loop. The black numbers (in kcal/mol) are the interaction energies of two neighboring molecules, and the red number (in kcal/mol) is the total interaction energy of the tetramer.

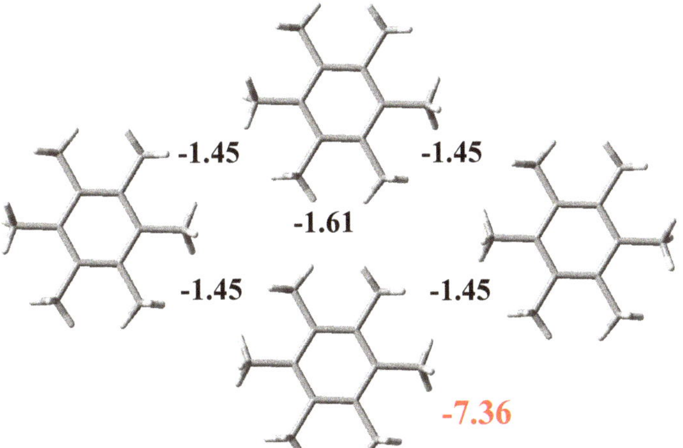

Figure 6. The four HMB molecules involved in a dispersion-bonded loop. The black numbers (in kcal/mol) are the interaction energies of two neighboring molecules, and the red number (in kcal/mol) is the total interaction energy of the tetramer.

Figures 5 and 6 also list the interaction energies for two neighboring monomers and the total interaction energies for the 1,3-DITFB tetramer and HMB tetramer. In the 1,3-DITFB tetramer, the interaction energy of one C–I···F halogen bond is −1.65 kcal/mol, and the interaction energy for the dimer without a C–I···F halogen bond is only −0.51 kcal/mol. The four-body interaction term for the 1,3-DITFB tetramer is about 0.05 kcal/mol, which means that the cooperativity of the noncovalent interactions in the 1,3-DITFB tetramer is negligible. The case for the HMB tetramer is quite similar. The interaction energy of two neighboring HMB molecules is a little smaller than that of two C–I···F halogen-bonded 1,3-DITFB molecules. The four-body interaction term of the HMB tetramer is also about 0.05 kcal/mol and can also be neglected.

Figure 7 lists the interaction energies for two neighboring monomers and the total interaction energies for the four-body complex formed by two HMB molecules and two 1,3-DITFB molecules.

Different from the complexes in Figures 4–6, the complex in Figure 7 is formed via mixed noncovalent interactions, which include a π···π stacking interaction, a C–I···F halogen bond, and a dispersion-dominated interaction. The four-body interaction term of this complex is about 0.09 kcal/mol, which is a little larger than that of the 1,3-DITFB tetramer and HMB tetramer. However, the absolute value of the total interaction energy of this complex is over three times larger than that of the 1,3-DITFB tetramer or HMB tetramer. Again, it is reasonable to estimate the total interaction energy of a large complex simply from the sum of the two-body interaction energies.

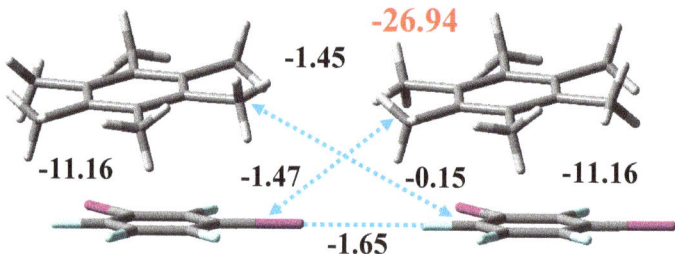

Figure 7. The loop formed by two HMB molecules and two 1,3-DITFB molecules. The black numbers (in kcal/mol) are the interaction energies of two neighboring molecules, and the red number (in kcal/mol) is the total interaction energy of the four molecules.

4. Conclusions

In this study, the cocrystal formed by HMB with 1,3-DITFB was successfully synthesized, and the noncovalent interactions in the crystal structure were calculated at the PBE0-D3(BJ)/def2-TZVPP level of theory. Unexpectedly, the cocrystal has a sandwiched-layer structure with alternating HMB layers and 1,3-DITFB layers. In the corrugated HMB layer, the HMB molecules attract each other mainly via the dispersion forces. In the 1,3-DITFB layer, the 1,3-DITFB molecules form a 2D sheet via the C–I···F halogen bonds, and one 1,3-DITFB molecule can form four C–I···F halogen bonds with four neighboring 1,3-DITFB molecules. The alternating HMB layers and 1,3-DITFB layers are stacked together by strong π···π stacking interactions between HMB and 1,3-DITFB molecules. No C–I···π halogen bonds and π···π stacking interactions between the HMB molecules or between the 1,3-DITFB molecules were found in the crystal structure. It was also found that the cooperativity of the noncovalent interactions in each layer is not very obvious. However, the cooperativity of the π···π stacking interactions in the sequence of alternating HMB and 1,3-DITFB molecules is considerable.

The formation of the unexpected sandwiched-layer structure of the molecular cocrystal is attributed to the special geometry of 1,3-DITFB. Other perfluoroiodobenzenes such as the 1,2-diiodotetrafluorobenzene, 1,4-diiodotetrafluorobenzene, and 1,3,5-trifluoro-2,4,6-triiodobenzene do not have such geometries and cannot form 2D sheets via the weak C–I···F halogen bonds. Hence, we predict that the cocrystals formed by HMB with these molecules will not have such a sandwiched-layer structure. The controlled experiments are in progress in our laboratory. The preliminary results show that these predicted cocrystals are a little more difficult to be synthesized than the cocrystal reported in this study.

Author Contributions: Y.Z. grew the cocrystals and performed all the quantum chemical calculations; J.-G.W. carried out the X-ray single diffraction studies; W.W. designed and supervised this project; Y.Z. and W.W. jointly wrote and revised the paper. All authors have read and agreed to the published version of the manuscript.

Funding: This research was funded by the National Science Foundation of China, grant number 21773104.

Acknowledgments: We thank the National Science Foundation of China for the financial support. W.W. thanks the National Supercomputing Center in Shenzhen for the computational support.

Conflicts of Interest: The authors declare no conflicts of interest.

References

1. Desiraju, G.R. *Crystal Engineering: The design of Organic Solids*; Elsevier: Amsterdam, The Netherlands, 1989.
2. Pimentel, G.C.; McClellan, A.L. *The Hydrogen Bond*; W.H. Freeman & Co.: San Francisco, CA, USA, 1960.
3. Hobza, P.; Müller-Dethlefs, K. *Noncovalent Interactions. Theory and Experiment*; RSC Theoretical and Computational Chemistry Series; Royal Society of Chemistry: Cambridge, UK, 2010.
4. Schreiner, P.R.; Chernish, L.V.; Gunchenko, P.A.; Tikhonchuk, E.Y.; Hausmann, H.; Serafin, M.; Schlecht, S.; Dahl, J.E.P.; Carlson, R.M.K.; Fokin, A.A. Overcoming Lability of Extremely Long Alkane Carbon-Carbon Bonds Through Dispersion Forces. *Nature* **2011**, *477*, 308–312. [CrossRef] [PubMed]
5. Clark, T.; Hennemann, M.; Murray, J.S.; Politzer, P. Halogen Bonding: The σ-Hole. *J. Mol. Model.* **2007**, *13*, 291–296. [CrossRef] [PubMed]
6. Murray, J.S.; Lane, P.; Clark, T.; Riley, K.E.; Politzer, P. σ-Holes, π-Holes and Electrostatically-Driven Interactions. *J. Mol. Model.* **2012**, *18*, 541–548. [CrossRef] [PubMed]
7. Alkorta, I.; Elguero, J.; Frontera, A. Not Only Hydrogen Bonds: Other Noncovalent Interactions. *Crystals* **2020**, *10*, 180. [CrossRef]
8. Bader, R.F.W. *Atoms in Molecules: A Quantum Theory*; Clarendon: Oxford, UK, 1990.
9. Thakuria, R.; Nath, N.K.; Saha, B.K. The Nature and Applications of π-π Interactions: A Perspective. *Cryst. Growth Des.* **2019**, *19*, 523–528. [CrossRef]
10. Hunter, C.A.; Sanders, J.K.M. The Nature of π-π Interactions. *J. Am. Chem. Soc.* **1990**, *112*, 5525–5534. [CrossRef]
11. Hernández-Trujillo, J.; Costas, M.; Vela, A. Quadrupole Interactions in Pure Non-dipolar Fluorinated or Methylated Benzenes and their Binary Mixtures. *J. Chem. Soc. Faraday Trans.* **1993**, *89*, 2441–2443. [CrossRef]
12. Dahl, T. Crystal Structure of the Trigonal Form of the 1:1 Complex Between Hexamethylbenzene and Hexafluorobenzene. *Acta Chem. Scand.* **1972**, *26*, 1569–1575. [CrossRef]
13. Cavallo, G.; Metrangolo, P.; Milani, R.; Pilati, T.; Priimagi, A.; Resnati, G.; Terraneo, G. The Halogen Bond. *Chem. Rev.* **2016**, *116*, 2478–2601. [CrossRef]
14. Wang, H.; Wang, W.; Jin, W.J. σ-Hole Bond vs π-Hole Bond: A Comparison Based on Halogen Bond. *Chem. Rev.* **2016**, *116*, 5072–5104. [CrossRef]
15. Pang, X.; Wang, H.; Wang, W.; Jin, W.J. Phosphorescent π-Hole···π Bonding Cocrystals of Pyrene with Haloperfluorobenzenes (F, Cl, Br, I). *Cryst. Growth Des.* **2015**, *15*, 4938–4945. [CrossRef]
16. Adamo, C.; Barone, V. Toward Reliable Density Functional Methods without Adjustable Parameters: The PBE0 Model. *J. Chem. Phys.* **1999**, *110*, 6158–6169. [CrossRef]
17. Grimme, S.; Antony, J.; Ehrlich, S.; Krieg, H. A Consistent and Accurate Ab Initio Parametrization of Density Functional Dispersion Correction (DFT-D) for the 94 Elements H-Pu. *J. Chem. Phys.* **2010**, *132*, 154104. [CrossRef]
18. Grimme, S.; Ehrlich, S.; Goerigk, L. Effect of the Damping Function in Dispersion Corrected Density Functional Theory. *J. Comput. Chem.* **2011**, *32*, 1456–1465. [CrossRef]
19. Weigend, F.; Ahlrichs, R. Balanced Basis Sets of Split Valence, Triple Zeta Valence and Quadruple Zeta Valence Quality for H to Rn: Design and Assessment of Accuracy. *Phys. Chem. Chem. Phys.* **2005**, *7*, 3297–3305. [CrossRef]
20. Boys, S.F.; Bernardi, F. The Calculation of Small Molecular Interactions by the Difference of Separate Total Energies. Some Procedures with Reduced Errors. *Mol. Phys.* **1970**, *19*, 553–566. [CrossRef]
21. Frisch, M.J.; Trucks, G.W.; Schlegel, H.B.; Scuseria, G.E.; Robb, M.A.; Cheeseman, J.R.; Scalmani, G.; Barone, V.; Mennucci, B.; Petersson, G.A.; et al. *Gaussian 09, Revision C.01*; Gaussian, Inc.: Wallingford, CT, USA, 2010.
22. Wang, W.; Zhang, Y.; Wang, Y.B. Highly Accurate Benchmark Calculations of the Interaction Energies in the Complexes $C_6H_6···C_6X_6$ (X = F, Cl, Br, and I). *Int. J. Quantum Chem.* **2017**, *117*, e25345. [CrossRef]
23. Wang, W.; Sun, T.; Zhang, Y.; Wang, Y.B. The Benzene···Naphthalene Complex: A more Challenging System than the Benzene Dimer for newly Developed Computational Methods. *J. Chem. Phys.* **2015**, *143*, 114312. [CrossRef]
24. Li, M.M.; Wang, Y.B.; Zhang, Y.; Wang, W. The Nature of the Noncovalent Interactions between Benzene and C_{60} Fullerene. *J. Phys. Chem. A* **2016**, *120*, 5766–5772. [CrossRef]
25. Sheldrick, G.M. SHELXT–Integrated Space-Group and Crystal-Structure Determination. *Acta Crystallogr.* **2015**, *A71*, 3–8. [CrossRef]

26. Dolomanov, O.V.; Bourhis, L.J.; Gildea, R.J.; Howard, J.A.K.; Puschmann, H. OLEX2: A Complete Structure Solution, Refinement and Analysis Program. *J. Appl. Cryst.* **2009**, *42*, 339–341. [CrossRef]
27. Pitoňák, M.; Neogrády, P.; Řezáč, J.; Jurečka, P.; Urban, M.; Hobza, P. Benzene Dimer: High-Level Wave Function and Density Functional Theory Calculations. *J. Chem. Theory Comput.* **2008**, *4*, 1829–1834.
28. Groom, C.R.; Bruno, I.J.; Lightfoot, M.P.; Ward, S.C. The Cambridge structural database. *Acta Crystallogr. Sect. B Struct. Sci. Cryst. Eng. Mater.* **2016**, *72*, 171–179. [CrossRef] [PubMed]
29. Wang, W.; Zhang, Y.; Wang, Y.B. The π···π Stacking Interactions between Homogeneous Dimers of $C_6F_xI_{(6-x)}$ (x = 0, 1, 2, 3, 4, and 5): A Comparative Study with the Halogen Bond. *J. Phys. Chem. A* **2012**, *116*, 12486–12491. [CrossRef]
30. Tsuzuki, S.; Uchimaru, T.; Wakisaka, A.; Ono, T. Magnitude and Directionality of Halogen Bond of Benzene with C_6F_5X, C_6H_5X, and CF_3X (X = I, Br, Cl, and F). *J. Phys. Chem. A* **2016**, *120*, 7020–7029. [CrossRef]
31. Nwachukwu, C.I.; Bowling, N.P.; Bosch, E. C–I···N and C–I···π Halogen Bonding in the Structures of 1-Benzyliodoimidazole Derivatives. *Acta Cryst.* **2017**, *C73*, 2–8. [CrossRef]
32. Tauer, T.P.; Sherrill, C.D. Beyond the Benzene Dimer: An Investigation of the Additivity of π-π Interactions. *J. Phys. Chem. A* **2005**, *109*, 10475–10478. [CrossRef]
33. Saraswatula, V.G.; Sharada, D.; Saha, B.K. Stronger π···π Interaction Leads to a Smaller Thermal Expansion in Some Charge Transfer Complexes. *Cryst. Growth Des.* **2018**, *18*, 52–56. [CrossRef]

© 2020 by the authors. Licensee MDPI, Basel, Switzerland. This article is an open access article distributed under the terms and conditions of the Creative Commons Attribution (CC BY) license (http://creativecommons.org/licenses/by/4.0/).

Article

N/N Bridge Type and Substituent Effects on Chemical and Crystallographic Properties of Schiff-Base (*Salen/Salphen*) Ni^{ii} Complexes

Cynthia S. Novoa-Ramírez [1,†], Areli Silva-Becerril [1,†], Fiorella L. Olivera-Venturo [1,2], Juan Carlos García-Ramos [1,3], Marcos Flores-Alamo [4] and Lena Ruiz-Azuara [1,*]

1. Department of Inorganic Chemistry, Faculty of Chemistry, Universidad Nacional Autónoma de México, Av. Universidad 3000, Circuito Exterior S/N, Coyoacán, C.P. 04510 Ciudad de México, Mexico; cynthiasinai@comunidad.unam.mx (C.S.N.-R.); aresilbec@comunidad.unam.mx (A.S.-B.); fiorella.olivera@upch.pe (F.L.O.-V.); juan.carlos.garcia.ramos@uabc.edu.mx (J.C.G.-R.)
2. Department of Science, Faculty of Science and Philosophy, Av. Honorio Delgado 430, 15102 San Martín de Porres, Peru
3. School of Health Sciences, Universidad Autónoma de Baja California, Campus Ensenada, Ensenada, 21100 Baja California, Mexico
4. USAII, Faculty of Chemistry, Universidad Nacional Autónoma de México, Av. Universidad 3000, Circuito Exterior S/N, Coyoacán, C.P. 04510 Ciudad de México, Mexico; mfa@unam.mx
* Correspondence: lena@unam.mx
† Equal participation in this work.

Received: 9 June 2020; Accepted: 13 July 2020; Published: 15 July 2020

Abstract: In total, 13 ligands R-*salen* (N,N'-bis(5-R-salicylidene)ethylenediamine (where R = MeO, Me, OH, H, Cl, Br, NO$_2$) and R-*salphen* (N,N'-bis(5-R-salicylidene)-1,2-phenylenediamine (where R = MeO, Me, OH, H, Cl, Br) and their 13 nickel complexes NiR*salen* and NiR*salphen* were synthesized and characterized using IR (infrared) spectroscopy, mass spectrometry, elemental analysis, magnetic susceptibility, NMR (nuclear magnetic resonance), UV-vis (ultraviolet-visible) spectroscopy, cyclic voltammetry, and X-ray crystal diffraction. Previous studies have shown that all complexes have presented a square planar geometry in a solid state and as a solution (DMSO). In electrochemical studies, it was observed that in N/N aliphatic bridge complexes, the Ni^{II} underwent two redox reactions, which were quasi-reversible process, and the half-wave potential followed a trend depending on the ligand substituent in the 5,5'-R position. The electron-donor substituent—as -OH, and -CH$_3$ decreased the $E_{1/2}$ potential—favored the reductor ability of nickel. The crystals of the complexes NiMe*salen*, NiMeO*salen*, NiMeO*salphen*, and N*isalphen* were obtained. It was shown that the crystal packaging corresponded to monoclinic systems in the first three cases, as well as the triclinic for N*isalphen*. The Hirshfeld surface analysis showed that the packaging was favored by H···H and C···H/H···C interactions, and C-H···O hydrogen bridges when the substituent was -MeO and π-stacking was added to an aromatic bridge. Replacing the N/N bridge with an aromatic ring decreased distortion in square-planar geometry where the angles O-Ni-N formed a perfect square-planar.

Keywords: nickel; Schiff bases; crystallography; Hirshfeld surface analysis

1. Introduction

Salen type ligands, derived from the condensation between salicylaldehyde and a primary diamine, are considered versatile ligands in coordination chemistry because the steric and electronic properties can be modulated by different amine aldehydes precursors. The metal *salen* complexes have been widely studied in diverse areas of chemistry. The interest lays in its easy synthesis, versatility, and kinetic and thermodynamic stability provided by the chelating capacity of the tetradentate ligand with N$_2$O$_2$

donor atoms. For the aforementioned reasons, such complexes not only play an important role in coordination chemistry, but in various areas such as asymmetric catalysis [1], epoxide formation [2–4], olefin hydrogenation [5,6], or in polymerization reactions [7,8]. Moreover, they have been extensively studied in connection with metalloprotein models and, more recently, in bioinorganic chemistry. A complex of Ni^{II}, Fe^{II}, and Cu^{II}-*salen* has been synthesized as a biomimetic compound for the study of metalloenzyme active sites and their catalytic mechanism [9–12].

It is well known that the steric and electronic effect plays an important role in the reactivity of M-*salen* compounds. The incorporation of electron-donor/withdrawing groups on *salen* skeleton allows redox potential modulation [9,13] and favors certain geometries [14,15] that impact the interaction with the substrates or specific recognition sites. In recent years, attention has focused on the biological properties of Schiff bases and their metal-compounds [16], showing that ligands by themselves can inhibit a carbonic anhydrase enzyme. The efficiency of these enzymes depend on the properties of the bridge N/N [16]. Several metal-*salen* complexes have presented different biological activities, such as an antibacterial [17,18] and antiproliferative against different tumor lines [19–22]. One of the proposed mechanisms of action of these complexes with *salen*-type ligands is the specific interactions with DNA and RNA. Different types of damage can occur depending on the chemical reactivity of the metal complex. Mn^{III}-*salen* in the presence of an oxidant induces DNA cleavages [23,24], whereas Co^{III}-*salen* can cleave DNA under aerobic conditions [25]. Co^{II}-*salen* and Fe^{III}-*salen* bind DNA at the minor groove [26], while Ni^{II}-*salen* causes damage to nucleic acids, specifically causing divisions in guanine residues in the hairpin region of a single chain [27]. On the other hand, it has been reported that square coordination compounds with conjugated systems in their coordination spheres present stacking interactions with DNA [28] and G-quadruplexes. The binding affinity and selectivity of interaction with the latter is modulated by changing the substituents on the *salen* skeleton and modifying the nature of the N/N-bridge for the Schiff base [29,30]. The correct choice of the N/N bridge plays an important role in the geometry of these compounds, as it lengthens the chain and favors octahedral geometry [31] while adding aromatic rings that favor planar-square geometry. Moreover, it increases π-π interactions [32]. These type of complexes also present a square-planar geometry [33] and multiple studies have shown how the reactivity of nickel-*salen* complexes can modulate by the design and control of the nature of the ligands by the choice of the N/N bridge or its substituents. Therefore, in this work, we study the effect the N/N bridge has on geometry and how the half-wave potential ($E_{1/2}$) changes when a substitute is added. We report the structure of four Nickel complexes with tetradentate N_2O_2 ligands, analyzing the intermolecular interactions favored by the substituent and the N/N bridge, which modify the distance of Ni-Ni interaction found between dimers.

2. Materials and Methods

The experiments were carried out in ambient conditions. Nickel acetate tetrahydrate, salicylaldehyde, 2-hydroxi-5-metoxybenzaldehyde, 2-hydroxi-5-metylbenzaldehyde, 2-hydroxi-5-nitrobenzaldehyde, 2-hydroxi-5-clorobenzaldehyde, 2-hydroxi-5-bromorobenzaldehyde, and o-phenylenediamine were acquired from Sigma-Aldrich (Saint Louis, MO, USA), whereas 1,2-diaminoethane was acquired from Merck (Darmstadt, Germany.). The solvents used were acquired from Quimica Alvi (Ciudad de México, Mexico). All reactive materials were used without further purification. Elemental analysis was carried out in the Reach and Industry Support Services Unit (USAII for its Spanish abbreviations), using an EAGER 200 elemental analyzer (EAGER 200 CHNS/method, Ciudad de México, Mexico). IR (infrared) spectra were performed on a Nicolet AVATAR 320 FT-IR (Ciudad de México, Mexico) in an interval of 4000–400 cm^{-1}. The samples incorporated onto a KBr disk in the range of 3500–500. UV-VIS (ultraviolet-visible) spectra were obtained on a Hewlett Packard 845× UV-visible system diode array spectrophotometer in a range from 250 nm to 800 nm in dimethylsulfoxide (DMSO) solutions Sigma-Aldrich (Saint Louis, MO, USA). The ^{1}H-NMR ^{13}C-NMR (nuclear magnetic resonance) (Ciudad de México, Mexico), COSY (correlation spectroscopy), and HSQC (heteronuclear single quantum correlation) (Tables S7 and S8, Figures S21–S39, Supplementary Materials), were performed with the USAII, collected by a VARIAN VNMRS 400 MHz. Chemical shifts were reported in ppm relative to the internal TMS (tetramethylsilane) standard. The solvents used were CDCl$_3$, Acetone-d$_6$, DMF-d$_7$, and DMSO-d$_6$, all of which were acquired from Sigma-Aldrich (Saint Louis, MO, USA). Mass spectrometers were acquired in the USAII (Ciudad de México, Mexico). All of the ligands and Ni*Rsalen* were obtained using FAB$^+$ in a LECO PEGASUSIII. NMR and mass spectrometry were not obtained for NiCl*salphen* and NiBr*salphen* due to their low solubility. Cyclic voltammetry was performed using PAR27 potentiostat/galvanostat (Ciudad de México, Mexico) with a conventional three-electrode array. Tetrabutylammonium hexafluorophosphate (Sigma-Aldrich, Saint Louis, MO, USA), served as a support electrolyte and DMSO (99.995, Sigma-Aldrich, Misuri, United State). Voltammogram were referenced with an internal adjustment using ferrocene (99.9%, Sigma-Aldrich, Saint Louis, MO, USA).

2.1. Abbreviation

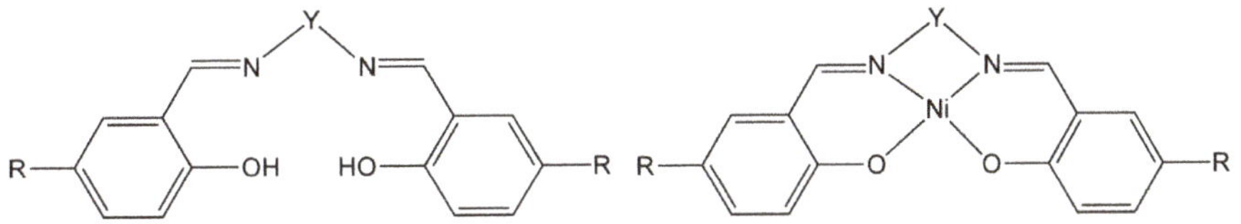

		CH$_3$O	MeO*salen*	NiMeO*salen*
		CH$_3$	Me*salen*	NiMe*salen*
salen	R =	OH	OH*salen*	NiOH*salen*
		H	*Salen*	Ni*salen*
		Cl	Cl*salen*	NiCl*salen*
		Br	Br*salen*	NiBr*salen*
		NO$_2$	NO$_2$*salen*	NiNO$_2$*salen*

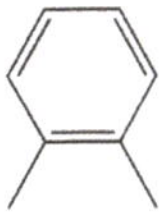		CH$_3$O	MeO*salphen*	NiMeO*salphen*
		CH$_3$	Me*salphen*	NiMe*salphen*
salphen	R =	OH	OH*salphen*	NiOH*salphen*
		H	*Salphen*	Ni*salphen*
		Cl	Cl*salphen*	NiCl*salphen*
		Br	Br*salphen*	NiBr*salphen*

2.2. Synthesis of Schiff Base Ligands

Next, 2 mmol of appropriate salicylaldehyde was dissolved in acetonitrile, then 1 mmol of 1,2-diaminoethane or *o*-phenylenediamine was dissolved in acetonitrile and added slowly. The dissolution was stirred for 20 min. The volume was reduced and a solid precipitated (range of yellows in the case of 1,2-diaminoethane bridge and range of orange for *o*-phenylenediamine derivatives), which was vacuum filtered and recrystallized in methanol. The ligands were characterized by elemental analysis, FT-IR spectroscopy, NMR (^{1}H and ^{13}C), and mass spectrometry:

MeO*salen* C$_{18}$H$_{20}$N$_2$O$_4$ analysis (%Calculated (Found)): C, 65.86 (66.07); H, 6.09 (6.02); N, 8.54 (8.84). FT-IR (cm^{-1}): υC=N, 1639; υC-O, 1276. FT-IR values were comparable to those reported [34]. NMR ^{1}H (ppm, CDCl$_3$): 8.30 (s, CH=N), 12.63 (s, C$_{Ar}$-O-H); ^{13}C (ppm): 166 (C=N), 155 (C$_{Ar}$-O). M$^+$ (m/z): 328 (328).

Me*salen* C$_{18}$H$_{20}$N$_2$O$_2$ analysis (%C(F)): C, 72.97 (73.10); H, 6.76 (6.38); N, 9.46 (10.05). FT-IR (cm^{-1}): υC=N, 1637; υC-O, 1282. FT-IR values were comparable to those reported [35]. NMR ^{1}H (ppm, CDCl$_3$): 8.29 (s, CH=N), 12.95 (s, C$_{Ar}$-O-H); ^{13}C (ppm): 166 (C=N), 159 (C$_{Ar}$-O). M$^+$ (m/z): 296 (296).

OHsalen $C_{16}H_{16}N_2O_4$; Analysis (%C(F)): C,63.55 (63.85); H, 5,37 (5.03); N, 9.32 (9.47). FT-IR (cm^{-1}): υC=N, 1640; υC-O, 1258. FT-IR values were slightly lower than those reported [36]. NMR ^{1}H (ppm, Acet-d$_6$): 8.46 (s, CH=N), 12.40 (s, C$_{Ar}$-O-H). Insoluble for NMR and mass spectrometry.

Salen $C_{16}H_{16}N_2O_2$ analysis (%C(F)): C,71.69 (71.62); H, 5.64 (6.01); N, 10.79 (10.44). FT-IR (cm^{-1}): υC=N, 1636; υC-O, 1284. FT-IR values were comparable to those reported [37]. ^{1}H NMR (ppm, CDCl$_3$): 8.36 (s, CH=N), 13.19 (s, C$_{Ar}$-O-H); ^{13}C (ppm): 166 (C=N), 161 (C$_{Ar}$-O). M$^+$ (m/z): 268 (268).

Clsalen $C_{16}H_{14}N_2Cl_2O_2$ analysis (%C(F)): C, 56.96 (57.13); H, 4.15 (3.88); N, 8.30 (8.52). FT-IR (cm^{-1}): υC=N, 1631; υC-O, 1274. FT-IR values were slightly lower than those reported [38]. ^{1}H NMR (ppm, CDCl$_3$): 8.29 (s, CH=N), 13.08 (s, C$_{Ar}$-O-H); ^{13}C (ppm): 165 (C=N), 159 (C$_{Ar}$-O). M$^+$ (m/z): 336 (336).

Brsalen $C_{16}H_{14}Br_2N_2O_2$ analysis (%C(F)): C, 45.07 (45.18); H, 3.29 (2.80); N, 6.57(7.22). FT-IR (cm^{-1}): υC=N, 1635; υC-O, 1273. FT-IR values were comparable to those reported [34]. ^{1}H NMR (ppm, CDCl$_3$): 8.28 (s, CH=N), 13.10 (s, C$_{Ar}$-O-H); ^{13}C (ppm): 165 (C=N), 160 (C$_{Ar}$-O). M$^+$ (m/z): 426 (426).

NO$_2$salen $C_{16}H_{14}N_4O_6$ analysis (%C(F)): C, 53.63 (53.69); H, 3.93 (3.59); N, 15.63 (15.76). FT-IR (cm^{-1}): υC=N, 1647; υC-O, 1326. FT-IR values were comparable to those reported [39]. ^{1}H NMR (ppm, DMSO-d6): 8.77 (s, CH=N). M$^+$ (m/z): 358 (358).

MeOsalphen $C_{22}H_{20}N_2O_4$ analysis (%C(F)): C, 70.19 (70.30); H, 5.35 (5.02); N, 7.44 (7.95). FT-IR (cm^{-1}): υC=N, 1616; υC-O, 1275. NMR ^{1}H (ppm, Acet-d6): 8.85 (s, CH=N), 12.46 (s, C$_{Ar}$-O-H); ^{13}C NMR (ppm): 166 (C=N), 155 (C$_{Ar}$-O). M + H$^+$ (m/z): 376 (376). The structure of this ligand was already reported [40].

Mesalphen $C_{22}H_{20}N_2O_2$ analysis (%C(F)): C 76.72 (76.83); H, 5.85 (5.63); N, 8.13 (8.68). FT-IR (cm^{-1}): υC=N, 1618; υC-O, 1283. FT-IR values were comparable to those reported [40] NMR ^{1}H (ppm, CD$_3$CN): 8.70 (s, CH=N), 12.86 (s, C$_{Ar}$-O-H); ^{13}C NMR (ppm): 165 (C=N), 159 (C$_{Ar}$-O). M + H$^+$ (m/z): 344 (345).

OHsalphen $C_{20}H_{16}N_2O_4$ analysis (%C(F)): C, 68.95 (68.96); H, 4.62 (4.34); N, 8.04 (8.26). FT-IR (cm^{-1}): υC=N, 1614; υC-O, 1277. NMR ^{1}H (ppm, Acet-d6): 8.76 (s, CH=N), 12.30 (s, C$_{Ar}$-O-H); ^{13}C NMR (ppm): 165 (C=N), 155 (C$_{Ar}$-O). M + H$^+$ (m/z): 348 (349).

Salphen $C_{20}H_{16}N_2O_2$ analysis (%C(F)): C,75.89 (75.93); H,5.0 (5.1); N,8.91 (8.85). FT-IR (cm^{-1}): υC=N, 1612; υC-O, 1276. IR values were comparable to those reported [41]. M + H$^+$ (m/z): 316 (317). Insoluble for NMR

Clsalphen $C_{20}H_{14}N_2Cl_2O_2$ analysis (%C(F)): C, 62.45 (62.25); H, 3.66 (3.19); N, 7.27(8.04). FT-IR (cm^{-1}): υC=N, 1614; υC-O, 1273. M + H$^+$ (m/z): 384 (385). Insoluble for NMR. The structure of this ligand has already been reported [42].

Brsalphen $C_{20}H_{14}N_2Br_2O_2$ analysis (%C(F)): C, 50.66 (50.74); H, 2.97 (2.57); N, 5.90 (6.42). FT-IR (cm^{-1}): υC=N, 1612; υC-O, 1373. FT-IR values were comparable to those reported [18]. NMR ^{1}H (ppm, Acet-d6): 8.91 (s, CH=N), 13.09 (s, C$_{Ar}$-O-H); ^{13}C NMR (ppm): 164 (C=N), 160 (C$_{Ar}$-O). M + H$^+$ (m/z): 474 (474).

2.3. Synthesis of Nickel Complexes

In total, 1 mmol of nickel acetate was dissolved in methanol and the ligand, previously dissolved in methanol/chloroform, was added the dropwise to the nickel acetate solution, in case the ligand NO$_2$salen was dissolved in the DMF (dimethylformamide). The mixture of the reaction was stirred for 15 min and a solid compound precipitated. In the case of the NiRsalen complex, a brown-orange precipitate was obtained. For the NiRsalphen complexes, a red solid was precipitated. The solid compound was vacuum filtered and washed with methanol and chloroform. NiNO$_2$salen was washed with cold DMF [43]. The compounds were characterized by elemental analysis, FT-IR spectroscopy, NMR (^{1}H and ^{13}C), and mass spectrometry:

NiMeOsalen, N,N'-bis(5-metoxisalicylidene)ethylenediamine, nickel(II), NiC$_{18}$H$_{18}$N$_2$O$_4$· H$_2$O analysis (%C(F)): C,53.63 (53.12); H, 5.00 (4.96); N, 6.95 (6.98). FT-IR (cm^{-1}): υC=N, 1626; υC-O, 1328 FT-IR values were slightly lower than those reported [34]. NMR ^{1}H (ppm, DMSO-d6): 7.80 s, CH=N); ^{13}C (ppm): 160 (C=N), 162 (C$_{Ar}$-O). M + H$^+$ (m/z): 384 (385).

NiMe*salen*, N,N'-bis(5-metylsalicylidene)ethylenediamine, nickel(II), NiC$_{18}$H$_{18}$N$_2$O$_2$ analysis (%C(F)): C,61.19 (61.23); H, 5.10 (5.13); N, 7.93 (7.93). FT-IR (cm^{-1}): υC=N, 1624; υC-O, 1316 FT-IR values were comparable to those reported [35]. NMR ^{1}H (ppm, CDCl$_3$): 7.25 (s, CH=N); ^{13}C (ppm): 161 (C=N), 163 (C$_{Ar}$-O). M + H$^+$ (m/z): 352 (353).

NiOH*salen*, N,N'-bis(5-hidroxisalicylidene)ethylenediamine, nickel(II), NiC$_{16}$H$_{14}$N$_2$O$_4$· 2H$_2$O analysis (%C(F)): C, 48.94 (48.49); H, 4.15 (4.61); N, 7.39 (7.12). FT-IR (cm^{-1}): υC=N, 1614; υC-O, 1301. NMR ^{1}H (ppm, DMSO-d6): 7.75 (s, CH=N); ^{13}C (ppm): 159 (C=N), 162 (C$_{Ar}$-O). M + H$^+$ (m/z): 356 (357). The structure of this complex was already reported [44].

Ni*salen*, N,N'-bis(salicylaldehyde)ethylenediamine, nickel(II), NiC$_{16}$H$_{14}$N$_2$O$_2$ analysis (%C(F)): C, 58.77 (59.51); H, 3.89 (3.92); N, 8.57 (8.92). FT-IR (cm^{-1}): υC=N, 1624; υC-O, 1320. NMR ^{1}H (ppm, CDCl$_3$): 7.38 (s, CH=N); ^{13}C (ppm): 162 (C=N), 165 (C$_{Ar}$-O). M + H$^+$ (m/z): 324 (325). The structure of this complex was already reported [33].

NiCl*salen*, N,N'-bis(5-chlorosalicylidene)ethylenediamine, nickel(II), NiC$_{16}$H$_{12}$N$_2$Cl$_2$O$_2$· H$_2$O analysis (%C(F)): C, 46.22 (46.22); H, 3.42 (2.75); N, 6.80 (6.89). FT-IR (cm^{-1}): υC=N, 1624; υC-O, 1312. FT-IR values were comparable to those reported [45]. NMR ^{1}H (ppm, CDCl$_3$): 7.44 (s, CH=N). M+H$^+$ (m/z): 394 (394).

NiBr*salen*, N,N'-bis(5-bromosalicylidene)ethylenediamine, nickel(II) NiC$_{16}$H$_{12}$Br$_2$N$_2$O$_2$· 2H$_2$O analysis (%C(F)): C, 37.04 (36.85); H, 3.10 (3.05); N, 5.39 (5.46). FT-IR (cm^{-1}): υC=N, 1626; υC-O, 1309 FT-IR values are comparable to those reported [34]. NMR ^{1}H (ppm, CDCl$_3$): 7.45 (s, CH=N). M + H$^+$ (m/z): 482 (483).

NiNO$_2$*salen*, N,N'-bis(5-nitrosalicylidene)ethylenediamine), nickel(II), NiC$_{16}$H$_{12}$N$_4$O$_6$· 1.6H$_2$O analysis (%C(F)): C, 42.98 (43.18); H, 3.11 (3.46); N, 12.74 (12.59). FT-IR (cm^{-1}): υC=N, 1639; υC-O, 1321. FT-IR values were comparable to those reported [39]. NMR ^{1}H (ppm, DMSO-d6): 7.95 (s, CH=N). Insoluble for mass spectrometry.

NiMeO*salphen* (N,N'-bis(5-metoxisalicylidene)-1,2-phenylenediamine, nickel(II) NiC$_{22}$H$_{18}$N$_2$O$_4$ analysis (%C(F)): C, 61.01 (61.50); H, 4.18 (3.71); N, 6.46 (7.24). FT-IR (cm^{-1}): υC=N, 1616; υC-O, 1213. NMR ^{1}H (ppm, CDCl$_3$): 8.23 (s, CH=N). M + H$^+$ (m/z): 432 (433).

NiMe*salphen*, N,N'-bis(5-metylsalicylidene)-1,2-phenylenediamine, nickel(II) NiC$_{22}$H$_{18}$N$_2$O$_2$, analysis (%C(F)): C, 65.88 (66.30); H, 4.52 (4.02); N, 6.98 (7.31). FT-IR (cm^{-1}): υC=N, 1624; υC-O, 1213. M + H$^+$ (m/z): 400 (401). Insoluble for NRM. The structure of this complex was already reported [35].

NiOH*salphen*, N,N'-bis(5-hidroxisalicylidene)-1,2-phenylenediamine, nickel(II) NiC$_{20}$H$_{14}$N$_2$O$_4$ CH$_3$OH analysis (%C(F)): C, 57.70 (57.63); H, 4.15 (3.66); N, 6.40 (6.62). FT-IR (cm^{-1}): υC=N, 1610; υC-O, 1220. NMR ^{1}H (ppm, DMF-7): 8.00 (s, CH=N); ^{13}C (ppm): 162 (C=N), 155 (C$_{Ar}$-O). M+H$^+$ (m/z): 404 (405).

Ni*salphen*, N,N'-bis(salicylaldehyde) -1,2-phenylenediamine, nickel(II) NiC$_{20}$H$_{14}$N$_2$O$_2$ analysis (%C(F)): C, 64.8 (64.4); H, 3.70(3.78); N, 7,32 (7.51). FT-IR (cm^{-1}): υC=N, 1604; υC-O, 1295 FT-IR values were comparable to those reported [41]. M + H$^+$ (m/z):372 (373). Insoluble for NRM.

NiCl*salphen*, N,N'-bis(5-chlorosalicylidene)-1,2-phenylenediamine, nickel(II), NiC$_{20}$H$_{12}$N$_2$Cl$_2$O$_2$; analysis (%C(F)): C, 54.35 (55.59); H, 2.73 (2.28); N, 6.33 (7.16). FT-IR (cm^{-1}): υC=N, 1608; υC-O, 1290. FT-IR values were comparable to those reported [38]. Insoluble for NRM and mass spectrometry.

NiBr*salphen*, N,N'-bis(5-bromosalicylidene)-1,2-phenylenediamine, nickel(II), NiC$_{20}$H$_{12}$N$_2$Br$_2$O$_2$; analysis (%C(F)): C, 45.25 (45.77); H, 2.27 (2.16); N, 5.27 (5.87). FT-IR (cm^{-1}): υC=N, 1606; υC-O, 1328. FT-IR values were comparable to those reported [18]. Insoluble for NRM and mass spectrometry.

2.4. X-Ray Crystallography

Suitable single crystals for compounds NiMe*salen*, NiMeO*salen*, NiMeO*salphen*, and Ni*salphen* were mounted on a glass fiber. Crystallographic data were collected with an Oxford Diffraction Gemini "A" diffractometer with a CCD area detector, with $\lambda_{MoK\alpha}$ = 0.71073 Å for NiMe*salen*, NiMeO*salen*, NiMeO*salphen*, and $\lambda_{CuK\alpha}$ = 1.54184 Å for Ni*salphen* at 130 K. Unit cell parameters were determined with a set of three runs of 15 frames (1° in ω). The double pass method of scanning was used to exclude any noise [46]. The collected frames were integrated by using an orientation matrix determined from the narrow frame scans. Final cell constants were determined by a global refinement. Collected data were corrected for absorbance by using an analytical numeric absorption correction with a multifaceted crystal model based on expressions upon the Laue symmetry with equivalent reflections [47]. Structure solutions and refinement were carried out with the SHELXS-2014 [48] and SHELXL-2014 [49] packages. WinGX v2018.3 [50] software was used to prepare material for publication. Full-matrix least-squares refinement was carried out by minimizing $(Fo^2 - Fc^2)^2$. All non-hydrogen atoms were refined anisotropically. H atoms attached to C atoms were placed in geometrically idealized positions and refined as riding on their parent atoms, with C-H = 0.95 – 0.99 Å and with $U_{iso}(H) = 1.2U_{eq}(C)$ for aromatic and methylene groups, and $1.5U_{eq}(C)$ for methyl groups. On the other hand, for the compound NiMe*salen*, the solvent molecules were significantly disordered and could not be modeled properly (i.e., SQUEEZE [51]). Part of the PLATON package of crystallographic software was used to calculate the solvent disorder area and remove contributions to the overall intensity data. The disordered solvents area was centered around the 0.500–0.034 position and showed an estimated total of 60 electrons and a void volume of 180 Å^3. Crystallographic data for all complexes are presented in Table 1. The crystallographic data for the structures reported in this paper was deposited with the Cambridge Crystallographic Data Centre as supplementary publication no. CCDC 2006691–2006694. Copies of the data can be obtained free of charge on application to CCDC, 12 Union Road, Cambridge, CB2 1EZ, UK. (fax: (+44) 1223-336-033, e-mail: deposit@ccdc.cam.ac.uk).

Table 1. X-ray diffraction data collection and refinement parameters for the compounds NiMes*alen*, NiMeOs*alen*, NiMeOs*alphen*, and Nis*alphen*.

Compound	NiMes*alen*	NiMeOs*alen*	NiMeOs*alphen*	Nis*alphen*
Empirical formula	$C_{18} H_{18} N_2 Ni O_2$	$C_{18} H_{18} N_2 Ni O_4$	$C_{22} H_{18} N_2 Ni O_4$	$C_{43} H_{31} C_{l9} N_4 Ni_2 O_4$
Formula weight	353.05	385.05	433.09	1104.19
Crystal system	Monoclinic	Monoclinic	Monoclinic	Triclinic
Space group	C 2/c	P 2₁/c	P 2₁/n	P -1
Unit cell dimensions				
a (Å)	a = 21.422(2)	a = 16.5157(9)	a = 13.5517(4)	a = 13.7180(10)
b (Å)	b = 13.1534(18)	b = 7.2835(3)	b = 7.7890(2)	b = 14.030(2)
c (Å)	c = 6.5394(6)	c = 13.7886(7)	c = 17.01984(4)	c = 14.908(2)
α (°)	α = 90.	α = 90	α = 90.	α = 115.285(12)
β (°)	β = 96.280(9).	β = 104.809(5)	β = 100.197(3).	β = 116.477(11)
γ (°)	γ = 90	γ = 90	γ = 90.	γ = 92.967(10)
Volume(Å³)	1831.6(4)	1603.56(14)	1768.14(8)	2215.8(5)
Z	4	4	4	2
Density (calculated) (mg/m³)	1.280	1.595	1.627	1.655
Absorption coefficient (mm⁻¹)	1.068	1.237	1.880	1.441
F(000)	736	800	896	1116
Crystal size(mm³)	0.470 × 0.090 × 0.070	0.550 × 0.160 × 0.130	0.470 × 0.230 × 0.200	0.520 × 0.170 × 0.160
Theta range for data collection	3.641 to 29.388°.	3.445 to 29.394°.	3.854 to 73.599°.	3.380 to 29.513°.
Index ranges	−27 <= h <= 25, −16 <= k <= 16, −5 <= l <= 9	−20 <= h <= 17, −9 <= k <= 10, −18 <= l <= 19	−16 <= h <= 16, −7 <= k <= 9, −20 <= l <= 19	−18 <= h <= 18, −18 <= k <= 18, −19 <= l <= 16
Reflections collected	4413	11931	11435	18883
Independent reflections	2132 [R(int) = 0.0441]	3921 [R(int) = 0.0215]	3514 [R(int) = 0.0177]	10314 [R(int) = 0.0282]
Completeness to theta = 25.242°	99.5 %	99.8 %	100.0%	99.7 %
Absorption correction	Analytical	Analytical	Analytical	Analytical
Max. and min. transmission	0.929 and 0.750	0.854 and 0.646	0.725 and 0.556	0.827 and 0.682
Refinement method	Full-matrix least-squares on F²	Full-matrix least-squares on F²	Full-matrix least-squares on F²	Full-matrix least-squares on F²
Data/restraints/parameters	2132/0/106	3921/0/228	3514/0/264	10314/0/559
Goodness-of-fit on F²	1.019	1.047	1.058	1.027
Final R indices [I > 2sigma(I)]	R1 = 0.0506, wR2 = 0.0944	R1 = 0.0247, wR2 = 0.0644	R1 = 0.0313, wR2 = 0.0854	R1 = 0.0394, wR2 = 0.0796
R indices (all data)	R1 = 0.0802, wR2 = 0.1062	R1 = 0.0287, wR2 = 0.0671	R1 = 0.0332, wR2 = 0.0870	R1 = 0.0561, wR2 = 0.0892
Largest diff. peak and hole	0.601 and −0.625 e.Å⁻³	0.373 and −0.344 e.Å⁻³	0.238 and −0.567 e.Å⁻³	0.914 and −0.815 e.Å⁻³

2.5. Cyclic Voltammetry

The cyclic voltammetry was carried out with a conventional arrangement of three electrodes: a vitreous carbon working electrode, a platinum counter electrode, and a silver pseud-electrode. The potentials were referenced to the saturated calomel electrode (SCE) with ferrocene as an internal standard (E°$_{Fc+/Fc}$ = +0.46 V vs. SCE). The experiments were collected in 0.001 M DMSO solutions under nitrogen atmosphere. The supporting electrolyte was 0.1 M of tetrabutylammonium hexafluorophosphate.

3. Results and Discussion

3.1. Electronic Spectra

The electronic spectra of the R*salen* ligand showed three absorption maxima. The first was the 255–270 nm region and the second was the 315–350 nm region. These two bands were attributed to π → π* transitions, with a high molar absorptivity coefficient. They also had a third band between 420–430 nm due to n → π* of the group C=N. Electronic spectra of R*salphen* were similar to R*salen* spectra. The difference was that R*salphen* spectra showed a small shoulder next to the 260 nm and R*salphen* ligands had another band in the 270–400 region due to the π → π* transition for the third aromatic ring in N/N bridge. All R*salphen* transitions were shifted to a major wavelength values and had bigger values of molar extinction coefficients, because the higher aromaticity of the ligands favored the delocalization of electron density.

NiR*salen* showed four characteristics bands (Figures S9–S20, Supplementary Materials): the first two were in the 260–268 nm and 320–380 nm regions, with a high molar absorptivity coefficient, both due to π → π* transitions of the ligand. The third ws the 405–518 nm due to a ligand-metal charge transfer (LMCT) transition, from the phenolate to M due to $^1A_{1g}$ → ^1E_g transition [52,53]. The last band in the 500–680 nm region was owed to the d-d transition [52,54]. These bands could not be characterized with precision because of the low solubility of the compounds, since they presented a very low molar absorptivity coefficient. These bands were attributed to $^1A_{1g}$ → $^1A_{2g}$, which is characteristic for a square planar geometry. These electronic transitions were confirmed measuring the magnetic moment (μeff ≈ 0.5, Table S6, Supplementary Materials), meaning that the nickel complexes presented a diamagnetic property, consequences of the eight paired electrons. NiR*salphen* had the same trend that their ligands and had one more π → π* transition due to the third aromatic ring and higher molar extinction coefficients (Data Table 2). All maxima shifted to major wavelength values [52,54].

Table 2. Electronic spectral data of the Schiff bases and their complexes.

Compound	λ$_{max}$ (ε, L mol^{-1} cm^{-1}) in DMSO	Compound	λ$_{max}$ (ε, L mol^{-1} cm^{-1}) in DMSO
MeOsalen	260, (21145), 345(14109)	NiMeOsalen	258(39143), 330(7681), 431(7096)
Mesalen	260(17555), 326(8568), 427(176)	NiMesalen	260(47643), 334(7975), 417(6478)
OHsalen	260(13265), 350(8593)	NiOHsalen	260(40311), 333(7338), 438(6053)
Salen	260(8747), 327(18295), 410(299)	NiSalen	260(58958), 324(8061), 407(5906)
Clsalen	258(14461), 327(7794), 420(552)	NiClsalen	258(44337), 324(8216), 415(6263)
Brsalen	260(16272), 327(7479), 419(653)	NiBrsalen	258(48523), 326(8863), 414(6419)
NO$_2$salen	258(17745), 370(20301), 422(30053)	NiNO$_2$salen	263(18147), 340(11715), 405(22660)
MeOsalphen	276(42299), 348(18694)	NiMeOsalphen	268(50128), 296(2969), 382(26461), 511(2355)
Mesalphen	274(18573), 341(14755), 450(790)	NiMesalphen	260(55055),292(22231), 380(26821), 487(8851), 679(5.6)
OHsalphen	274(20172), 370(13706)	NiOHsalphen	260(39286), 298(19113), 386(22873), 518(8406)
Salphen	269(22051), 332(18444), 448(1418)	NiSalphen	260(37709), 298(14534), 377(22632), 475(7305)
Clsalphen	262(18058), 275(16555), 338(8525), 398(1779)	NiClsalphen	262(59148), 380(27100), 476(10739), 580(179)
Brsalphen	274(18364), 339(2102), 404(3323), 451(3341)	NiBrsalphen	264(36210), 314(13762), 378(21675) 478(23408), 674(750)

The last band involved with the d-d transition provided an approximation of the intensity of the complex field, since the energy of this electronic transition was associated with 10 Dq. This band could not be observed for complexes with the imine aliphatic bridge since it was masked by high intensity transitions. On the other hand, in the compounds with the aromatic bridge, we observed that those with a substituted electron-withdrawn (-Br, -Cl) had a greater wavelength value, thus decreasing the energy necessary to carry out this transition, especially when compared to the substituted electron-donor (-MeO, -OH) [43].

3.2. X-Ray

From the single crystal X-ray diffraction analysis we found that the compounds NiMe*salen* and NiMeO*salen* of Ni^{II} were polymorphs of (2,2'-(ethane-1,2-diylbis((nitrilo)methylylidene)) bis(4-methylphenolato))-nickel(II) methanol solvate [35] and dinuclear bis(2,2'-(ethane-1,2-diylbis ((nitrilo)methylylidene))bis(4-methoxyphenolato))-di-nickel(II) methanol solvate [35], respectively. Polymorphism was found in the crystalline arrangement, since the compound NiMe*salen* (Figure 1) crystallized in the monoclinic crystal system with a space group of C2/c, while in literature it was found that methanol solvated compound crystallized in the triclinic crystal system with a space group of P-1. On the other hand, NiMeO*salen* crystallized in the monoclinic crystal system with a space group of P2$_1$/c, while the previously reported methanol solvated compound was space group P2$_1$/n and monoclinic crystal system [35].

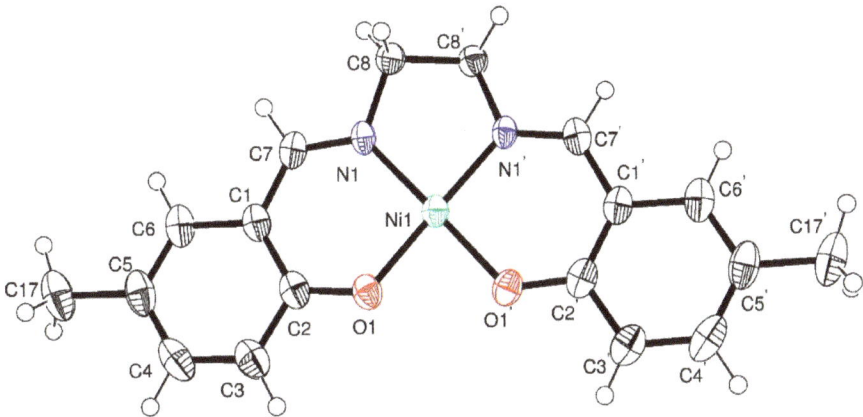

Figure 1. View on the perspective of the compound NiMe*salen*, with a displacement ellipsoid at a 50% probability level for non-H atoms.

Discrete unit NiMe*salen* contained one central Ni^{II} ion and one unit of deprotonated 2,2'-(ethane-1,2-diylbis((nitrilo)methylylidene))bis(4-methylphenolato) tetradentate ligand. Figure 1 shows the tetracoordinated metal center of Ni^{II} despites having a N$_2$O$_2$ coordination environment. Selected bond and angle parameters are given in Table 3.

The Ni^{II} center had almost a perfect square-planar geometry, which was defined by two N and two O atoms with τ_4 of 0.02 and torsion angles O1-Ni1-N1-C8 (169.73(17)°), N1'-Ni1-N1-C7 (173.6(2)°), C2-O1-Ni1-N1 (7.8(2)°), and C2-O1-Ni1-O1' (171.2(2)°). In fact, the Ni (II) atom was 0.016 Å out of the plane and formed by O1/O1'/N1'/N1. Each discrete molecule coplanar had an rms (root-mean squeare) of 0.030. There were intermolecular interactions of type C-H···O, a hydrogen bond, and π···π contacts that stabilized the crystal packing (Figure 2). Intermolecular interactions were established as follows: the hydrogen atom (C) carbon donor atom interacted with the (O) oxygen acceptor atom C8-H8A···O1 (2.45 Å), thus forming a $R_1^2(4)$ motif along the c axes. In this same crystallographic direction, the interaction of type π···π was represented by the centroid Cg4 and six membered ring

C1/C6. The intermolecular contacts of the hydrogen bond and π-stacking array formed a bidimensional complex array along the *a-b* plane.

Table 3. Selected bond lengths (Å) and angles (°) for compounds NiMe*salen*, NiMeO*salen*, NiMeO*salphen*, and Ni*salphen*.

NiMesalen		NiMeOsalen		NiMeOsalphen	
Bond *	Lengths	Bond	Lengths	Bond	Lengths
C2—O1	1.323(3)	C2—O1	1.3125(17)	C2—O1	1.3077(18)
C5—C17	1.525(4)	C5—O3	1.3770(17)	C5—O3	1.3765(18)
C7—N1	1.297(3)	C7—N1	1.2892(19)	C7—N1	1.298(2)
C8—N1	1.469(4)	C8—N1	1.4774(17)	C8—C9	1.397(2)
C8—C8#1	1.515(5)	C8—C9	1.504(2)	C8—N1	1.4233(19)
O1—Ni1	1.852(2)	C17—O3	1.4256(18)	C17—O3	1.4219(19)
Ni1—N1#1	1.844(2)	O1—Ni1	1.8505(10)	N1—Ni1	1.8626(12)
Ni1—N1	1.844(2)	O2—Ni1	1.8544(10)	N2—Ni1	1.8643(13)
		Ni1—N2	1.8476(12)	Ni1—O2	1.8394(11)
		Ni1—N1	1.8520(12)	Ni1—O1	1.8536(11)
Bond	Angles	Bond	Angles	Bond	Angles
N1#1—Ni1—N1	86.32(15)	N1—C8—C9	107.28(11)	C9—C8—N1	113.44(13)
N1#1—Ni1—O1#1	94.95(9)	N2—C9—C8	107.04(11)	C8—C9—N2	114.08(13)
N1—Ni1—O1#1	178.40(10)	N2—Ni1—O1	179.52(5)	C8—N1—Ni1	113.37(10)
N1#1—Ni1—O1	178.40(10)	N2—Ni1—N1	86.03(5)	C9—N2—Ni1	113.04(10)
N1—Ni1—O1	94.95(9)	O1—Ni1—N1	94.40(5)	O2—Ni1—O1	84.12(5)
O1#1—Ni1—O1	83.79(13)	N2—Ni1—O2	93.77(5)	O2—Ni1—N1	179.34(5)
		O1—Ni1—O2	85.80(4)	O1—Ni1—N1	95.22(5)
		N1—Ni1—O2	179.60(5)	O2—Ni1—N2	94.60(5)
		C8—N1—Ni1	114.60(9)	O1—Ni1—N2	177.72(5)
		C9—N2—Ni1	113.22(9)	N1—Ni1—N2	86.05(6)

Nisalphen			
Molecule A		Molecule B	
Bond *	Lengths	Bond	Lengths
C2A-O1A	1.306(3)	C2B-O1B	1.306(3)
C7A-N1A	1.308(3)	C7B-N1B	1.299(3)
C8A-C9A	1.395(3)	C8B-C9B	1.389(4)
C8A-N1A	1.417(3)	C8B-N1B	1.427(3)
C12A-O2A	1.312(3)	C12B-O2B	1.309(3)
O1A-Ni1A	1.8398(16)	O1B-Ni1B	1.8370(17)
O2A-Ni1A	1.8346(18)	O2B-Ni1B	1.8360(16)
N1A-Ni1A	1.857(2)	N1B-Ni1B	1.8567(19)
N2A-Ni1A	1.8564(19)	N2B-Ni1B	1.856(2)
Bond	Angles	Bond	Angles
C9A-C8A-N1A	113.8(2)	C9B-C8B-N1B	113.8(2)
C8A-C9A-N2A	113.8(2)	C8B-C9B-N2B	113.8(2)
C8A-N1A-Ni1A	113.13(16)	C8B-N1B-Ni1B	112.87(16)
C9A-N2A-Ni1A	112.97(16)	C9B-N2B-Ni1B	113.22(16)
O2A-Ni1A-O1A	83.05(8)	O2B-Ni1B-O1B	83.03(7)
O2A-Ni1A-N2A	95.17(8)	O2B-Ni1B-N2B	95.47(8)
O1A-Ni1A-N2A	178.13(9)	O1B-Ni1B-N2B	178.47(8)
O2A-Ni1A-N1A	178.42(8)	O2B-Ni1B-N1B	178.24(9)
O1A-Ni1A-N1A	95.46(8)	O1B-Ni1B-N1B	95.23(8)
N2A-Ni1A-N1A	86.32(9)	N2B-Ni1B-N1B	86.27(9)

* Operators for generating equivalent atoms: −x + 1, y, −z + 1/2#1.

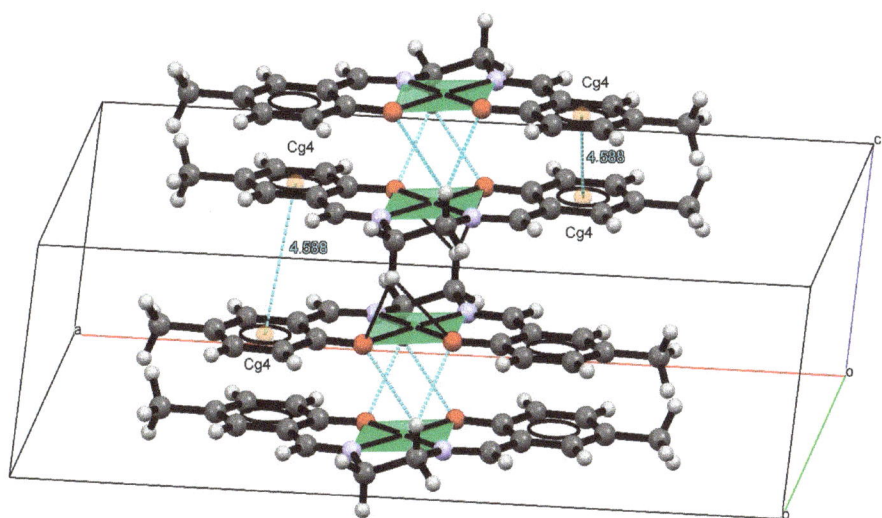

Figure 2. Crystal array of compound NiMe*salen* shows the view along the *b*-axis and with perspective to plane formed by *a-c* axes emphasizing the π-stacking and R^1_2 (4) motif.

Figure 3 shows the perspective view of the molecular structure of polymorphic compound NiMeO*salen*. The NiMeO*salen* discrete unit contained one central Ni^{II} ion and one 2,2'-(ethane-1,2-diylbis((nitrilo)methylylidene))bis(4-methoxyphenolato) tetradentate ligand. The Ni^{II} ion was tetracoordinated with an N_2O_2 coordination environment. All bond length and angles corresponded to those reported in the literature [43]. Selected bond and angles parameters are given in Table 3.

The Ni^{II} center had a perfect square-planar geometry, defined by two N and two O atoms with rms of 0.004, τ_4 of 0.006, and torsion angles O1-Ni1-N1-C8 (176.01(2)°), N2-Ni1-N1-C7 (176.33(2)°), C2-O1-Ni1-N1 (10.03 (2)°), and C2-O1-Ni1-O2 (170.32(2)°). In fact, the Ni^{II} atom was 0.005 Å out of the plane and formed by O1/O2/N2/N1. Nonetheless, it was observed that the six-membered rings deviated slightly from the coplanarity, thus finding an angle of 2.60(1)° between the aromatic rings.

Similarly with NiMe*salen*, in the crystalline arrangement for the compound NiMeO*salen*, there were intermolecular interactions for the C-H⋯O hydrogen bonding and intermolecular contacts of type π⋯π (Figure 4). For the no classical hydrogen bond, these interactions were formed between the (C) carbon donor atom and two (O) oxygen acceptor atoms (C8-H8B⋯O1 (2.61 Å) and C8-H8B⋯O1(2.45 Å)), thus forming an $R_1^2(4)$ motif along the *b* axes. Additionally, there were C17-H17C⋯O4 (2.75 Å) and C18-H18C⋯O3 (2.82 Å), which formed an $R_2^2(6)$ motif along the *b-c* plane. The weak interaction π⋯π had a distance of 3.95(8) Å between Cg4 and Cg5. Cg4 represent the six membered ring C1/C6 and Cg5 correspond to the C11/C16 ring. Finally, the intermolecular contacts of no classical hydrogen bond and π-stacking formed a tridimensional supramolecular array.

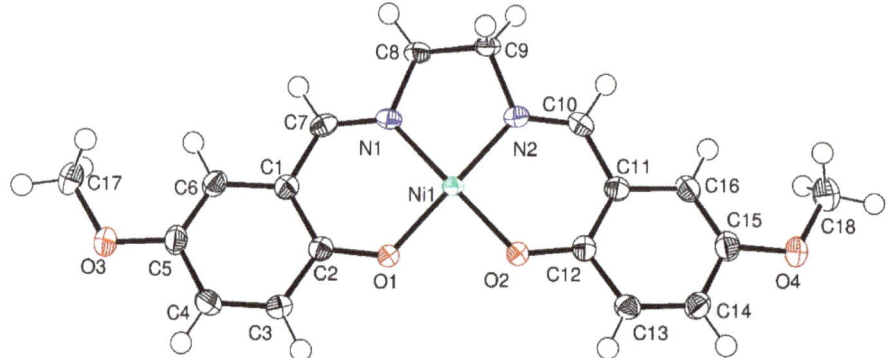

Figure 3. Perspective of the compound NiMeO*salen* with a displacement ellipsoid at a 50% probability level for non-H atoms.

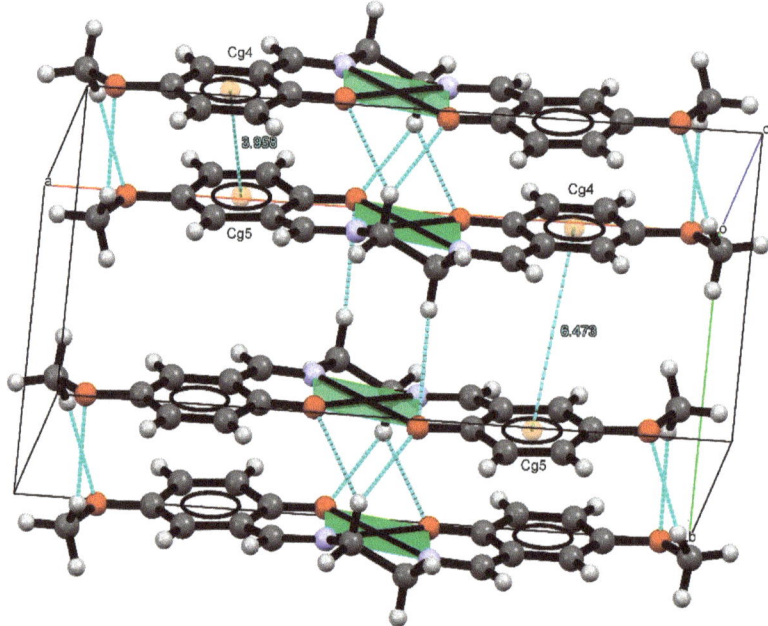

Figure 4. Crystal array of compound NiMeO*salen*. View along the *c* axes and with a perspective of a plane formed by the *a-b* axes, which emphasizes the $R_1^2(4)$ and $R_2^2(6)$ motifs, as well as π-stacking.

The compound NiMeO*salphen* crystallized in a monoclinic crystal system with the space group P2$_1$/n. The asymmetric unit consisted of one nickel(II) center and one 2,2′-{1,2-phenylenebis[(azanylylidene)methylylidene]}bis(4-methoxyphenolato) ligand. The ORTEP diagram is shown in Figure 5. The squared plane of Ni^{II} center was chelated by two oxygen and two nitrogen atoms that derived from a *salen* ligand, with Ni-O and Ni-N bond distances ranging from 1.8394 to 1.8643 (13) Å. O-Ni-O, N-Ni-N, and O-Ni-N bond angles of 84.12 to 179.34°. The length distance of Ni-N was, on average, 1.8634(12) Å (Table 3), which was slightly higher than that observed in compounds NiMe*salen* and NiMeO*salen*. Nevertheless, the tetracoordinate Ni^{II} in compound NiMeO*salphen* had a square plane geometry with a rms of 0.0147 and a τ$_4$ of 0.020. An analysis of the coplanarity shows that there were angles of 3.07 (7) and 5.06 (7)° between the square plane N$_2$O$_2$

at the metal center. The planes formed by the six-membered rings C1/C6 and C11/C16, respectively. Additionally, there was a perfect coplanarity between the square plane N_2O_2 at the metal center and the ring formed by the C8-C9-C1 /C22 atoms with an angle of 0.52 (7)°.

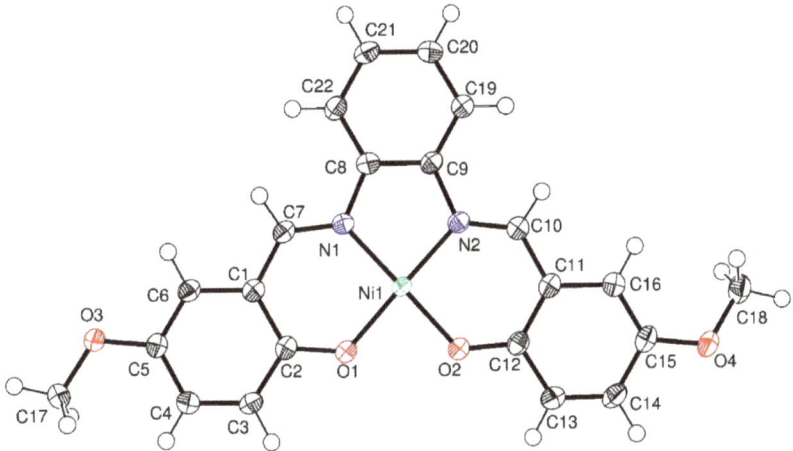

Figure 5. Perspective of the compound NiMeO*salphen* with a displacement ellipsoids at a 65% probability level for non-H atoms.

In the crystal packing, there were C-H···O no classic hydrogen bonds and π···π intermolecular contacts (Figure 6).

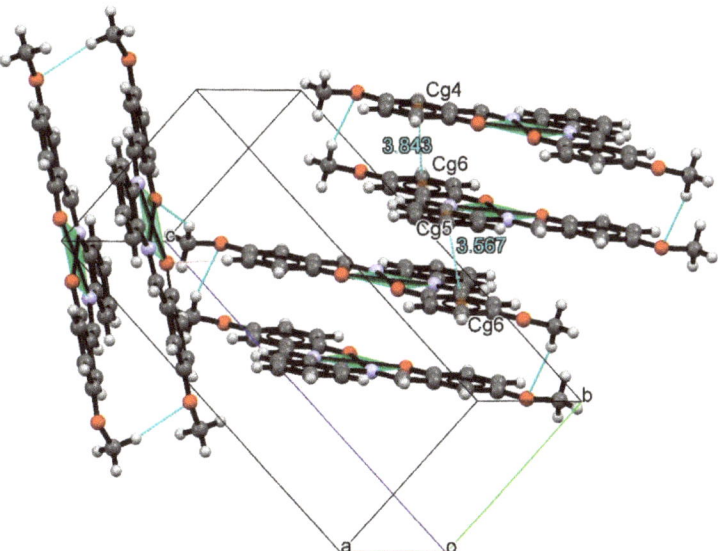

Figure 6. Crystal array of compound NiMeO*salphen*. View is a tridimensional perspective emphasizing $R_2^2(30)$ and $R_2^2(11)$ *motifs* and π-stacking.

The interactions of type hydrogen bond were observed between C17-H17C···O2 (2.43 Å), C18-H18C···O3 (2.50 Å), and C21-H21···O3 (2.42 Å). These intermolecular contacts formed R22(30) and $R_2^2(11)$ motifs along the a-c plane. Moreover, there were weak π-π interactions with a distance of centroids Cg4-Cg6 (3.84 Å) and Cg5-Cg6 (3.56 Å). Cg4 represented the six membered ring C1/C6, while Cg5 corresponded to the C8-C9-C19/C22. Cg6 was formed by the C11/C16 ring. Finally, all intermolecular contacts formed a tridimensional supramolecular array.

Unlike previously mentioned compounds NiMe*salen*, NiMeO*salen*, and NiMeO*salphen*, the single crystal X-ray diffraction analysis revealed that compound Ni*salphen* crystalized in the triclinic space group P-1. The asymmetric unit of Ni*salphen* contained two molecules of the nickel coordination compound and three molecules of the chloroform, which was then used as a solvent (Figure 7). Each metal central of Ni^{II} ion was tetracoordinated with one unit of a deprotonated *salphen* ligand with *salphen* = N,N'-o-phenylenebis(salicylideneimine). Selected bond and angles parameters are given in Table 3.

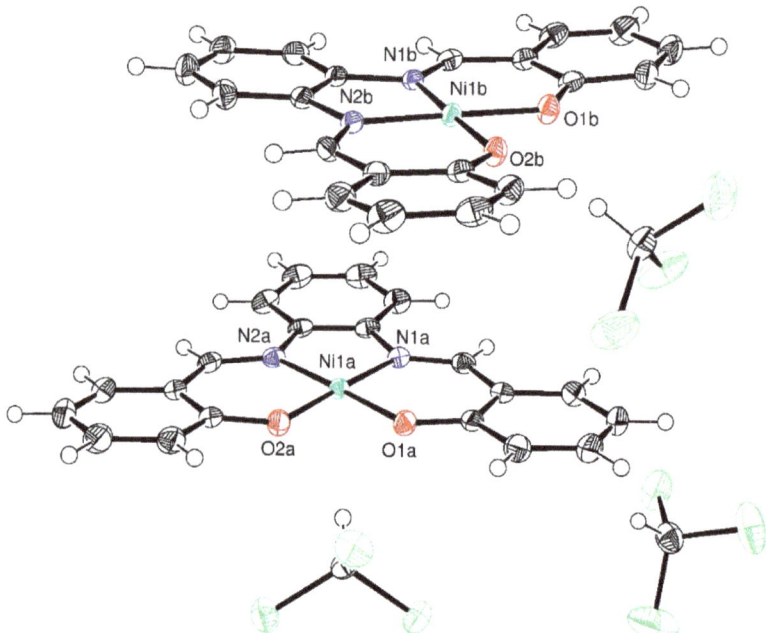

Figure 7. Perspective of the compound Ni*salphen* with a displacement ellipsoids at a 55% probability level for non-H atoms.

While investigating the plane formed by three aromatic rings and the square symmetry N_2O_2 at the metallic center of Ni^{II}, we found a coplanarity in each molecule with rms of 0.060 and 0.019 Å for molecules A and B, respectively. Furthermore, these molecules had a parallel arrangement between them, with an angle of 1.4°.

In molecule A, there was a perfect square planar geometry with a τ_4 de 0.026 with torsion angles O1A-Ni1A-N1A-C8A (179.72(15)°), N2A-Ni1A-N1A-C7A (177.8(2)°), C2A-O1A-Ni1A-N1A (0.90 (2)°), and C2A-O1A-Ni1A-O2A (179.6(2)°). Similarly, molecule B had a perfect square planar geometry (τ_4 de 0.024) with torsion angles O1B-Ni1B-N1B-C8B (179.26(15)°), N2B-Ni1B-N1B-C7B (178.8(2)°), C2B-O1B-Ni1B-N1B (6.0(2)°), and C2B-O1B-Ni1B-O2B (173.7(2)°).

In the crystalline arrangement of compound Ni*salphen*, a Ni-Ni distance of 3.26 Å was observed. The short distance found between both metal centers was favored by the interaction of the π···π. There were electronic densities in the coplanar and parllel A-A and A-B molecules.

This system obtained dinuclear structural arrangements with possible applications in molecular modeling and bioinorganic systems. Additionally, there were intermolecular interactions for C-H···O hydrogen bonding. Figure 8 shows the crystalline array with intermolecular contacts.

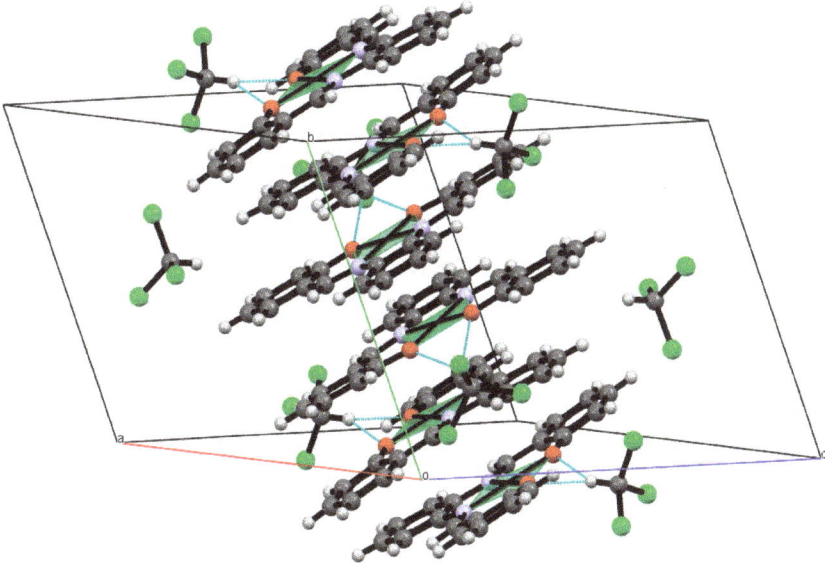

Figure 8. Crystal array of compound Ni*salphen*, view a tridimensional perspective emphasizing the π-stacking.

For the no classical hydrogen bond, interactions were formed between the carbon donor atom of the chloroform molecule solvent and the two oxygen acceptor atoms of the *salphen* ligand: C41-H41···O1A (2.26 Å), C41-H41···O2A (2.26 Å), C51-H51···O1B (2.29 Å), and C51-H51···O2B (2.19 Å). These show a linear, bifurcated, and trifurcated form for two, three, and four centers, respectively, in the intermolecular interaction. Additionally, there were C51···H16A (2.99 Å) and C52···H14B (3.01 Å) intermolecular contacts.

Despite the small differences on the Ni^{II}-donor atoms length, the nature of the N/N bridge and electron-donor/withdrawn character of the substituents in the 5- and 5′-position of the Schiff base play a key role in packing Ni^{II} coordination compounds. This can be observed in the Ni-Ni distance found in the different crystal structures obtained here and those previously reported.

Another important factor that influenced crystalline packing was the solvent. The Ni*salen* solvate reported by Siegler et al. showed a crystalline arrangement; the interactions stabilizing the crystal depended on it. It favored dimers when the Ni-Ni distances were modified according to acetone, 3.16 Å, $CHCl_3$, 3.13 Å (system monoclinic); $CHCl_3$, 3.19 Å (system orthorhombic); CH_2Cl_2, 3.28 Å; $C_2H_4O_2$, 3.37 Å; DMF, 3.3901 Å; or the favor 1-D chain, as was the case for the methanol solvate Ni-Ni 3.44 Å, wherein the solvent joined the monomers through C-H···O interactions in one direction [33].

Comparing the intermolecular interactions found on Ni*salen* and Ni*salphen*, the incorporation of an extra aromatic ring in the ligand structure increased the number on the π···π and C-H···π interactions. In the two Ni*salen* structures reported with the same crystalline system (triclinic), the π···π interaction found a length of 3.63 [55] and 4.43 Å [33]. Meanwhile, the two C-H···π were observed. On the other hand, the Ni*salphen* structure presented two π···π interactions with lengths of 3.89 and 4.55 and four C-H···π interactions of 3.22, 3.39, 3.65, and 3.68 Å. The sum of all interactions led to a Ni-Ni distance of 3.26 Å, which was slightly smaller than the length found in both Ni*salen* with 3.63 and 3.36 Å, respectively.

Substituents also played an important role in crystal packing. *NiMeOsalphen* presented three π···π interactions (i.e., 3.56, 3.65, and 3.84 Å), three C-H··· π interactions (3.24, 3.33 and 3.38 Å), and a C-H···O interaction with a length of 2.50 Å. For *NiMeOsalen*, only one π···π interaction of 3.95 Å was observed. There were two C-H···π interactions with lengths of 3.61 and 3.77 Å and two C-H···O interactions with lengths of 2.75 and 2.82 Å. The Ni-Ni distance observed in these examples could be closely related with the C-H···O and π···π interactions from the methoxy groups and the extra aromatic ring for *NiMeOsalphen*. The Ni-Ni distance on *NiMeOsalphen* could be longer than *NiMeOsalen* due to the π···π interaction found between the two dimeric units.

For NiMe*salen*, the π···π interaction was retained but the main contribution for the crystal stabilization relied on the C-H···π interaction with distance values of 3.64 and 3.66 Å. These interactions kept the two units close enough to establish a Ni-Ni distance of 3.39 Å. The C-H···O interactions elicited by the methoxy groups contributed a shorter Ni-Ni distance for NiMeO*salen* (3.18 Å) than NiMe*salen* (3.39 Å). The same was observed for the compounds NiMeO*salphen* and NiMe*salphen* [35].

In the crystalline structure NiOH*salen*, two interactions of π-π were shown. However, the -OH groups in the structure stabilized the crystalline packing mainly by the interaction of the hydrogen bridges for the $O_{solvent}$-H···O_{salen} and O_{salen}-H···$O_{solvent}$ with methanol molecules [44]. The Ni-Ni distance was 3.61 Å, which, when compared to NiMeO*salen* (3.18 Å), increased because of the sovlent's role in the packing. One methanol molecule formed a hydrogen bridge interaction with two neighboring molecules, $O_{solvent}$-H···O_{salen} and O_{salen}-H···$O_{solvent}$ [44]. These solvent interactions also occurred in the Ni*salen* structure when methanol was the solvate [33].

3.3. Hirshfeld Surface Analysis

Hirshfeld's surface (HS) analysis provided detailed information regarding intermolecular interactions. A better understanding of the problem may help address the challenge of quantitatively understanding intermolecular contacts using visual information on color and shadow on surfaces [56].

The Crystal Explorer 17 program [57] was used to generate the HS and 2D fingerprint plots of the complexes (i.e., NiMe*salen*, NiMeO*salen*, NiMeO*salphen*, and Ni*salphen*). The d_{norm} HS was obtained, which combined the normalized distances from the closer atom inside the surface (d_i) and outside the surface (d_e) to the HS, showing all contacts of the crystal structure. The red regions indicate the contacts were shorter than the sum of the van der Waals radii of the involved atoms. The blue and white regions indicated that the contacts were longer and closer to the van der Waals limit. Figure 9 shows the HS and all compound interactions.

The d_{norm} HS of the compounds showed red spots, which indicated close-contacts in the crystal structure, i.e., non-classical hydrogen bonds C-H···O and π···π, as well as intermolecular interactions between centroids of six-membered rings in phenyl groups. The shape index was a function of HS and very helpful when investigating the π···π stacking interaction. The blue and red zone indicated a region with a stacking arrangement. Figure 10 presents the shape index mapped on the compounds' HS. The blue zone indicated the presence of π···π stacking interactions in the crystal structure. The π···π interaction in compound Ni*salphen* stabilized and favored the 3.26 Å distance between the Ni^{II} metal centers, due to the presence of molecules A and B in the asymmetric unit of Ni*salphen*. Figures S1–S8 present the details of the fingerprint plots for each compound. In them, they describe the intermolecular interactions around the HS.

Figure 11 shows the contributions of contacts obtained from the decomposition of the fingerprint plots. The fingerprint plots of NiMe*salen*, NiMeO*salen*, and NiMeO*salphen* were similar, indicating that the H···H and C···H/H···C were the most important contributors for crystal packing. H···H contacts contributed 64.4% (NiMe*salen*), 46.4% (NiMeO*salen*), and 32.4% (NiMeO*salphen*), while C···H/H···C contacts contributed 16% (NiMe*salen*), 21.2% (NiMeO*salen*), and 20% (NiMeO*salphen*). A similar trend was observed in the fingerprint plot for Ni*salphen*, where the H···H and X···H/H···X contacts had greater contributors for stabilizing interactions, with H···H contacts contributing 32.5% and 27.5% in molecules A and B, respectively. The contributions for C···H/H···C, O···H, and C···C contacts were approximately of 20%, 5%, and 8% for molecules A and B, while the Cl···H/H···Cl contact contributed 19.9% and 23.8% in molecules A and B, respectively.

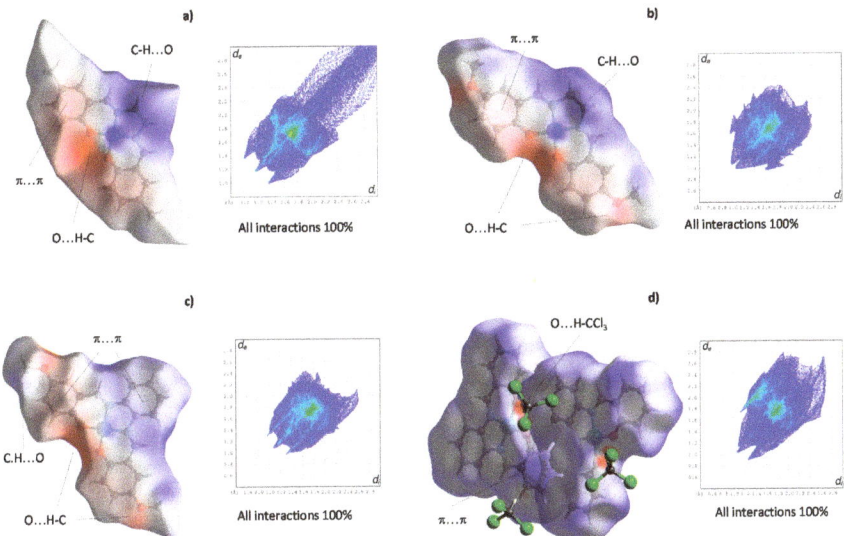

Figure 9. Hirshfeld surface (HS) with d_{norm} mapped and fingerprinted plots of the compounds NiMe*salen* (**a**), NiMeO*salen* (**b**), NiMeO*salphen* (**c**), and Ni*salphen* (**d**) for all interactions.

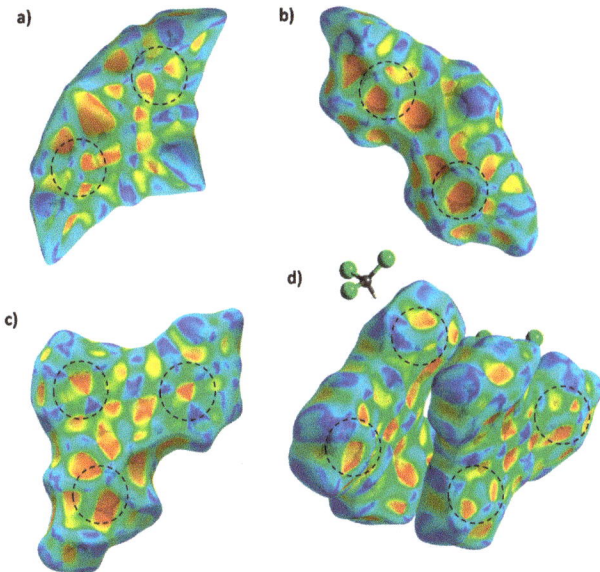

Figure 10. HS of the compounds NiMe*salen* (**a**), NiMeO*salen* (**b**), NiMeO*salphen* (**c**), and Ni*salphen* (**d**), mapped with shape index.

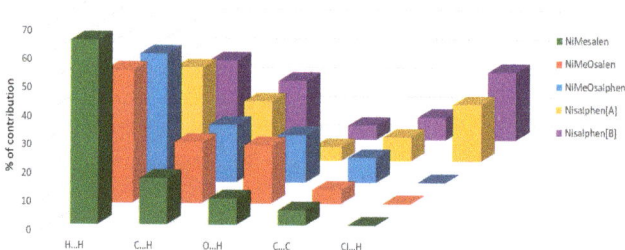

Figure 11. Contribution of some intermolecular contacts for HS of the compounds NiMe*salen*, NiMeO*salen*, and NiMeO*salphen*, as well as for molecules A and B of compound Ni*salphen*.

3.4. Cyclic Voltammetry

Nickel^{II}-*salen* compounds have a neutral charge and show low solubility. Adding an extra aromatic ring in the ligand structure (i.e., *salphen*-type ligands) causes the solubility to decrease even more. When the NiR*salphen* solution bubbled with nitrogen, it started to precipitate. Because the low solubility of the compounds, it was only possible to characterize NiR*salen* compounds in the electrochemical study.

We performed voltammetry of the ligands (Figure 12b). The *salen* ligand ran in the direction of the positive potential in an interval of −3.2 to 1.0 V. In an inversion study, reduction signals 3a and 3b were associated with C=N reduction and an irreversible oxidation signal, 4a [58]. Figure 12a shows the Ni*salen* voltammogram under the same condition, caused by the nickel oxidation process ([Ni^{II}L] → [Ni^{III}L] + 1 ē) and 2a and 2b due to nickel reduction process ([Ni^{II}L] + 1 ē → [Ni^{I}L]) [14,59]. Signal 3a and 3b was also observed to shift lower potential values. The other NiR*salen* complexes showed a similar behavior with the signals that shifted to different potentials due to the substituent in the 5,5′position (Figures S40–S46, Supplementary Materials). In this work, only the processes associated with the reduction and oxidation of nickel were reported. The voltammograms were run in an interval of −2.4 to 1.0 V (Figure 12c,d).

The cathodic and anodic peak current were plotted in the square root function of the sweep speed ($v^{1/2}$). Only the complexes NiMeOH*salen*, NiOH*salen*, NiCl*salen*, and NiBr*salen* presented a linear dependence, which means that the oxidation of nickel was a diffusion-controlled process. A coupled reaction was suggested to impact the reversible process, thus confirming that Ni*salen* and NiMe*salen* via plotting i_{pc}/i_{pa} vs. logV. The oxidation process for the complexes were irreversible due to the ΔE being too big. The electron transference was a slow process, as is shown in Table 4. The oxidation process involved an EC mechanism and the Ni^{III}-*salen* complex coordinates solvents, such as DMSO, in their axial position to stabilize the Ni^{III} oxidations for electronic density [14,60].

With regard to the oxidation process, the reductions were a quasi-reversible process and we found that all nickel complex reductions were diffusion-controlled processes, except for NiOH*salen*, which presented coupled reactions. In comparison with oxidation reactions, the reduction of Ni^{II} was a more quantitative process. ΔE values were close to 59 mV and the i_{pc}/i_{pa} ratio was closer to 1 (Table 4).

For both processes, we found a trend between $E_{1/2}$ and the effect of the substituent. Correlations were made with the Hammett sigma in the para-position. The metal center's acidity was influenced by the effect of the substituent. Therefore, the oxidative and reductive capacity of nickel modulated with the correct use of these substituents [13,61–63]. Electron-donor substituents shifted the $E_{1/2}$ to a lower potential value and the electron-withdrawn groups shifted toward a more positive potential value. Thus, an electron-donor group improved the reductive capacity and electron-withdrawn groups improved the oxidative capacity of nickel, as shown in Figure 13.

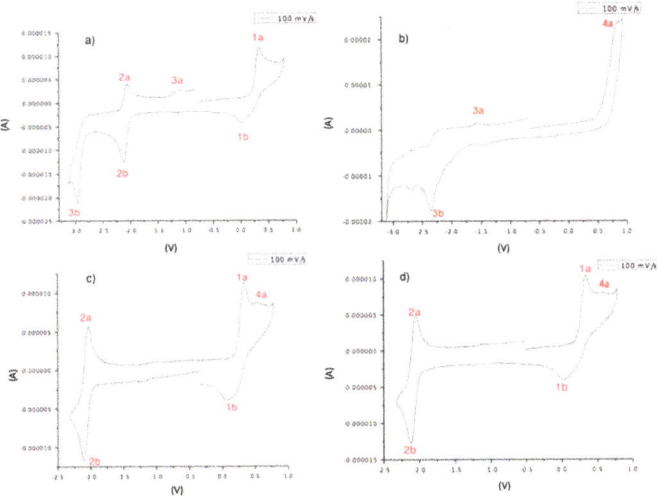

Figure 12. Voltammogram of Ni*salen* 1 mM (DMSO). (**a**) Ni*salen* voltammogram from −3.2 to 1.0 V; (**b**) *salen* voltammogram from −3.2 to 1.0 V; (**c**) Ni*salen* voltammogram from negative potential to 2.4 to 1.0 V; (**d**) Ni*salen* voltammogram from positive potential to −2.4 to 1.0 V. All the experiments were referenced to the pair Fc$^+$/Fc.

Table 4. Cyclic voltammetric parameter for NiR*salen* complexes, referenced to the pair Fc$^+$/Fc.

Process Compound	Ni^{II}L→Ni^{III}L + 1 ē			Ni^{II}L + 1 ē →Ni^{I}L		
	ΔE (mV)	i_{pc}/i_{pa}	$E_{1/2}$ (V)	ΔE (mV)	i_{pc}/i_{pa}	$E_{1/2}$ (V)
NiMeO*salen*	110	0.19	0.19	110	0.50	−2.08
NiMe*salen*	240	0.31	0.17	77	0.61	−2.13
NiOH*salen*	200	0.12	0.073	92	0.30	−2.17
Ni*salen*	410	0.16	0.16	76	0.72	−2.09
NiCl*salen*	390	0.16	0.22	73	0.83	−1.98
NiBr*salen*	400	0.23	0.21	60	0.45	−1.96
NiNO$_2$*salen*	200	0.41	0.33	120	0.57	−1.72

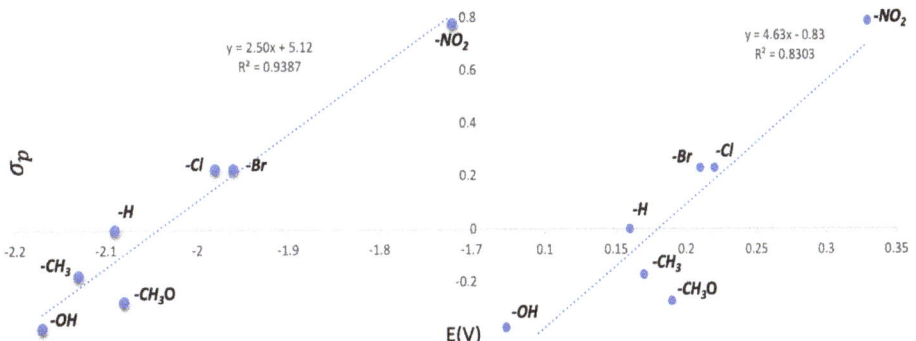

Figure 13. Left: correlation between $E_{1/2}$ of Ni^{II}/Ni^{I} and the σ_p constant for the NiR*salen* complexes. Right: correlation between $E_{1/2}$ of Ni^{III}/Ni^{II} and the σ_p constant for the NiR*salen* complexes.

4. Conclusions

In this study, various Schiff bases and their Ni^{II} complexes were synthesized. All the prepared ligand and complexes were analyzed via C, H, and N analyses. They were assigned molecular structures and geometries using information obtained from UV-Vis, magnetic susceptibility, and X-ray crystallography, all of which corresponded to square-planar geometry in the solid state. The Hirshfeld surface analysis was used to study intermolecular interactions. This analysis revealed that the O···H, H···H and π···π contacts were the most significant in the crystal array of the compounds *NiMesalen*, *NiMeOsalen*, and *NiMeOsalphen*, and O···H, H···H, Cl···H and π···π contacts in the crystal array of the compound *Nisalphen*. The no classical hydrogen bonding and π···π stacking information conveyed by Hirshfeld surface analysis were consistent with the crystal structure analysis. The substituents and the N/N bridge affected the crystal packing and electronic properties of nickel. According to the structures obtained for *Nisalphen*, *NiMeOsalen*, and *NiMeOsalphen*, it was possible to observe that the addition of an aromatic ring in the N/N bridge increased the number on π···π and C-H···π interactions and decreased their length. Substituents also played an important role in crystal packing for *NiMeOsalen* and *NiMeOsalphen*. i.e., a higher contribution for the O···H interaction. Due to this contribution, the length of π···π interactions were minor in both complexes. In N/N aliphatic bridge complexes, the substituents also had an important role. The $E_{1/2}$ depended on the electron-withdrawn or electron-donor nature of the R (R') substituent, which followed a correlation with the σ_p of Hammet in such a way that, for the electron-donor substituent, -OH and -CH$_3$ decreased the half-wave potential, instead favoring nickel's reductor ability.

Supplementary Materials: The following are available online at http://www.mdpi.com/2073-4352/10/7/616/s1, Table S1. Atomic coordinates (x 104) and equivalent isotropic displacement parameters (Å2 103) for compound NiMesalen. U(eq) is defined as one third of the trace of the orthogonalized Uij tensor. Table S2. Atomic coordinates (×104) and equivalent isotropic displacement parameters (Å2 × 103) for compound NiMeOsalen. U(eq) is defined as one third of the trace of the orthogonalized Uij tensor. Table S3. Atomic coordinates (x 104) and equivalent isotropic displacement parameters (Å2 × 103) for compound NiMeOsalphen. U(eq) is defined as one third of the trace of the orthogonalized Uij tensor. Table S4. Atomic coordinates (×104) and equivalent isotropic displacement parameters (Å2 × 103) for compound Nisalphen. U(eq) is defined as one third of the trace of the orthogonalized Uij tensor. Table S5. Bond lengths [Å] and angles [°] for compound Nisalphen. Table S6. Effective magnetic moment and number of unpaired electrons of Ni^{II} complexes. Table S7. ^{1}H-NMR values for the ligands and nickel complexes. Table S8. ^{13}C-NMR values for the ligands and nickel complexes. Figure S1. Normalized contact distance (d$_{norm}$, defined in terms of d$_e$, d$_i$, and the van der Waals radii of the atoms) mapped on the Hirshfeld surface of the compound NiMeOsalen, represented with one surrounding moiety to visualize the intermolecular interaction. Figure S2. Hirshfeld surface with d$_{norm}$ mapped and fingerprint plots for compound NiMesalen, with C···H interaction (first row) and H···H, O···H interactions (row 2–3). The color ranges from dark blue to red with increasing frequency (relative area of the surface), corresponding to each kind of interaction. Figure S3. Normalized contact distance (d$_{norm}$, defined in terms of d$_e$, d$_i$, and the van der Waals radii of the atoms) mapped on the Hirshfeld surface of the compound NiMeOsalen, represented with one surrounding moiety to visualize the intermolecular interaction. Figure S4. Hirshfeld surface with d$_{norm}$ mapped and fingerprint plots for NiMeOsalen, with C···H interaction (first row) and H···H, O···H interactions (row 2–3). The color ranges from dark blue to red with increasing frequency (relative area of the surface) corresponding to each kind of interaction. Figure S5. Normalized contact distance (d$_{norm}$, defined in terms of d$_e$, d$_i$, and the van der Waals radii of the atoms) mapped on the Hirshfeld surface of the compound NiMeOsalphen, represented together with one surrounding moiety to visualize the intermolecular interaction. Figure S6. Hirshfeld surface with d$_{norm}$ mapped and fingerprint plots for compound NiMeOsalphen, with C···H interaction (first row) and H···H, O···H interactions (row 2–3). The color ranges from dark blue to red with increasing frequency (relative area of the surface) corresponding to each kind of interaction. Figure S7. Normalized contact distance (d$_{norm}$, defined in terms of d$_e$, d$_i$, and the van der Waals radii of the atoms) mapped on the Hirshfeld surface of the compound Nisalphen, represented together with one surrounding moiety to visualize the intermolecular interaction. Figure S8. Hirshfeld surface with d$_{norm}$ mapped and fingerprint plots of the two molecules name A and B in compound Nisalphen for C···H interaction (first row) and H···H, Cl···H interactions (row 2–3). The color ranges from dark blue to red with increasing frequency (relative area of the surface) corresponding to each kind of interaction. Figure S9. UV-vis of NiMeOsalen in DMSO solution. Figure S10. UV-vis NiMesalen in DMSO solution. Figure S11. UV-vis NiOHsalen in DMSO solution. Figure S12. UV-vis of Nisalen in DMSO solution. Figure S13. UV-vis of NiClsalen in DMSO solution. Figure S14. UV-vis of NiBrsalen in DMSO solution. Figure S16. UV-vis NiMeOsalphen in DMSO solution. Figure S17. UV-vis NiOHsalphen in DMSO solution. Figure S18. UV-vis Nisalphen in DMSO solution. Figure S19. UV-vis NiClsalphen in DMSO solution. Figure S20. UV-vis NiBrsalphen in DMSO solution. Figure S21. ^{1}H-NMR of NiMeOsalen in

DMSO-d6. Figure S22. ^{13}C-NMR of NiMeO*salen* in DMSO-d6. Figure S23. ^{1}H-NMR of NiMe*salen* in chloroform. Figure S24. ^{13}C-NMR of NiMe*salen* in chloroform. Figure S25. ^{1}H-NMR of NiOH*salen* in DMSO-d6. Figure S26. ^{13}C-NMR of NiOH*salen* in DMSO-d6. Figure S27. ^{1}H-NMR of Ni*salen* in chloroform. Figure S28. ^{13}C-NMR of Ni*salen* in chloroform. Figure S29. COSY spectrum of Ni*salen* in chloroform. Figure S30. HSQC spectrum of Ni*salen* in choloform. Figure S31. ^{1}H-NMR of NaCl*salen* in chloroform. Figure S32. ^{1}H-NMR of NiBr*salen* in chloroform. Figure S33. ^{1}H-NMR of NiNO$_2$*salen* in DMSO-d6. Figure S34. ^{1}H-NMR of NiOH*salphen* in DMF-d7. Figure S35. ^{13}C-NMR of NiOH*salphen* in DMF-d7. Figure S36. COSY spectrum of NiOH*salphen* in DMF-d7. Figure S37. HSQC spectrum of NiOH*salphen* in DMF-d7. Figure S38. ^{1}H-NMR of NiMeO*salphen* in CDCl$_3$. Figure S39. COSY spectrum of NiMeO*salphen* in CDCl$_3$. Figure S40. Voltammogram of NiMeO*salen* 1 mM in DMSO. Figure S41. Voltammogram of NiMe*salen* 1 mM in DMSO. Figure S42. Voltammogram of NiOH*salen* 1 mM in DMSO. Figure S43. Voltammogram of Ni*salen* 1 mM in DMSO. Figure S44. Voltammogram of NiCl*salen* 1 mM in DMSO. Figure S45. Voltammogram of NiBr*salen* 1 mM in DMSO. Figure S46. Voltammogram of NiNO$_2$*salen* 1 mM in DMSO.

Author Contributions: Conceptualization, C.S.N.-R. and A.S.-B.; methodology, C.S.N.-R., A.S.-B., and F.L.O.-V.; software, M.F.-A.; formal analysis, C.S.N.-R., A.S.-B., J.C.G.-R., and M.F.-A.; investigation, L.R.-A.; direction and resources, L.R.-A.; data curation, M.F.-A.; writing—original draft preparation, C.S.N.-R., A.S.-B., and M.F.-A.; writing—review and editing, L.R.-A. and J.C.G.-R.; supervision,. L.R.-A.; project administration, L.R.-A.; funding acquisition, L.R.-A. All authors have read and agreed to the published version of the manuscript.

Funding: This research received no external funding.

Acknowledgments: We thank USAII for elemental analysis, mass spectrometry, nuclear magnetic resonance tests and chromatography coupled to mass spectrometry tests; PAPIIT IN230020 for financial support and PAPIIT IN217613 for scholarships (CSNR, and ASB).

Conflicts of Interest: The authors declare no conflict of interest.

References

1. Ito, Y.N.; Katsuki, T. Asymmetric Catalysis of New Generation Chiral Metallo*salen* Complexes. *Bull. Chem. Soc. Jpn.* **1999**, *72*, 603–619. [CrossRef]
2. Naeimi, H.; Karshenas, A. Highly regioselective conversion of epoxides to β-hydroxy nitriles using metal(II) Schiff base complexes as new catalysts under mild conditions. *Polyhedron* **2013**, *49*, 234–238. [CrossRef]
3. Dalton, C.T.; Ryan, K.M.; Langan, I.J.; Coyne, É.J.; Gilheany, D.G. Asymmetric alkene epoxidation with chromium oxo *salen* complexes: Effect of π-rich and other types of additives. *J. Mol. Catal. A Chem.* **2002**, *187*, 179–187. [CrossRef]
4. Yang, H.; Zhang, L.; Su, W.; Yang, Q.; Li, C. Asymmetric ring-opening of epoxides on chiral Co(*Salen*) catalyst synthesized in SBA-16 through the "ship in a bottle" strategy. *J. Catal.* **2007**, *248*, 204–212. [CrossRef]
5. Bordoloi, A.; Amrute, A.P.; Halligudi, S.B. [Ru(*salen*)(NO)] complex encapsulated in mesoporous SBA-16 as catalyst for hydrogenation of ketones. *Catal. Commun.* **2008**, *10*, 45–48. [CrossRef]
6. Chatterjee, D.; Bajaj, H.C.; Das, A.; Bhatt, K. First report on highly efficient alkene hydrogenation catalysed by Ni(*salen*) complex encapsulated in zeolite. *J. Mol. Cat.* **1994**, *92*, L235–L238. [CrossRef]
7. Hamdan, H.; Navijanti, V.; Nur, H.; Muhid, M.N.M. Fe(III)-*salen* encapsulated Al-MCM-41 as a catalyst in the polymerisation of bisphenol-A. *J. Solid State Sci.* **2005**, *7*, 239–244. [CrossRef]
8. Ding, L.; Liang, S.; Zhang, J.; Ding, C.; Chen, Y.; Lü, X. Cu^{2+}-templated self-assembly of an asymmetric *Salen*-Cu(II) complex and its application in catalytic polymerization of methyl methacrylate (MMA). *Inorg. Chem. Commun.* **2014**, *44*, 173–176. [CrossRef]
9. Butsch, K.; Günther, T.; Klein, A.; Stirnat, K.; Berkessel, A.; Neudörfl, J. Redox chemistry of copper complexes with various *salen* type ligands. *Inorganica Chimi. Acta* **2013**, *394*, 237–246. [CrossRef]
10. Pratt, R.C.; Stack, T.D.P. Mechanistic insights from reactions between copper(II)-phenoxyl complexes and substrates with activated C-H bonds. *Inorg. Chem.* **2005**, *44*, 2367–2375. [CrossRef]
11. Thomas, F. Ten years of a biomimetic approach to the copper(II) radical site of galactose oxidase. *Eur. J. Inorg. Chem.* **2007**, 2379–2404. [CrossRef]
12. Mahapatra, P.; Drew, M.G.B.; Ghosh, A. Ni(II) Complex of N$_2$O$_3$ Donor Unsymmetrical Ligand and Its Use for the Synthesis of NiII-MnII Complexes of Diverse Nuclearity: Structures, Magnetic Properties, and Catalytic Oxidase Activities. *Inorg. Chem.* **2018**, *57*, 8338–8353. [CrossRef] [PubMed]
13. Kianfar, A.H.; Paliz, M.; Roushani, M.; Shamsipur, M. Synthesis, spectroscopy, electrochemistry and thermal study of vanadyl tridentate Schiff base complexes. *Spectrochim Acta A* **2011**, *82*, 44–48. [CrossRef] [PubMed]

14. Santos, I.C.; Vilas-boas, M.; Piedade, M.F.M.; Freire, C.; Duarte, M.T.; Castro, B. De Electrochemical and X-ray studies of nickel (II) Schiff base complexes derived from salicylaldehyde. Structural effects of bridge substituents on the stabilisation of the +3 oxidation state. *Polyhedron* **2000**, *19*, 655–664. [CrossRef]
15. Shehata, E.E.; Masoud, M.S.; Khalil, E.A.; Abdel-Gaber, A.M. Synthesis, spectral and electrochemical studies of some Schiff base N_2O_2 complexes. *J. Mol. Liq.* **2014**, *194*, 149–158. [CrossRef]
16. Carradori, S.; De Monte, C.; D'Ascenzio, M.; Secci, D.; Celik, G.; Ceruso, M.; Vullo, D.; Scozzafava, A.; Supuran, C.T. *Salen* and tetrahydro*salen* derivatives act as effective inhibitors of the tumor-associated carbonic anhydrase XII—A new scaffold for designing isoform-selective inhibitors. *Bioorg. Med. Chem. Lett.* **2013**, *23*, 6759–6763. [CrossRef]
17. Hang, Z.X.; Dong, B.; Wang, X.W. Synthesis, crystal structures, and antibacterial activity of zinc(II) complexes with bis-schiff bases. *Synth. React. Inorg. Met. Org. Chem.* **2012**, *42*, 1345–1350. [CrossRef]
18. Fasina, T.M.; Ogundele, O.; Ejiah, F.N.; Dueke-Eze, C.U. Biological Activity of Copper (II), Cobalt (II) and Nickel (II) Complexes of Schiff Base Derived from O-phenylenediamine and 5-bromosalicylaldehyde. *Int. J. Biol. Chem.* **2012**, *6*, 24–30. [CrossRef]
19. Woldemariam, G.A.; Mandal, S.S. Iron(III)-*salen* damages DNA and induces apoptosis in human cell via mitochondrial pathway. *J. Inorg. Biochem.* **2008**, *102*, 740–747. [CrossRef]
20. Ansari, K.I.; Grant, J.D.; Kasiri, S.; Woldemariam, G.; Shrestha, B.; Mandal, S.S. Manganese(III)-*salens* induce tumor selective apoptosis in human cells. *J. Inorg. Biochem.* **2009**, *103*, 818–826. [CrossRef]
21. Meshkini, A.; Yazdanparast, R. Chemosensitization of human leukemia K562 cells to taxol by a Vanadium-*salen* complex. *Exp. Mol. Pathol.* **2010**, *89*, 334–342. [CrossRef]
22. Immel, T.A.; Grützke, M.; Batroff, E.; Groth, U.; Huhn, T. Cytotoxic dinuclear titanium-salan complexes: Structural and biological characterization. *J. Inorg. Biochem.* **2012**, *106*, 68–75. [CrossRef] [PubMed]
23. Gravert, D.J.; Griffin, J.H. Specific DNA cleavage mediated by manganese complex [Salen Mn(III)]+. *J. Org. Chem.* **1993**, *58*, 820–822. [CrossRef]
24. Peng, B.; Zhou, W.H.; Yan, L.; Liu, H.W.; Zhu, L. DNA-binding and cleavage studies of chiral Mn(III) *salen* complexes. *Transit. Metal. Chem.* **2009**, *34*, 231–237. [CrossRef]
25. Bhattacharya, S.; Mandal, S.S. Ambient oxygen activating water soluble Cobalt-*Salen* complex for DNA cleavage. *J. Chem. Soc. Chem. Commun.* **1995**, 2489–2490. [CrossRef]
26. Ali, A.; Kamra, M.; Bhan, A.; Mandal, S.S.; Bhattacharya, S. New Fe(III) and Co(II) *salen* complexes with pendant distamycins: Selective targeting of cancer cells by DNA damage and mitochondrial pathways. *Dalton Trans.* **2016**, *45*, 9345–9353. [CrossRef]
27. Muller, J.G.; Paikoff, S.J.; Rokita, S.E.; Burrows, C.J. DNA modification promoted by water-soluble nickel(II) *salen* complexes: A switch to DNA alkylation. *J. Inorg. Biochem.* **1994**, *54*, 199–206. [CrossRef]
28. Mariappan, M.; Suenaga, M.; Mukhopadhyay, A.; Maiya, B.G. Synthesis, structure, DNA binding and photonuclease activity of a nickel(II) complex with a N,N'-Bis(salicylidene)-9-(3,4-diaminophenyl) acridine ligand. *Inorg. Chim. Acta* **2012**, *390*, 95–104. [CrossRef]
29. Zhou, C.Q.; Liao, T.C.; Li, Z.Q.; Gonzalez-Garcia, J.; Reynolds, M.; Zou, M.; Vilar, R. Dinickel–*salphen* complexes as binders of human telomeric dimeric G-quadruplexes. *Chem. Eur. J.* **2017**, *23*, 4713–4722. [CrossRef]
30. Reed, J.E.; Arnal, A.A.; Neidle, S.; Vilar, R. Stabilization of G-quadruplex DNA and inhibition of telomerase activity by square-planar nickel(II) complexes. *J. Am. Chem. Soc.* **2006**, *128*, 5992–5993. [CrossRef]
31. Mukherjee, P.; Biswas, C.; Drew, M.G.B.; Ghosh, A. Structural variations in Ni(II) complexes of *salen* type di-Schiff base ligands. *Polyhedron* **2007**, *26*, 3121–3128. [CrossRef]
32. Ghaemi, A.; Fayyazi, K.; Keyvani, B.; Ng, S.W.; Tiekink, E.R.T. {2,2'-[o-Phenylenebis(nitrilomethanylylidene)] diphenolato- κ4 O,N,N',O'}nickel(II) monohydrate. *Acta Crystallogr. Sect. E Struct. Rep. Online* **2011**, *E67*, m1481–m1482. [CrossRef]
33. Siegler, M.A.; Lutz, M. Ni(*salen*): A system that forms many solvates with interacting Ni atoms. *Cryst. Growth Des.* **2009**, *9*, 1194–1200. [CrossRef]
34. Choudhary, A.; Das, B.; Ray, S. Enhanced catalytic activity and magnetization of encapsulated nickel Schiff-base complexes in zeolite-Y: A correlation with the adopted non-planar geometry. *Dalton Trans.* **2016**, *45*, 18967–18976. [CrossRef]
35. Xu, Y.; Xue, L.; Wang, Z.G. Synthesis, X-ray crystal structures, and antibacterial activities of Schiff base nickel(II) complexes with similar tetradentate Schiff bases. *Russ. J. Coord. Chem.* **2017**, *43*, 314–319. [CrossRef]

36. Lamour, E.; Routier, S.; Bernier, J.L.; Catteau, J.P.; Bailly, C.; Vezin, H. Oxidation of Cu(II) to Cu(III), free radical production, and DNA cleavage by hydroxy-*salen*-copper complexes. Isomeric effects studied by ESR and electrochemistry. *J. Am. Chem. Soc.* **1999**, *121*, 1862–1869. [CrossRef]
37. Aein Jamshid, K.; Asadi, M.; Hossein Kianfar, A. Synthesis, characterization and thermal studies of dinuclear adducts of diorganotin(IV) dichlorides with nickel(II) Schiff-base complexes in chloroform. *J. Coord. Chem.* **2009**, *62*, 1187–1198. [CrossRef]
38. Biradar, N.S.; Karajagi, G.V.; Roddabasanagoudar, V.L.; Aminabhavi, T.M. Geometrical transformations around nickel(ii) with silicon(iv) tetrachloride. *Synth. React. Inorg. Met.-Org. Chem.* **1984**, *14*, 773–783. [CrossRef]
39. Naeimi, H.; Moradian, M. Efficient synthesis and characterization of some novel nitro-schiff bases and their complexes of nickel(II) and copper(II). *J. Chem.* **2013**, *2013*. [CrossRef]
40. Eltayeb, N.E.; Teoh, S.G.; Chantrapromma, S.; Fun, H.K.; Ibrahim, K. 4,4′-Dimethoxy-2,2′-[1,2-phenylene-bis(nitrilomethylidyne)] diphenol. *Acta Crystallogr. Sect. E Struct. Rep. Online* **2007**, *63*. [CrossRef]
41. Wang, J.; Bei, F.; Xu, X.; Yang, X.; Wang, X. Crystal structure and characterization of 1, 2-N, N-disallicydene-phenylamineato nickel (II) complex. *J. Chem. Crystallogr.* **2003**, *33*, 845–849. [CrossRef]
42. Elerman, Y.; Elmali, A.; Kabak, M.; Aydin, M.; Peder, M. Crystal structure of bis-N,N′-p-chloro-salicylideneamine-1,2.diaaminobenzene. *J. Chem. Cryst.* **1994**, *24*, 603–606. [CrossRef]
43. Batley, G.; Graddon, D. Nickel(II) hydrosalicylamide complexes. *Aust. J. Chem.* **1967**, *20*, 1749. [CrossRef]
44. Kondo, M.; Nabari, K.; Horiba, T.; Irie, Y.; Shimizu, Y.; Fuwa, Y. Synthesis and crystal structure of [Ni{bis(2,5-dihydroxysalicylidene)ethylenediaminato}]: A hydrogen bonded assembly of Ni (II)–*salen* complex. *Inorga Chem Commun* **2003**, *6*, 154–156. [CrossRef]
45. Asadi, M.; Jamshid, K.A.; Kyanfar, A.H. Synthesis, characterization and equilibrium study of the dinuclear adducts formation between nickel(II) *Salen*-type complexes with diorganotin(IV) dichlorides in chloroform. *Inorganica Chimica Acta* **2007**, *360*, 1725–1730. [CrossRef]
46. Oxford Diffraction Ltd. Available online: https://www.rigaku.com/products/smc/crysalis (accessed on 17 July 2019).
47. Clark, R.C.; Reid, J.S. The analytical calculation of absorption in multifaceted crystals. *Acta Crystallogr. A* **1995**, *51*, 887–897. [CrossRef]
48. Sheldrick, G.M. SHELXT—Integrated space-group and crystal-structure determination. *Acta Crystallogr. A Found. Crystallogr.* **2015**, *71*, 3–8. [CrossRef]
49. Sheldrick, G.M. Crystal structure refinement with SHELXL. *Acta Crystallogr. C* **2015**, *C71*, 3–8. [CrossRef]
50. Farrugia, L.J. WinGX and ORTEP for Windows: An update. *J. Appl. Crystallogr.* **2012**, *45*, 849–854. [CrossRef]
51. Spek, A.L. PLATON SQUEEZE: A tool for the calculation of the disordered solvent contribution to the calculated structure factors. *Acta Crystallogr. Sect. C Struct. Chem.* **2015**, *C71*, 9–18. [CrossRef]
52. Garg, B.S.; Nandan Kumar, D. Spectral studies of complexes of nickel(II) with tetradentate schiff bases having N2O2 donor groups. *Spectrochim Acta A Mol. Biomol. Spectrosc.* **2003**, *59*, 229–234. [CrossRef]
53. Lever, A.B.P. *Inorganic Electronic Spectroscopy*, 2nd ed.; Elsevier: New York, NY, USA, 1984.
54. Maki, G. Ligand field theory of Ni (II) complexes. I. Electronic energies and singlet ground-state conditions of Ni (II) complexes of different symmetries. *J. Chem. Phys.* **1958**, *28*, 651–662. [CrossRef]
55. Kianfar, A.H.; Dostani, M.; Mahmood, W.A.K. An unprecedented DDQ-nickel(II)*Salen* complex interaction and X-ray crystal structure of nickel(II)*Salen*.DDH co-crystal. *Polyhedron* **2015**, *85*, 488–492. [CrossRef]
56. Spackman, M.A.; Jayatilaka, D. Hirshfeld surface analysis. *Cryst. Eng. Comm.* **2009**, *11*, 19–32. [CrossRef]
57. Wolff, S.K.; Grimwood, D.J.; McKinnon, J.J.; Jayatilaka, D.; Spackman, M.A. Available online: http://hirshfeldsurface.net/CrystalExplorer (accessed on 17 July 2019).
58. Isse, A.A.; Gennaro, A.; Vianello, E. Electrochemical reduction of Schiff base ligands H2*salen* and H2salophen. *Electrochim. Acta* **1997**, *42*, 2065–2071. [CrossRef]
59. Pooyan, M.; Ghaffari, A.; Behzad, M.; Amiri Rudbari, H.; Bruno, G. Tetradentate N 2O 2 type Nickel(II) Schiff base complexes derived from meso -1,2-diphenyle-1,2-ethylenediamine: Synthesis, characterization, crystal structures, electrochemistry, and catalytic studies. *J. Coord. Chem.* **2013**, *66*, 4255–4267. [CrossRef]
60. Freire, C. Spectroscopic characterisation of electrogenerated nickel(III) species. Complexes with N_2O_2 Schiff-base ligands derived from salicylaldehyde. *J. Chem. Soc. Dalton Trans.* **1998**, 1491–1498. [CrossRef]

61. Jäger, E.G.; Schuhmann, K.; Görls, H. Syntheses, characterization, redox behavior and Lewis acidity of chiral nickel(II) and copper(II) Schiff base complexes. *Inorg. Chim. Acta* **1997**, *255*, 295–305. [CrossRef]
62. Zolezzi, S.; Spodine, E.; Decinti, A. Electrochemical studies of copper(II) complexes with Schiff-base ligands. *Polyhedron* **2002**, *21*, 55–59. [CrossRef]
63. Kianfar, A.H.; Sobhani, V.; Dostani, M.; Shamsipur, M.; Roushani, M. Synthesis, spectroscopy, electrochemistry and thermal study of vanadyl unsymmetrical Schiff base complexes. *Inorg. Chim. Acta* **2011**, *365*, 108–112. [CrossRef]

© 2020 by the authors. Licensee MDPI, Basel, Switzerland. This article is an open access article distributed under the terms and conditions of the Creative Commons Attribution (CC BY) license (http://creativecommons.org/licenses/by/4.0/).

Article

Anion–Cation Recognition Pattern, Thermal Stability and DFT-Calculations in the Crystal Structure of H_2dap[Cd(HEDTA)(H_2O)] Salt (H_2dap = H_2(N3,N7)-2,6-Diaminopurinium Cation)

Jeannette Carolina Belmont-Sánchez [1], Noelia Ruiz-González [1], Antonio Frontera [2], Antonio Matilla-Hernández [1], Alfonso Castiñeiras [3] and Juan Niclós-Gutiérrez [1,*]

1. Department of Inorganic Chemistry, Faculty of Pharmacy, University of Granada, 18071 Granada, Spain; carol.bs.quimic@hotmail.com (J.C.B.-S.); noeliarg13@correo.ugr.es (N.R.-G.); amatilla@ugr.es (A.M.-H.)
2. Departament de Química, Universitat de les Illes Balears, Crta. de Valldemossa km 7.5, 07122 Palma de Mallorca (Baleares), Spain; toni.frontera@uib.es
3. Department of Inorganic Chemistry, Faculty of Pharmacy, University of Santiago de Compostela, 15782 Santiago de Compostela, Spain; alfonso.castineiras@usc.es
* Correspondence: jniclos@ugr.es

Received: 24 March 2020; Accepted: 14 April 2020; Published: 15 April 2020

Abstract: The proton transfer between equimolar amounts of [Cd(H_2EDTA)(H_2O)] and 2,6-diaminopurine (Hdap) yielded crystals of the out-of-sphere metal complex H_2(N3,N7)dap [Cd(HEDTA)(H_2O)]·H_2O (**1**) that was studied by single-crystal X-ray diffraction, thermogravimetry, FT-IR spectroscopy, density functional theory (DFT) and quantum theory of "atoms-in-molecules" (QTAIM) methods. The crystal was mainly dominated by H-bonds, favored by the observed tautomer of the 2,6-diaminopurinium(1+) cation. Each chelate anion was H-bonded to three neighboring cations; two of them were also connected by a symmetry-related anti-parallel π,π-staking interaction. Our results are in clear contrast with that previously reported for H_2(N1,N9)ade [Cu(HEDTA)(H_2O)]·2H_2O (EGOWIG in Cambridge Structural Database (CSD), Hade = adenine), in which H-bonds and π,π-stacking played relevant roles in the anion–cation interaction and the recognition between two pairs of ions, respectively. Factors contributing in such remarkable differences are discussed on the basis of the additional presence of the exocyclic 2-amino group in 2,6-diaminopurinium(1+) ion.

Keywords: EDTA; 2,6-diaminopurine; cadmium; co-crystal; H-bonding; π–π stacking

1. Introduction

Nucleobase complexes with transition metals are continuously under investigation due to their applications as advanced functional materials, their biologic importance, structural diversity and use as molecular recognition models for nucleic acids [1–6]. The majority of structural information available in these systems is mainly dedicated to the adenine nucleobase [7–16] and a variety of N-alkylated derivatives as ligands [17–31]. In contrast, available structural information in the Cambridge Structural Database (CSD) on metal complexes, co-crystals and salts with 2,6-diaminopurine (Hdap) nucleobase is much more limited, despite the fact that Hdap is an analog of adenine. Interestingly, the Hdap nucleobase is able to form the same coordination bonds than adenine and, additionally, the extra exocyclic amino group of Hdap can further function as H-bond donor. Therefore, Hdap can generate novel metal complexes, coordination polymers and supramolecular assemblies.

This study reports the synthesis, X-ray structure and density functional theory study of a new metal complex of formula H_2(N3,N7)dap[Cd(HEDTA)(H_2O)]·H_2O (**1**). A comparison with the previously reported analog of adenine, [Cu(HEDTA)(H_2O)]·2H_2O [5,32], was also performed. The H-bonding networks that are established at both faces of H_2dap were also studied using DFT calculations and the relative strength of each H-bond was estimated using the QTAIM theory. The antiparallel π,π-stacking interactions that were formed between the cations were also studied, focusing on the effect of the counter-ions.

2. Materials and Methods

2.1. Reagents

H_4EDTA acid (TCI), Hdap (Alfa Aesar) and $CdCO_3$ (Alfa Aesar) were used as received.

2.2. Crystallography

A colorless needle crystal of H_2dap[Cd(HEDTA)(H_2O)]·H_2O (1) was mounted on a glass fiber and used for data collection. Crystal data were collected at 100(2) K, using a Bruker D8 VENTURE PHOTON III-14 diffractometer. Graphite-monochromated MoK(α) radiation (λ = 0.71073 Å) was used throughout. The data were processed with APEX2 [33] and corrected for absorption using SADABS (transmissions factors: 1.000–0.962) [34]. The structure was solved by direct methods using the program SHELXS-2013 [35] and refined by full-matrix least-squares techniques against F^2 using SHELXL-2013 [35]. Positional and anisotropic atomic displacement parameters were refined for all non-hydrogen atoms. Hydrogen atoms were located in difference maps and included as fixed contributions riding on attached atoms with isotropic thermal parameters 1.2/1.5 times those of their carrier atoms. Criteria of a satisfactory complete analysis were the ratios of 'rms' shift to standard deviation less than 0.001 and no significant features in final difference maps. Atomic scattering factors were taken from the International Tables for Crystallography [36]. Molecular graphics were plotted with PLATON [37]. A summary of the crystal data, experimental details and refinement results are listed in Table 1. Crystallographic data for 1 has been deposited in the Cambridge Crystallographic Data Center with the CCDC number 1992206.

2.3. Other Physical Measurements

Analytical data (CHN) were obtained in a Fisons–Carlo Erba EA 1108 elemental micro-analyzer. The cadmium content was cheeked as CdO by the weight of final residue in the thermogravimetric analysis (TGA) within 1% of assumed experimental error. FT-IR spectrum was recorded (KBr pellet) on a Jasco FT-IR 6300 spectrometer. TGA was carried out (r.t. to 950 °C) in air flow (100 mL/min) by a Shimadzu Thermobalance TGA–DTG–50H instrument and a series of 35 time-spaced FT-IR spectra of evolved gasses were recorded with a coupled FT-IR Nicolet Magna 550 spectrometer.

2.4. Synthesis and Relevant IR Spectrum Data

Compound 1 was obtained in a two-step process. First, $CdCO_3$ (1 mmol, 0.17 g) and H_4EDTA (1 mmol, 0.29 g) were reacted in water (100 mL) inside an open Kitasato flask at 50–70 °C, with permanent stirring until a clear solution was observed. The heat was ceased and then small portions of Hdap (1 mmol, 0.15 g) were added to the Cd-H_2EDTA chelate. The reaction mixture was filtered without vacuum (to remove any insoluble material) on a crystallization flask. The slow evaporation of the solution (two-three weeks at r.t.) produces needle crystals of 1. Yield: ~70%. Elemental analysis (%): Calc. for $C_{15}H_{24}CdN_8O_{10}$: C 30.60, H 4.11, N 19.03, Cd (as CdO) 21.81; Found: C 30.57, H 4.08, N 18.87, Cd (as CdO, final residue at 675 °C, in the TGA curve) 22.46. FT–IR data [cm^{-1}]: 3500–3100 vbr $ν_{as}/ν_s(H_2O) + ν_{as}/ν_s(NH_2) + ν_{as}(NH)$, 3411s, br, $ν(OH)$, 2931w $ν_{as}(CH_2)$, 1674s, $ν(C=O)$, 1596 vs $δ(NH_2) + δ(H_2O) + ν_{as}(COO)$, 1400 m $ν_s(COO)$, 923 w, 849 w $π(C–H)$.

Table 1. Crystal data and structure refinement for $H_2(N3,N7)dap[Cd(HEDTA)(H_2O)]\cdot H_2O$.

Empirical Formula	$C_{15}H_{24}CdN_8O_{10}$
Formula weight	588.82
Temperature	100(2) K
Wavelength	0.71073 Å
Crystal system, space group	Triclinic, P-1
Unit cell dimensions	a = 7.4924(3) Å, α = 81.9310(10)°
	b = 9.0078(4) Å, β = 78.0170(10)°
	c = 17.2884(6) Å, γ = 70.545(2)°
Volume	1072.99(8) Å^3
Z, Calculated density	2, 1.822 Mg/m^3
Absorption coefficient	1.090 mm^{-1}
F(000)	596
Crystal size	0.160 × 0.030 × 0.020 mm
Theta range for data collection	2.405 to 30.507°
Limiting indices	−10 ≤ h ≤ 10, −12 ≤ k ≤ 12, −24 ≤ l ≤ 24
Reflections collected / unique	88812 / 6551 [R(int) = 0.0556]
Completeness to θ = 25.242	99.9%
Absorption correction	Semi-empirical from equivalents
Max. and min. transmission	1.000 and 0.962
Refinement method	Full-matrix least-squares on F^2
Data / restraints / parameters	6551 / 0 / 307
Goodness-of-fit on F^2	1.073
Final R indices [I > 2σ(I)]	R_1 = 0.0222, wR_2 = 0.0454
R indices (all data)	R_1 = 0.0273, wR_2 = 0.0477
Largest diff. peak and hole	0.588 and −0.469 e.Å^{-3}
CCSD refcode	1992206

2.5. Theoretical Methods

All DFT calculations were carried out using the Gaussian-16 program [38] at the PBE1PBE-D3/def2-TZVP level of theory and using the crystallographic coordinates. The formation energies of the assemblies have been evaluated by calculating the difference between the total energy of the assembly and the sum of the monomers that constitute the assembly, which have been maintained frozen. This methodology has been used by us [39,40] and others [41–45] to analyze supramolecular assemblies in crystal structures. The molecular electrostatic potential was computed at the same level of theory and plotted onto the 0.001 a.u. isosurface. The quantum theory of atoms-in-molecules (QTAIM) [46] analysis was carried out at the same level of theory by means of the AIMAll program [47]. The Cartesian coordinates of the theoretical models are given in the Supplementary Materials.

3. Results and Discussion

3.1. Thermal Stability

Under air-dry flow, the weight loss versus temperature TGA behavior consists of five steps (Figure 1). The experimental results and assignations are summarized in Table 2.

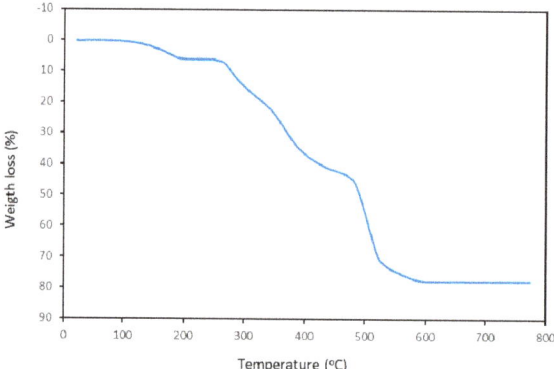

Figure 1. Weight loss versus temperature (in the range r.t. to 775 °C) in the thermogravimetric analysis of compound 1 (sample: 12.29 mg).

Table 2. Summary of the results and assignations in the thermogravimetric analysis of compound 1.

Step or R	Temperature (°C)	Time (min)	Weight (%) Experimental	Weight (%) Calulated	Evolved Gases or Residue (R)
1	55–220	2.5–21	6.056	6.159 *	2 H_2O, CO_2 (t)
2	220–315	21–31	12.071	-	CO_2, H_2O, CO,
3	315–450	31–43	23.569	-	CO_2, H_2O, CO, NH_3, N_2O, NO, NO_2, CH_4
4	450–560	43–53	33.071	-	CO_2, H_2O, CO, NH_3, N_2O, NO, NO_2, CH_4
5	560–600	53–70	2.676	-	CO_2, H_2O, NH_3, N_2O, NO, NO_2
R	600	-	22.557	21.808	CdO
R	675	-	22.462	21.808	CdO

* Calculated only for the loss of 2 H_2O. t = trace amounts.

First of all, compound **1** overlaps the loss of uncoordinated water and aqua ligand content (with small amounts of CO_2) in a consistent wide range of temperature (55–220 °C, experimental lost 6.056%, calculated for 2 H_2O molecules 6.159%). In the second step (200–315 °C, with a weight loss of 12.071%) only CO_2, CO and H_2O were evolved, strongly suggesting that the combustion of organic ligands begins by the HEDTA^{3-} chelator. Third and fourth steps (315–450 and 450–560 °C) produce (in addition to H_2O, CO_2 and CO) NH_3 and N-oxides (N_2O, NO and NO_2) plus amounts of CH_4. In the last fifth step (560–675 °C) the presence of CH_4 and CO were less relevant. The weight loss during the burning steps (under an air flow) of organic material cannot be attributed to specific fragments of HEDTA^{3-} or H_2dap$^+$. In contrast, the estimated residue (22.557% at 600 °C and 22.462% at 675 °C) reasonably agrees to the calculated weight for CdO (21.808%) within a reasonable experimental error (<1%).

3.2. Crystal Structure and Anion–Cation Recognition Pattern

This compound has an equimolar ratio of the tautomer H_2(N3,N7)dap$^+$ cation, the ternary anion [Cd(HEDTA)(H_2O)]$^-$ and an unbounded to the metal aqua molecule (Figure 2). Table 3 shows the coordination bond distances and angles in the novel Cd(II) 'out-sphere' complex. Table 4 reports data concerning H-bonding interactions in its crystal. The first structural insight was that the assumed most basic N9 donor atom of Hdap diamino–purine in such a tautomeric form of the cation was unable to remove the aqua ligand from the seven coordinated Cd(II) chelate anion. The [Kr]4d^{10} electronic configuration and the size of the Cd(II) center enables its rather common hepta-coordination as well as the inequality of its bond distances [2.267(1)–2.459(1) Å]. The Cd(II) coordination polyhedron in the chelate anions is best referred as a distorted mono-caped octahedron. The shortest bond is Cd-O(aqua) whereas the largest ones (<2.40 A) were Cd-N10<Cd-O(carboxyl)<Cd-N20. Interestingly the largest

Cd-N20 bond involves de N20-HEDTA atom supporting the N-(carboxymethyl) arm of the chelating ligand. Table 3 summarizes the H-bonding interactions in compound 1.

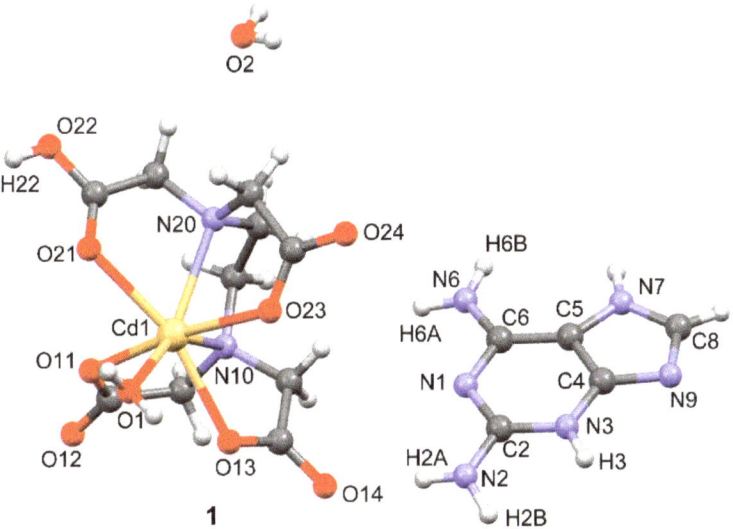

Figure 2. Asymmetric unit in the crystal of compound **1**, with relevant atom numbering scheme.

Table 3. Coordination bond lengths (Å) and angles (°) in the crystal of compound **1**, H$_2$(N3,N7) dap[Cd(HEDTA)(H$_2$O)]·H$_2$O. See Figure 1 for numbering scheme.

Atoms	Distance or Angle	Atoms	Distance or Angle
Cd(1)-O(1)	2.2672(11)	Cd(1)-N(10)	2.4111(13)
Cd(1)-O(11)	2.2984(11)	Cd(1)-O(21)	2.4400(11)
Cd(1)-O(23)	2.3010(11)	Cd(1)-N(20)	2.4585(13)
Cd(1)-O(13)	2.3748(11)	O(1)-Cd(1)-O(11)	94.13(4)
O(1)-Cd(1)-O(23)	91.28(4)	O(11)-Cd(1)-O(21)	81.61(4)
O(11)-Cd(1)-O(23)	168.52(4)	O(23)-Cd(1)-O(21)	109.06(4)
O(1)-Cd(1)-O(13)	79.59(4)	O(13)-Cd(1)-O(21)	161.48(4)
O(11)-Cd(1)-O(13)	91.09(4)	N(10)-Cd(1)-O(21)	123.95(4)
O(23)-Cd(1)-O(13)	79.93(4)	O(1)-Cd(1)-N(20)	138.89(4)
O(1)-Cd(1)-N(10)	145.66(4)	O(11)-Cd(1)-N(20)	111.24(4)
O(11)-Cd(1)-N(10)	73.31(4)	O(23)-Cd(1)-N(20)	70.22(4)
O(23)-Cd(1)-N(10)	96.63(4)	O(13)-Cd(1)-N(20)	129.28(4)
O(13)-Cd(1)-N(10)	69.10(4)	N(10)-Cd(1)-N(20)	74.65(4)
O(1)-Cd(1)-O(21)	83.96(4)	O(21)-Cd(1)-N(20)	69.17(4)

In the crystal, each anion is H-bonded to three independent neighboring cations, revealing that the anion–cation recognition of **1** is mainly featured by this kind of inter-molecular interaction (Figure 3). Deeping in this question, a H$_2$dap$^+$ cation links the complex anions by the H-bonds: N2-H2B···O24#1 (2.987(1) Å, 169.5°), N3-H3···O23#1 (2.712(1) Å, 179,5°) and (aqua)O1-H1WA···N9#1 (2.902(1) Å, 166.6°) with #1 = −x + 1, −y + 1, −z + 1. This recognition pattern involves both O-acceptors of the same HEDTA$^-$ carboxylate group and the most basic N9 atom of the purinium(1+) ion. Another H$_2$dap ion builds two H-bonds with O-carboxylate acceptors of the same HEDTA$^-$ carboxylate group: N6-H6B···O13#5 (2.831(1) Å, 1.67.7°) and N7-H7···O14#5 (2.675(1) Å, 177.2°) with #5 = x, −y, z. Figure 4 shows the way these two purinium(1+) cations were additionally related by a moderate anti-parallel π,π-stacking interaction between their 5- and 6-membered rings (inter-centroid distance d$_{c-c}$ 3.49 Å, interplanar distance d$_{\pi-\pi}$ 3.21 Å, dihedral interplanar angle 0°, slipping angles β = γ = 25.91°, slippage index

1.56). In this interaction the shortest interplanar distance would be related to the remarkable slippage. A third purinium(1+) ion is related with the chelate anion by the H-bond N6-H6A···O24 (2.990(1) Å, 144.6°). Thus, O24 atom acts as twice-acceptor for H-bonding interactions.

Table 4. Geometric features of the hydrogen bonds in the crystal structure of H_2(N3,N7)dap [Cd(HEDTA)(H_2O)]·H_2O (**1**). The distances were measured between the heavy atoms.

H-bond	D···A (Å)	Angle (°)
O(1)-H(1WA)···N(9)#1	2.9017(17)	166.6
O(1)-H(1WB)···O(12)#2	2.7398(16)	169.0
O(22)-H(22)···O(11)#3	2.5552(16)	175.5
N(2)-H(2A)···O(14)	2.8158(18)	165.9
N(2)-H(2B)···O(24)#1	2.9784(18)	169.5
N(3)-H(3)···O(23)#1	2.7123(17)	179.5
N(6)-H(6A)···O(24)#4	2.9898(18)	144.6
N(6)-H(6B)···O(13)#5	2.8307(17)	167.7
N(7)-H(7)···O(14)#5	2.6746(18)	177.2
O(2)-H(2WA)···O(12)#6	2.7517(17)	163.4
O(2)-H(2WB)···O(11)#7	2.9970(18)	131.5

Symmetry transformations to generate equivalent atoms: #1 − x + 1, − y + 1, − z + 1, #2 x + 1, y, z, #3 − x + 1, − y + 2, − z, #4 x − 1, y, z, #5 x, y − 1, z, #6 x + 1, y − 1, z, #7 − x + 1, −y + 1, − z.

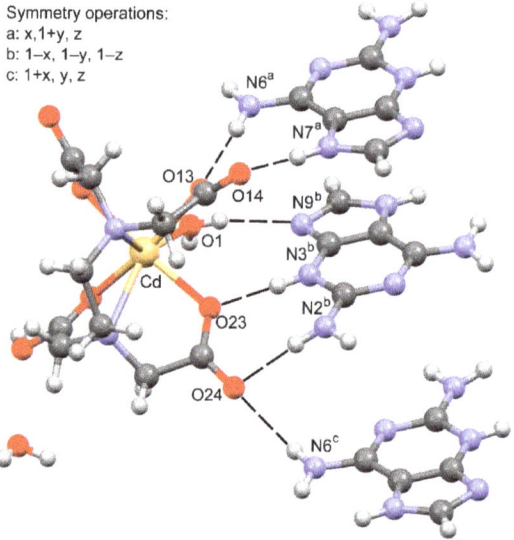

Figure 3. Molecular recognition pattern showing the cooperation of H-bonds between the [Cd(HEDTA)(H_2O)]$^-$ chelate anion and three neighboring H_2(N3,N7)dap$^+$ ions.

In this compound all N–H and O–H bonds were involved in N-H···O or O-H···O interactions excepting for the above mentioned (aqua)O(1)-H(1WA)···N(9)#1 one (Table 3). In this manner the packing was essentially dominated by the H-bonding array that forms bilayers with Cd(II) chelate anions and unboned water molecules whereas H_2(N3,N7)dap$^+$ ions fall oriented towards both external surfaces. These 2D-frameworks lie parallel to the ab crystal plane and were H-bonded pillared along the c axis in the 3D-network (Figure 5).

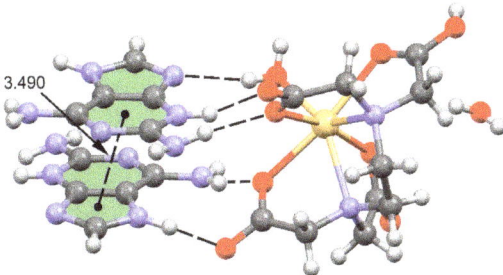

Figure 4. Molecular recognition pattern showing the cooperation of H-bonds and π,π-stacking interactions connecting the [Cd(HEDTA)(H$_2$O)]$^-$ chelate anion with two of the neighboring H$_2$(N3,N7)dap$^+$ ions.

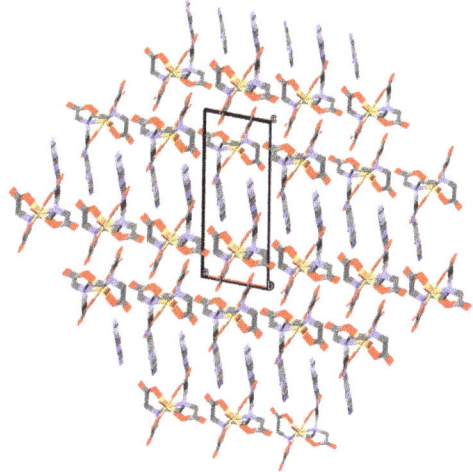

Figure 5. In the 3D H-bonded crystal of compound **1**, 2,6-diaminopurinium(1+) cations oriented towards the external faces of 2D H-bonded frameworks build by complex anions and unbounded to the cadmium(II) water molecules. All H atoms and H-bonding interactions are omitted for clarity.

3.3. DFT Calculations

The DFT study was focused to analyze the interesting supramolecular assemblies and H-bonding networks described above. First of all, the molecular electrostatic potential (MEP) surfaces of the anion and cation have been calculated in order to evaluate the best complementary dimer in terms of the electrostatic attraction between electron rich and electron poor regions of both molecules. The pure Coulombic attraction between the counter-ions is not directional; however weaker interaction like H-bonds or π,π-stacking interactions were able to nicely tune the final geometry of the supramolecular assembly. Evidence for the possibly structure-directing nature of these contacts was supported through an examination of MEP surfaces represented in Figure 6a. These reveal strong electropositive region (blue) at the NH groups of the Hdap$^+$ cation and at the H-atoms of the Cd-coordinated water molecule. Moreover, the surfaces show excess of negative charge (red) at the O-atoms of the Cd-coordinated carboxylate group and at the N-atom of the five-membered ring of H$_2$dap$^+$ thus affording potentially favorable O-H···N and N-H···O interactions between the counter-ions. The MEP surface of Hdap$^+$ also evidences that the N1-atom was less basic than N9, thus it was a worse H-bond acceptor. The MEP surface of the complex represented in Figure 6b shows how the charge density was significantly redistributed upon complexation.

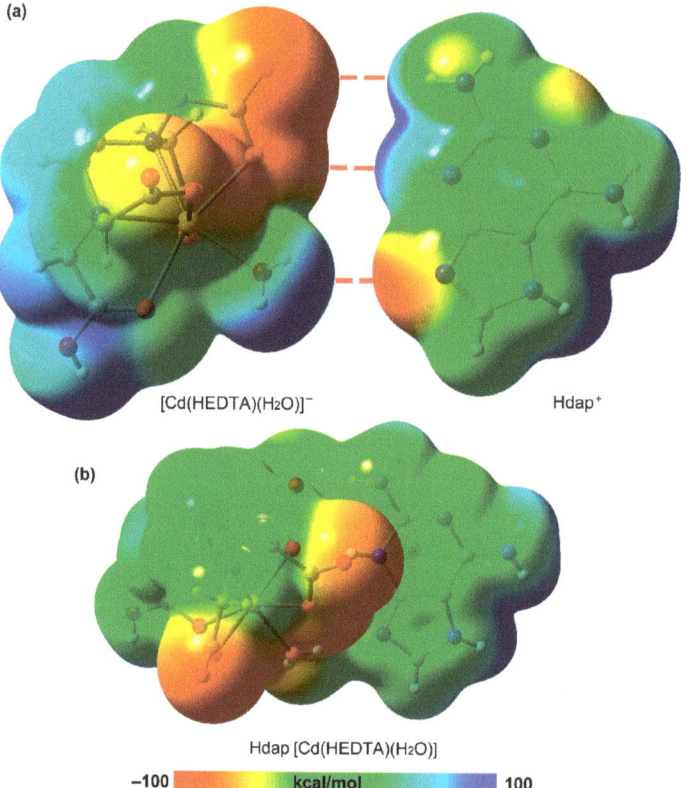

Figure 6. (a) MEP surfaces (0.001 a.u.) for [Cd(HEDTA)(H$_2$O)]$^-$ and H$_2$dap$^+$ highlighting the electropositive (blue) and electronegative (red) regions of each molecule. The dashed lines highlight a favorable electrostatic interaction between these two species. (b) MEP surface (0.001 a.u.) of the assembly at the PBE1PBE/def2-TZVP level of theory.

We have selected the supramolecular assembly commented above in Figure 7a–d to analyze the energetic features of the H-bonds and π,π-stacking interactions in **1**. Figure 7a shows a partial view of the solid state of **1** where these interactions are highlighted. From this quaternary assembly, we have first analyzed two H-bonded dimers (see Figure 7b,c), which present very large dimerization energies due to the strong contribution of the electrostatic attraction between counter-ions. Curiously the dimer with two H-bonds (Figure 7c) was stronger than that with three H-bonds (Figure 7b), likely due to the shorter H-bond distances. This aspect is further analyzed below. Regarding the π,π-stacked dimer, it presents a positive (repulsive) binding energy because it occurs against the Coulombic repulsion between both H$_2$dap$^+$ cations (ΔE_3 = +44.8 kcal/mol). However, if the counter-ions were taken into consideration, the interaction becomes favorable, ΔE_4 = −92.8 kcal/mol.

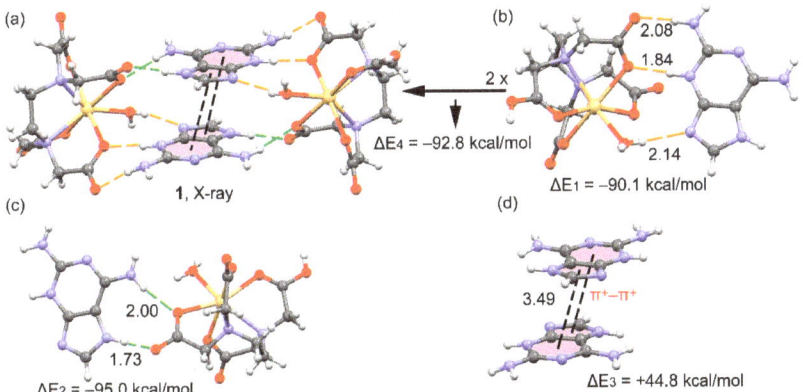

Figure 7. (a) Partial view of the X-ray solid state of **1** showing the self-assembled tetramer. (**b**,**c**) H-bonded dimers extracted from the assembly dimers. (**d**) Isolated π,π-stacked dimer of H2dap$^+$ moieties. The distances are given in Å.

As commented above, the interaction energies were strongly dominated by the Coulombic attraction between the counter-ions and it was difficult to evaluate the real effect of the H-bonding interactions. In order to better analyze the H-bonding network, we have used the QTAIM method to estimate the contribution of each H-bond. The existence of a bond path (lines of maximum density) and bond critical point (CP) connecting two atoms is a universal indication of interaction [48]. The distribution of bond CPs and bond paths in the two H-bonded dimers of compound **1** are given in Figure 8. Each H-bond interaction was characterized by a bond CP (green sphere) and bond path interconnecting the H-atom to the N/O-atoms and confirming the interaction. The energy of each contact has been evaluated according to the approach suggested by Espinosa et al. [49] and Vener et al. [50]. The energy predictors were developed specifically for HBs and were based on the kinetic energy density (V_r) of the Lagrangian energy density (G_r). These values along with the charge density (ρ_r) are summarized in Table 5 for the CPs indicated in Figure 8. It can be observed that both energy predictors show that the N9-H···O H-bond (CP4) was the strongest one, even stronger than N3$^+$-H3···O (the second strongest HB) that bears the positive charge, in line with the shortest distance (1.73 Å) and larger electron density (ρ_r) of CP4, see Table 5. The dissociation energies obtained for the other H-bonds were in the typical range of moderately strong H-bonds. There is an acceptable agreement between both energy predictors thus giving reliability to the study. It is worth mentioning that the sum of the dissociation energies of the two H-bonds of the dimer shown in Figure 8b (18.62 kcal/mol, using the V_r predictor) was larger than the sum of the three H-bonds in the dimer shown in Figure 8a (16.60 kcal/mol), in good agreement with the DFT energies computed for the assemblies shown in Figure 7a,b. This result confirms the fact that the H-bonds were stronger in the dimer where only two H-bonds were formed. Finally, it is interesting to note that the total energy density ($H_r = V_r + G_r$) was negative in CP4 thus evidencing partial covalent character for the N9-H9···O H-bond, in agreement with its large dissociation energy. The rest of CPs exhibit positive Hr values, evidencing their negligible covalent character.

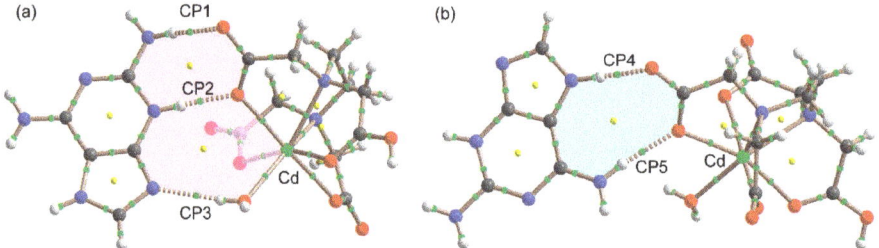

Figure 8. (**a,b**). Distribution of bond and ring critical points (green and yellow spheres, respectively) and bond paths in two dimers of complex 1. The QTAIM data at the bond CPs denoted as CP1 to CP5 are given in Table 3.

Table 5. Values of ρ_r, V_r and G_r (in a.u.) for CP1 to CP5 as indicated in Figure 7. The dissociation energy (E_{dis}) of each H-bond based on V_r and G_r parameters are also indicated in kcal/mol.

CP#	ρ_r	V_r	G_r	E_{dis} ($-0.5 \times V_r$)	E_{dis} ($0.429 \times G_r$)
1	0.0186	−0.0122	0.0155	3.83	4.17
2	0.0328	−0.0297	0.0308	9.32	8.29
3	0.0171	−0.0110	0.0148	3.45	3.98
4	0.0429	−0.0421	0.0380	12.2	10.2
5	0.0224	−0.0172	0.0207	6.49	5.57

3.4. Structural Insides on N(heterocyclic)-Proton Affinities, H-Tautomerism and Metal Binding Patterns from Hdap and Its Cationic Forms in Salts and Their Metal Complexes.

In a rather comprehensive review [51] we have look at the molecular recognition patterns between metal complexes and adenine or a variety of deaza- and aza-adenines (such as Hdap) on the basis of the cooperation between coordination bonds and intra-molecular interligand H-bonding interactions. This review emphasizes the relevance of the N(heterocyclic)-H tautomeric possibilities in neutral and protonated forms of such kinds of natural or synthetic closely related N-heterocyclic ligands. Recent reports from our groups extend these points of view to the guanine-synthetic acyclovir as a ligand [52–54]. Now we have the opportunity to deep into the relevance of these factors on the basis of the available crystallographic results related to cationic forms of Hdap, its salts and inner- or out-sphere metal complexes.

It is generally assumed that the proton affinity of hardly versatile ligand adenine (Hade) follows the order N9>N1>N7>N3>>N6(exocyclic amino) [51]. In a private communication to the CSD basis [55] the structure of the salt H$_3$(N1,N7,N9)dapCl$_2$·H$_2$O (see reference code NULCOO in CSD Database) revealed the lesser proton affinity of the N3 atom of Hdap. That seems also agree with the depleted proton affinity found for the N3-atom of acyclovir, a well-known guanine-synthetic nucleoside [52]. The tautomers H$_2$(N1,N3)dap$^+$, H$_2$(N3,N9)dap$^+$ and H$_2$(N7,N9)dap$^+$ do not have received crystallographic support. The H$_2$(N1,N7)dap$^+$ tautomer acts as N9-donor ligand in two isomorphous compounds having all-trans octahedral complex molecules [M^{II}(H$_2$dap)$_2$(hpt)$_2$ (H$_2$O)$_2$]·4H$_2$O (M = Co or Ni, htp = homophthalato(2-) ligand) [56] (see Scheme 1). The κN9-H$_2$(N1,N7)dap+ coordinating role was consistent not only with the highest proton affinity of the donor atom but also to its less steric hindrance. Interestingly in these complexes there was a cooperation of each M-N9 bond with an intra-molecular (aqua)O-H···N3 interligand interaction. The H$_2$(N1,N9)dap$^+$ tautomer was the counter cation of two rather distinct salts, with a dicarboxylate [57] or a dodecafluoro-*closo*-dodecaborate(2-) anion [58]. This tautomer binds metal ions by its N7-donor, in two Cd(II)-dicarboxylate coordination polymers [59,60] and a mononuclear Co(II) complex [61] displaying the appropriate cooperation between the metal-N7 coordination bond and an N6-H···O interligand interaction. This tautomeric form was in agreement of the N-proton affinity assumed for the free base Hdap (N9>N1>N7>N3) which is also consistent by the crystal structure of H(N9)dap·H_2O [59].

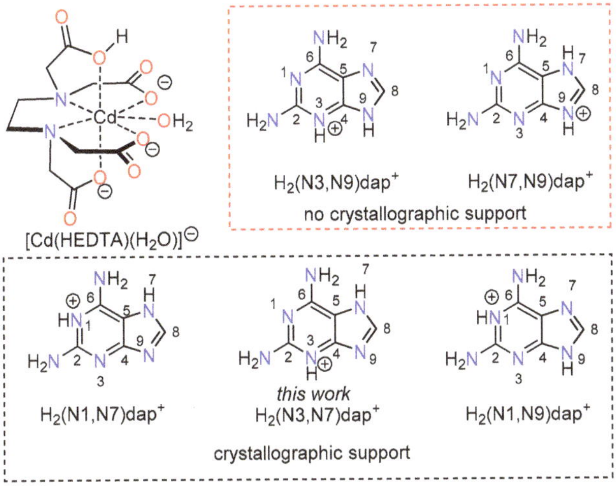

Scheme 1. Structure of [Cd(HEDTA)(H_2O)]$^-$ and different tautomeric forms of Hdap$^+$ with the atom numbering scheme.

The H_2(N3,N7)dap+ tautomer, also here reported, is previously document in three rather distinct compounds. The out-sphere complex (H_2(N3,N7)dap)$_2$[Nd(μ_2-croco) (croco)(H_2O)$_4$]$_2$ (croco = croconate(2-) ion). This compound also builds a sophisticated H-bonded network, carefully describe by R. Baggio et al. [62] where any relevant π,π-stacking interactions appears precluded by coordination of the croco ligands. Why the H_2(N3,N7)dap$^+$ ions does not bind to Nd(III) centers can by explained on the basis of the Pearson's border-line basis of the Hdap and its cation whereas trivalent lanthanide cations were typical hard Pearson's acids. The two other compounds exhibit the κN9-H_2(N3,N7)dap+ ligand mode in presence of benzene-polycarboxylate anions. In the complex cation of trans-[Co^{II}(H_2O)$_4$(H_2(N3,N7)dap)$_2$] (btec)·4H_2O] (btec = bezene-1,2,4,5-tetracarboxylate) [63], aqua ligands cannot acts as H-acceptor for the N3-H bond of the H_2(N3,N7)dap$^+$ ions. Consequently, the Co-N9[H_2(N3,N7)dap] coordination bond does not cooperate with an interligand N3-H···O interaction. In clear contrast the polymeric compound {[Zn(btc)(H_2O) (H_2(N3,N7)dap)]·4H_2O}$_n$ (btc = benzene-1,2,3-tricarboxylate(3-) ion) exhibits the cooperation between the Zn-N9 coordination bond and an interligand (H_2dap) N3-H···O(carboxy, btc) interaction (2.587(4) Å, 157°) [59]. Curiously the O-carboxylate(btc) acceptor involved in such interligand H-bonding interaction implies an un-bonded to the Zn(II) O atom. This is certainly a relevant fact because of the common cooperation of metal-N(purine-like) bonds with (purine-like)N-H···O(carboxylate) intra-molecular interligand interactions was built with a metal-O(coordinated) H-acceptor atom [53].

4. Concluding Remarks

In summary, the proton transfer between 2,6-diaminopurine and [Cd(H_2EDTA)(H_2O)] yields the outer sphere complex reported herein. The geometric features of the nucleobase in the solid state have been discussed in terms of binding pattern, protonation degree and proton tautomer as well as the hydrogen-bonding. Significantly, the solid-state structure was tuned by the synergistic formation of H-bonds and π$^+$–π$^+$ interactions that have been described in detail. Moreover, the interaction energies of several supramolecular assemblies observed in the solid state have been evaluated and discussed by using MEP surfaces and DFT calculations. Finally, the individual H-bonding dissociation energies have been computed using two available energy predictors by means of the QTAIM method.

On the basis of our results and other above referred, it seems clear that the tautomerism plays a relevant role in the crystal having H$_2$dap+ ions. The lack of literature concerning H$_2$(N1,N3)dap$^+$, H$_2$(N3,N9)dap$^+$ and H$_2$(N7,N9)dap$^+$ could be related to one of the following factors: The steric hindrance on N1, the depleted proton affinity of N3 and the suitability of the highest basic of N7 and N9 to metal binding. In the here reported compound, the use of the H$_2$(N3,N7)dap$^+$ favors the extensive H-bonding of its crystal, at the same time that precludes its coordination to the Cd(II) center instead of the aqua ligand.

Supplementary Materials: The following are available online at http://www.mdpi.com/2073-4352/10/4/304/s1, Cartesian coordinates of the theoretical models shown in Figure 7.

Author Contributions: Conceptualization, J.N-G. and A.M.-H.; methodology, all authors; software, A.M.-H., A.F. and A.C.; investigation, J.C.B.-S. and N.R.-G.; writing—original draft preparation, A.M.-H., A.C., A.F. and J.N.-G.; writing—review and editing, all authors, visualization, A.M.-H., A.C., A.F. project administration, J.N.-G.; funding acquisition, A.M.-H, A.F. and J.N-G. All authors have read and agreed to the published version of the manuscript.

Funding: This research was funded by the Excellence Network 'Metal Ions in Biological Systems' MetalBio CTQ2017-90802-REDT, the Research group FQM-283 (Junta de Andalucía) and MICIU/AEI of Spain (project CTQ2017-85821-R FEDER funds).

Acknowledgments: We thank the Centre de Tecnologies de la Informació (CTI), Universitat de les Illes Balears for computational facilities. We also thank all projects for financial support.

Conflicts of Interest: The authors declare no conflict of interest.

References

1. Terrón, A.; Fiol, J.J.; García-Raso, A.; Barceló-Oliver, M.; Moreno, V. Biological recognition patterns implicated by the formation and stability of ternary metal ion complexes of low-molecular-weight formed with amino acid/peptides and nucleobases/nucleosides. *Coord. Chem. Rev.* **2007**, *251*, 1973–1986. [CrossRef]
2. Sivakova, S.; Rowan, S.J. Nucleobases as supramolecular motifs. *Chem. Soc. Rev.* **2005**, *34*, 9–21. [CrossRef]
3. Lippert, B. Multiplicity of metal ion binding patterns to nucleobases. *Coord. Chem. Rev.* **2000**, *200–202*, 487–516. [CrossRef]
4. Navarro, J.A.R.; Lippert, B. Simple 1:1 and 1:2 complexes of metal ions with heterocycles as building blocks for discrete molecular as well as polymeric assemblies. *Coord. Chem. Rev.* **2001**, *222*, 219–250. [CrossRef]
5. Choquesillo-Lazarte, D.; Brandi-Blanco, M.P.; García-Santos, I.; González-Pérez, J.M.; Castiñeiras, A.; Niclós-Gutiérrez, J. Interligand interactions involved in the molecular recognition between copper (II) complexes and adenine or related purines. *Coord. Chem. Rev.* **2008**, *25*, 1241–1256. [CrossRef]
6. Olea, D.; Alexandre, S.S.; Amo-Ochoa, P.; Guijarro, A.; De Jesus, F.; Soler, J.M.; De Pablo, P.J.; Zamora, F.; Gomez-Herrero, J. From Coordination Polymer Macrocrystals to Nanometric Individual Chains. *Adv. Mater.* **2005**, *17*, 1761–1765. [CrossRef]
7. García-Terán, J.P.; Castillo, O.; Luque, A.; Garcíía-Couceiro, U.; Beobide, G.; Román, P. Molecular Recognition of Adeninium Cations on Anionic Metal−Oxalato Frameworks: An Experimental and Theoretical Analysis. *Inorg. Chem.* **2007**, *46*, 3593–3602. [CrossRef]
8. González-Pérez, J.M.; Alarcón-Payer, C.; Castiññiras, A.; Pivetta, T.; Lezama, L.; Choquesillo-Lazarte, D.; Crisponi, G.; Niclós-Gutiérrez, J. A Windmill-Shaped Hexacopper(II) Molecule Built Up by Template Core-Controlled Expansion of Diaquatetrakis(μ$_2$-adeninato-N3,N9)dicopper(II) with Aqua(oxydiacetato)copper(II). *Inorg. Chem.* **2006**, *45*, 877–882. [CrossRef]
9. Amo-Ochoa, P.; Rodríguez-Tapiador, M.I.; Castillo, O.; Olea, D.; Guijarro, A.; Alexandre, S.S.; Gómez-Herrero, J.; Zamora, F. Assembling of Dimeric Entities of Cd(II) with 6-Mercaptopurine to Afford One-Dimensional Coordination Polymers: Synthesis and Scanning Probe Microscopy Characterization. *Inorg. Chem.* **2006**, *45*, 7642–7650. [CrossRef]
10. García-Terán, J.P.; Castillo, O.; Luque, A.; García-Couceiro, U.; Román, P.; Lloret, F. One-Dimensional Oxalato-Bridged Cu(II), Co(II), and Zn(II) Complexes with Purine and Adenine as Terminal Ligands. *Inorg. Chem.* **2004**, *43*, 5761–5770. [CrossRef]
11. García-Terán, J.P.; Castillo, O.; Luque, A.; García-Couceiro, U.; Román, P.; Lezama, L. An Unusual 3D Coordination Polymer Based on Bridging Interactions of the Nucleobase Adenine. *Inorg. Chem.* **2004**, *43*, 4549–4551. [CrossRef]

12. Rojas-González, P.X.; Catiñeiras, A.; González-Pérez, J.M.; Choquesillo-Lazarte, D.; Niclós-Gutiérrez, J. Interligand Interactions Controlling the μ-N7,N9-Metal Bonding of Adenine (AdeH) to the N-Benzyliminodiacetato(2−) Copper(II) Chelate and Promoting the N9 versus N3 Tautomeric Proton Transfer: Molecular and Crystal Structure of [Cu$_2$(NBzIDA)$_2$(H$_2$O)$_2$(μ-N7,N9-Ade(N3)H)]·3H$_2$O. *Inorg. Chem.* **2002**, *41*, 6190–6192. [PubMed]
13. García-Terán, J.P.; Castillo, O.; Luque, A.; García-Couceiro, U.; Beobide, G.; Román, P. Supramolecular architectures assembled by the interaction of purine nucleobases with metal-oxalato frameworks. Non-covalent stabilization of the 7H-adenine tautomer in the solid-state. *Dalton Trans* **2006**, 902–911. [CrossRef]
14. Bugella-Altamirano, E.; Choquesillo-Lazarte, D.; González-Pérez, J.M.; Sánchez-Moreno, M.J.; Martín-Ramos, R.; Covelo, B.; Carballo, R.; Catiñeiras, A.; Niclós-Gutiérrez, J. Three new modes of adenine-copper(II) coordination: Interligand interactions controlling the selective N3-, N7- and bridging μ-N3,N7–metal-bonding of adenine to different N-substituted iminodiacetato-copper(II) chelates. *Inorg. Chim. Acta* **2002**, *339*, 160–170. [CrossRef]
15. Suzuki, T.; Hirai, Y.; Monjushiro, H.; Kaizaki, S. Cobalt(III) Complexes of Monodentate N(9)-Bound Adeninate (ade-), [Co(ade-κN9)Cl(en)$_2$]$^+$ (en = 1,2-Diaminoethane): Syntheses, Crystal Structures, and Protonation Behaviors of the Geometrical Isomers. *Inorg. Chem.* **2004**, *43*, 6435–6444. [CrossRef]
16. Yang, E.-C.; Zhao, H.-K.; Feng, Y.; Zhao, X.-J. A Tetranuclear Cu^{II}-Based 2D Aggregate with an Unprecedented Tetradentate μ$_4$-N1,N3,N7,N9-Adeninate Nucleobase. *Inorg. Chem.* **2009**, *48*, 3511–3513. [CrossRef]
17. García-Raso, A.; Terrón, A.; Ortega-Castro, J.; Barceló-Oliver, M.; Lorenzo, J.; Rodríguez-Calado, S.; Franconetti, A.; Frontera, A.; Vázquez-López, E.M.; Fiol, J.J. Iridium(III) coordination of N(6) modified adenine derivatives with aminoacid chains. *J. Inorg. Biochem.* **2020**, *205*, 111000. [CrossRef]
18. Ruiz-González, N.; García-Rubiño, M.E.; Domínguez-Martín, A.; Choquesillo-Lazarte, D.; Franconetti, A.; Frontera, A.; Catiñeiras, A.; González-Pérez, J.M.; Niclós-Gutiérrez, J. Molecular and supramolecular recognition patterns in ternary copper(II) or zinc(II) complexes with selected rigid-planar chelators and a synthetic adenine-nucleoside. *J. Inorg. Biochem.* **2020**, *203*, 110920. [CrossRef]
19. Martínez, D.; Pérez, A.; Cañellas, S.; Silió, I.; Lancho, A.; García-Raso, A.; Fiol, J.J.; Terrón, A.; Barceló-Oliver, M.; Ortega-Castro, J.; et al. Synthesis, reactivity, X-ray characterization and docking studies of N7/N9-(2-pyrimidyl)-adenine derivatives. *J. Inorg. Biochem.* **2020**, *203*, 110879. [CrossRef]
20. Roitzsch, M.; Lippert, B. Metal Coordination and Imine–Amine Hydrogen Bonding as the Source of Strongly Shifted Adenine pK$_a$ Values. *J. Am. Chem. Soc.* **2004**, *126*, 2421–2424. [CrossRef]
21. Añorbe, M.G.; Welzel, T.; Lippert, B. Migration of a cis-(NH3)2PtII Moiety along Two Adenine Nucleobases, from N1 to N6, is Markedly Facilitated by Additional Pt^{II} Entities Coordinated to N7. *Inorg. Chem.* **2007**, *46*, 8222–8227. [CrossRef]
22. Purohit, C.S.; Verma, S. A Luminescent Silver–Adenine Metallamacrocyclic Quartet. *J. Am. Chem. Soc.* **2006**, *128*, 400–401. [CrossRef]
23. Purohit, C.S.; Mishra, A.K.; Verma, S. Four-Stranded Coordination Helices Containing Silver–Adenine (Purine) Metallaquartets. *Inorg. Chem.* **2007**, *46*, 8493–8495. [CrossRef]
24. Purohit, C.S.; Verma, S. Patterned Deposition of a Mixed-Coordination Adenine–Silver Helicate, Containing a π-Stacked Metallacycle, on a Graphite Surface. *J. Am. Chem. Soc.* **2007**, *129*, 3488–3489. [CrossRef]
25. Shipman, M.A.; Price, C.; Gibson, A.E.; Elsegood, M.R.J.; Clegg, W.; Houlton, A. Monomer, Dimer, Tetramer, Polymer: Structural Diversity in Zinc and Cadmium Complexes of Chelate-Tethered Nucleobases. *Chem. Eur. J.* **2000**, *6*, 4371–4378. [CrossRef]
26. Kruger, T.; Ruffer, T.; Lang, H.; Wagner, C.; Steinborn, D. Synthesis, Characterization, and Reactivity of [Li(N^6,N^9-Me$_2$Ade$_{-H}$)]: A Structurally Characterized Lithiated Adenine. *Inorg. Chem.* **2008**, *47*, 1190–1195. [CrossRef]
27. Amantia, D.; Price, C.; Shipman, M.A.; Elsegood, M.R.J.; Clegg, W.; Houlton, A. Minor Groove Site Coordination of Adenine by Platinum Group Metal Ions: Effects on Basicity, Base Pairing, and Electronic Structure. *Inorg. Chem.* **2003**, *42*, 3047–3056. [CrossRef]
28. Zobi, F.; Spingler, B.; Alberto, R. Structure, Reactivity and Solution Behaviour of [Re(ser)(7-MeG)(CO)$_3$] and [Re(ser)(3-pic)(CO)$_3$]: 'Nucleoside-mimicking' Complexes Based on the fac-[Re(CO)$_3$]$^+$ Moiety. *Dalton Trans.* **2005**, 2859–2865. [CrossRef]

29. Jiang, Q.; Wu, Z.-Y.; Zhang, Y.-M.; Hotze, A.C.G.; Hannon, M.J.; Guo, Z.-J. Effect of Adenine Moiety on DNA Binding Property of Copper(II)–terpyridine Complexes. *Dalton Trans* **2008**, 3054–3060. [CrossRef]
30. Price, C.; Horrocks, B.R.; Mayeux, A.; Elsegood, M.R.J.; Clegg, W.; Houlton, A. Self-Complementary Metal Complexes Containing a DNA Base Pair. *Angew. Chem. Int. Ed.* **2002**, *41*, 1047–1049. [CrossRef]
31. Dobrzynska, D.; Jerzykiewicz, L.B. Adenine Ribbon with Watson– Crick and Hoogsteen Motifs as the 'Double-Sided Adhesive Tape' in the Supramolecular Structure of Adenine and Metal Carboxylate. *J. Am. Chem. Soc.* **2004**, *126*, 11118–11119. [CrossRef]
32. Serrano Padial, E.; Choquesillo-Lazarte, D.; Bugella Altamirano, E.; Castiñeiras, A.; Carballo, R.; Niclós Gutiérrez, J. New Copper(II) Compound having Protonated forms of Ethylenediamine-tetraacetate(4-) ion (EDTA) and Adenine (AdeH): Synthesis, Crystal Structure, Molecular Recognition and Physical Properties of (AdeH$_2$)[Cu(HEDTA)(H$_2$O)]·2H$_2$O. *Polyhedron* **2002**, *21*, 1451–1457. [CrossRef]
33. Bruker. *APEX3 Software*; v2018.7-2; Bruker AXS Inc.: Madison, WI, USA, 2018.
34. Sheldrick, G.M. *SADABS. Program for Empirical Absorption Correction of Area Detector Data*; University of Goettingen: Goettingen, Germany, 1997.
35. Sheldrick, G.M. A short history of SHELX. *Acta Crystallogr.* **2008**, *64*, 112–122. [CrossRef]
36. Wilson, A.J.C. *International Tables of Crystallography*; Kluwer Academic Publishers: Dordrecht, The Netherlands, 1995; Volume C.
37. Spek, A.L. PLATON. A multipurpose Crystallographic tool. *Acta Crystallogr.* **2009**, *65*, 148–155.
38. Frisch, M.J.; Trucks, G.W.; Schlegel, H.B.; Scuseria, G.E.; Robb, M.A.; Cheeseman, J.R.; Scalmani, G.; Barone, V.; Petersson, G.A.; Nakatsuji, H.; et al. *Gaussian 16, Revision A.01*; Gaussian, Inc.: Wallingford, CT, USA, 2016.
39. Hazari, A.; Das, L.K.; Kadam, R.M.; Bauza, A.; Frontera, A.; Ghosh, A. Unprecedented structural variations in trinuclear mixed valence Co (II/III) complexes: Theoretical studies, pnicogen bonding interactions and catecholase-like activities. *Dalton Trans.* **2015**, *44*, 3862–3876. [CrossRef]
40. Mitra, M.; Manna, P.; Bauzá, A.; Ballester, P.; Seth, S.K.; Ray Choudhury, S.; Frontera, A.; Mukhopadhyay, S. 3-Picoline Mediated Self-Assembly of M (II)–Malonate Complexes (M= Ni/Co/Mn/Mg/Zn/Cu) Assisted by Various Weak Forces Involving Lone Pair– π, π–π, and Anion··· π–Hole Interactions. *J. Phys. Chem. B* **2014**, *118*, 14713–14726. [CrossRef]
41. Kolaria, K.; Sahamies, J.; Kalenius, E.; Novikov, A.S.; Kukushkin, V.Y.; Haukka, M. Metallophilic interactions in polymeric group 11 thiols. *Solid State Sci.* **2016**, *60*, 92–98. [CrossRef]
42. Novikov, A.S.; Ivanov, D.M.; Bikbaeva, Z.M.; Bokach, N.A.; Kukushkin, V.Y. Noncovalent Interactions Involving Iodofluorobenzenes: The Interplay of Halogen Bonding and Weak lp(O)···π-Holearene Interactions. *Cryst. Growth Des.* **2018**, *18*, 7641–7654. [CrossRef]
43. Kinzhalov, M.A.; Novikov, A.S.; Chernyshev, A.N.; Suslonov, V.V. Intermolecular hydrogen bonding H···Cl– in the solid palladium(II)-diaminocarbene complexes. *Z. Kristallogr. Cryst. Mater.* **2017**, *232*, 299–305. [CrossRef]
44. Baykov, S.V.; Dabranskaya, U.; Ivanov, D.M.; Novikov, A.S.; Boyarskiy, V.P. Pt/Pd and I/Br Isostructural Exchange Provides Formation of C–I···Pd, C–Br···Pt, and C–Br···Pd Metal-Involving Halogen Bonding. *Cryst. Growth Des.* **2018**, *18*, 5973–5980. [CrossRef]
45. Usoltsev, A.N.; Adonin, S.A.; Novikov, A.S.; Samsonenko, D.G.; Sokolov, M.N.; Fedina, V.P. One-dimensional polymeric polybromotellurates(iv): Structural and theoretical insights into halogen···halogen contacts. *CrystEngComm* **2017**, *19*, 5934–5939. [CrossRef]
46. Bader, R.F.W. A quantum theory of molecular structure and its applications. *Chem. Rev.* **1991**, *91*, 893–928. [CrossRef]
47. Keith, T.A. *AIMAll (Version 13.05.06)*; TK Gristmill Software: Overland Park, KS, USA, 2013.
48. Bader, R.F.W. A Bond Path: A Universal Indicator of Bonded Interactions. *J. Phys. Chem. A* **1998**, *102*, 7314–7323. [CrossRef]
49. Espinosa, E.; Molins, E.; Lecomte, C. Hydrogen Bond Strengths Revealed by Topological Analyses of Experimentally Observed Electron Densities. *Chem. Phys. Lett.* **1998**, *285*, 170–173. [CrossRef]
50. Vener, M.V.; Egorova, A.N.; Churakov, A.V.; Tsirelson, V.G. Intermolecular Hydrogen Bond Energies in Crystals Evaluated Using Electron Density Properties: DFT Computations with Periodic Boundary Conditions. *J. Comput. Chem.* **2012**, *33*, 2303–2309. [CrossRef]

51. Domínguez-Martín, A.; Brandi-Blanco, M.P.; Matilla-Hernández, A.; El Bakkali, H.; Nurchi, V.M.; González-Pérez, J.M.; Castiñeiras, A.; Niclós-Gutiérrez, J. Unraveling the Versatile Metal Binding Modes of Adenine: Looking at Molecular Recognition Patterns of Deaza- and Aza-adenines in Mixed Ligand Metal Complexes. *Coord. Chem. Rev.* **2013**, *257*, 2814–2838.
52. Vílchez-Rodríguez, E.; Pérez-Toro, I.; Bauzá, A.; Matilla-Hernández, A. Structural and Theoretical Evidence of the Depleted Proton Affinity of the N3-Atom in Acyclovir. *Crystals* **2016**, *6*, 193.
53. Pérez-Toro, I.; Domínguez-Martín, A.; Choquesillo-Lazarte, D.; Vílchez-Rodríguez, E.; González-Pérez, J.M.; Castiñeiras, A.; Niclós-Gutiérrez, J. Lights and Shadows in the Challenge of Binding acyclovir, a synthetic Purine-like Nucleoside with Antiviral Activity, at an Apical-Distal Coordination Site in Copper-Polyamine Chelates. *J. Inorg. Biochem.* **2015**, *148*, 84–92. [CrossRef]
54. Pérez-Toro, I.; Domínguez-Martín, A.; Choquesillo-Lazarte, D.; González-Pérez, J.M.; Castiñeiras, A.; Niclós-Gutiérrez, J. Highest Reported Denticity of a Synthetic Nucleoside in the Unprecedented Tetradentate Mode of Acyclovir. *Cryst. Growth Des.* **2018**, *18*, 4282–4286. [CrossRef]
55. Bats, J.W.; Nasiri, H.R. CSD Database, reference code NULCO. Private Communication, 2015.
56. Atria, A.M.; Corsini, G.; Herrera, N.; Garland, M.T.; Baggio, R. Two Isomorphous Transition Metal Complexes Containing a Protonated Diaminopurine Ligand: Diaquabis(2,6-diamino-7H-purin-1-ium-κN^9) bis(homophthalato-κO)nickel(II) Tetrahydrate and the Cobalt(II) Analogue. *Acta Crystallogr.* **2011**, *67*, m169–m172. [CrossRef]
57. Atria, A.M.; Garland, M.T.; Baggio, R. 2,6-Diamino-9H-purine Monohydrate and Bis(2,6-diamino-9H-purin-1-ium) 2-(2-Carboxylatophenyl)acetate Heptahydrate: Two Simple Structures with Very Complex Hydrogen-bonding Schemes. *Acta Crystallogr.* **2010**, *66*, o547–o552. [CrossRef]
58. Belletire, J.L.; Schneider, S.; Shackelford, S.A.; Peryshkov, D.V.; Strauss, S.H. Pairing Hetherocyclic Cations with closo-Dodecafluorododecaborate(-). Synthesis of Binary Heterocyclium(1+) Salts and Ag$_4$(hetherocycle)$_8^{4+}$ Salt of B$_{11}$F$_{12}^{2-}$. *J. Fluor. Chem.* **2011**, *132*, 925–936. [CrossRef]
59. Yang, E.-C.; Chan, Y.-N.; Liu, H.; Wang, Z.-C.; Zhao, X.J. Unusual Polymeric Zn^{II}/Cd^{II} Complexes with 2,6-Diaminopurine by Synergistic Coordination of Nucleobases and Polycarboxylate Anions: Binding Behavior, Self-Assembled Pattern of the Nucleobase, and Luminescent Properties. *Cryst. Growth Des.* **2009**, *9*, 4933–4944. [CrossRef]
60. Liu, Z.-Y.; Dong, H.-M.; Wang, X.G.; Zhao, X.-J.; Yang, E.-C. Three Purine Containing Metal Complexes with Discrete Binuclear and Polymeric Chain Motifs: Synthesis, Crystal Structure and Luminescence. *Inorg. Chim. Acta* **2014**, *416*, 135–141. [CrossRef]
61. Atria, A.M.; Parada, J.; Moreno, Y.; Suárez, S.; Baggio, R.; Peña, O. Synthesis, Crystal Structure and Magnetic Properties of Diaquabis(2,6-diamino-7H-purin-1-ium-κN^9)bis(4,4'-oxydibenzoato-κO)cobalt(II) Dihydrate. *Acta Crystallogr.* **2018**, *74*, 37–44.
62. Atria, A.M.; Morel, M.; Garland, M.T.; Baggio, R. Bis(2,6-diamino-1H-purin-3-ium) Di-μ-croconato-κ^3O,O':O'';κ^3O:O',O''-bis[tetraaqua(croconato-κ^2O,O')-neodymium(III)]. *Acta Crystallogr.* **2011**, *67*, m17–m21. [CrossRef]
63. Atria, A.M.; Garland, M.T.; Baggio, R. Tetraaquabis(2,6-diamine-7H-κN^9)cobalt(II) benzene-1,2,4,5-tetracarboxylate tetrahydrate. *Acta Crystallogr.* **2011**, *67*, m275–m278. [CrossRef]

© 2020 by the authors. Licensee MDPI, Basel, Switzerland. This article is an open access article distributed under the terms and conditions of the Creative Commons Attribution (CC BY) license (http://creativecommons.org/licenses/by/4.0/).

Article

Local Vibrational Mode Analysis of π–Hole Interactions between Aryl Donors and Small Molecule Acceptors

Seth Yannacone [1], Marek Freindorf [1], Yunwen Tao [1], Wenli Zou [2] and Elfi Kraka [1,*]

1   Department of Chemistry, Southern Methodist University, 3215 Daniel Avenue, Dallas, TX 75275, USA; syannacone@smu.edu (S.Y.); mfreindorf@smu.edu (M.F.); yunwent@smu.edu (Y.T.)
2   Institute of Modern Physics, Northwest University, Xi'an 710127, China; zouwl@nwu.edu.cn
*   Correspondence: ekraka@smu.edu; Tel.: +1-214-768-2611

Received: 6 June 2020; Accepted: 24 June 2020; Published: 30 June 2020

Abstract: 11 aryl–lone pair and three aryl–anion π–hole interactions are investigated, along with the argon–benzene dimer and water dimer as reference compounds, utilizing the local vibrational mode theory, originally introduced by Konkoli and Cremer, to quantify the strength of the π–hole interaction in terms of a new local vibrational mode stretching force constant between the two engaged monomers, which can be conveniently used to compare different π–hole systems. Several factors have emerged which influence strength of the π–hole interactions, including aryl substituent effects, the chemical nature of atoms composing the aryl rings/π–hole acceptors, and secondary bonding interactions between donors/acceptors. Substituent effects indirectly affect the π–hole interaction strength, where electronegative aryl-substituents moderately increase π–hole interaction strength. N-aryl members significantly increase π–hole interaction strength, and anion acceptors bind more strongly with the π–hole compared to charge neutral acceptors (lone–pair donors). Secondary bonding interactions between the acceptor and the atoms in the aryl ring can increase π–hole interaction strength, while hydrogen bonding between the π–hole acceptor/donor can significantly increase or decrease strength of the π–hole interaction depending on the directionality of hydrogen bond donation. Work is in progress expanding this research on aryl π–hole interactions to a large number of systems, including halides, CO, and OCH_3^- as acceptors, in order to derive a general design protocol for new members of this interesting class of compounds.

Keywords: π–hole interaction; substituent effects; vibrational spectroscopy; local vibrational mode theory; direct measure for π–hole interaction strength; noncovalent interaction; hydrogen bonding

1. Introduction

The term 'π–hole interaction' was coined by Murray and Politzer [1–4], and is described as a noncovalent interaction (NCI) between a region of positive electrostatic potential (ESP) located on a π–bond (i.e., a 'π–hole') [5], and a lone–pair (lp) donor [6–8], anion [9,10], or other electron rich species [11,12]; where the π–hole is perpendicular to the molecular framework and electrons from the π–hole acceptor interact with an empty π* orbital of the donor. Some classic examples of π–hole interactions involving aryl groups include the benzene/hexafluorobenzene–water complexes, where an oxygen–lp interacts favorably with the center of the aromatic ring [13–20]. This special type of interaction has been identified in several important and highly relevant areas of modern chemical research, including drug targets [21,22], biological systems [23,24], and molecular crystals/solid state chemistry [25–30]. Interestingly, noble gases have recently been found capable of forming both σ– and π–hole interactions [31–33]. Ideal π–hole donors should contain heavier and more polarizable atoms, as these properties improve accessibility, size, and positive ESP of a π–hole [34–37].

Electron withdrawing π–hole acceptors can also increase the positive ESP of the π–hole [38,39]. The main interaction energy terms describing π–hole interactions are: ion induced polarization and a permanent quadrupole moment (Q_{zz}) from the electrostatic forces [40–42]. Though there have been several recent theoretical and experimental studies on π–hole interactions [43–56], often the strength of these interactions is discussed in terms of bond lengths (r) or binding energies (BE)/dissociation energies (DE). However, these properties are not necessarily qualified as bond strength descriptors. There is an ample number of examples in which the shorter bond is not the stronger bond [57–59]. It is often assumed that BE or DE provide a measure of the intrinsic bond strength of the NCI in question. However this might not even be true in a qualitative sense, as BE and DE are cumulative properties; i.e., they are the sum of all interactions between the monomers, including long–range electrostatic interactions which may even involve the more remote atoms of the monomer [60]. Therefore, it is difficult to single out a specific interaction between atoms or groups of monomers; even computationally this can only be done in a qualitative way via an energy decomposition scheme, which leads to model dependent results [61–65]. In this situation, vibrational spectroscopy provides an excellent alternative for the description of the interactions between the monomers of a complex, and offers a platform for deriving a spectroscopic measure of complex stability. However, as has been frequently pointed out [60,66–69], any description of bond strength based on vibrational modes has to consider that normal vibrational modes are generally delocalized due to the coupling of the motions of the atoms within a molecule or complex [70–74]. Therefore, only decoupled local vibrational modes can serve as bond strength measurements, as was realized in the Local Vibrational Mode (LVM) theory originally formulated by Konkoli and Cremer [75–82]. Local mode stretching force constants (k^a) are directly related to the intrinsic strength of a bond, and therefore provide a unique measure of bond strength based on vibrational spectroscopy [83]. The local mode procedure was inspired by the isotopic substitution of McKean [84]. McKean found that if an XH fragment in a molecule is replaced by XD, a local X–D stretching mode may be detected in the IR spectrum, and therefore the force constant of the X–H or X–D stretching may be measured. This technology has been used to measure the force constants of many X–H bonds, but it cannot be extended to other systems due to the weak isotope effect. However, theoretical calculations are not limited to natural isotopes, allowing for isotopes of any mass to be "invented." The local mode procedure treats all the atoms which are not involved in a particular local mode as massless particles, so that they can effortless follow the local motion. For each local mode associated with an internal coordinate such as a bond length, bond angle, dihedral angle or puckering coordinate a unique local mode force constant, associated local mode mass and frequency can be obtained. So far, the LVM analysis has been successfully applied to characterize covalent bonds [59,66,83,85–88] and weak chemical interactions such as halogen [89–92], chalcogen [58,93,94], pnicogen [95–97], and tetrel interactions [98]; as well as hydrogen bonding (HB) [67,69,99–102]. For a comprehensive review the reader is referred to Ref. [80].

In this work, LVM theory is utilized to obtain a more accurate measurement of strength and the intrinsic nature of interactions between various aryl systems as π–hole donors and a number of small electron rich π–hole acceptors; where the π–hole either interacts with lp–electrons from a charge neutral acceptor, or an anionic acceptor species. A special inter-monomer LVM stretching force constant is utilized, which directly assesses the strength of the π–hole⋯π–hole acceptor interaction. Based on this special inter-monomer k^a measure, recently and for the first time, the strength of metal–ring interactions in a series of actinide sandwich compounds was quantified [103], and a nonclassical HB involving a BH⋯π interaction was identified [104,105]. Burianova et al. concurrently verified this type of nonclassical HB involving a BH⋯π interaction both experimentally and theoretically while performing a mechanistic study involving the nucleophilic addition of hydrazines, hydrazides, and hydrazones to C≡N groups of boron–based clusters [106].

The current work investigates the interactions of π–hole acceptors H_2O, HCN, NH_3, and NO_3^-, with the following aromatic π–hole donors: C_6F_6, C_6F_5H, $C_6F_4H_2$, $C_6F_3H_3$, $N_3C_3H_3$, $N_3C_3F_3$, and $N_4C_2H_2$ (see Figure 1). Original theoretical works of similar nature date back to 1997, when Alkorta et al. investigated the effects of F–substitution on reactivity of the aromatic rings in systems where small electron-donating molecules interact with the π–clouds of benzene and hexafluorobenzene [107]. An extension of this work was reported in 2002, which included a larger array of aromatics and benzene derivatives and several negatively charged electron donors [108]. Simultaneously, a similar phenomenon was reported involving 1,3,5–triazine derivatives interacting with F$^-$, Cl$^-$, and azide (N$_3$) [109], and a computation study was combined with crystallographic evidence to confirm such interactions can favorably occur [110].

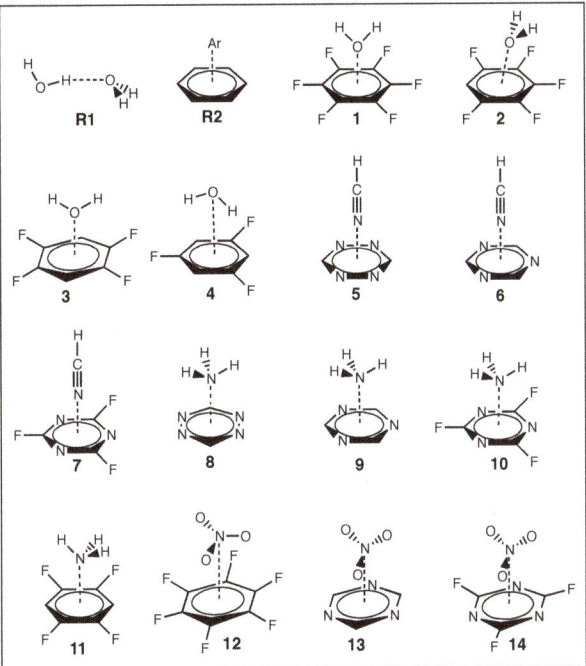

Figure 1. Schematic of the two references systems, **R1** and **R2**, and π–hole systems **1–14** studied in this work showing molecular geometries of each system; calculated at the ωB97X–D/aug–cc–pVTZ level of theory.

2. Computational Methods

DFT was utilized to optimize molecular geometries, calculate stationary point normal mode vibrational frequencies (ω_μ), LVM frequencies (ω^a), k^a [75,78,79], and Natural Bond Orbital (NBO) charges. Calculations were carried out at the ωB97X–D/aug–cc–pVTZ level of theory with tight convergence criteria and superfine integration grid [111–116]. All stationary points were confirmed to be minima by absence of imaginary ω_μ. Calculated and experimental vibrational frequencies of the H$_2$O$\cdots$C$_6$F$_6$ [117] system were used to gauge the accuracy of several model chemistries (see Tables 1 and 2). Theoretical vibrational spectroscopy was utilized to quantify the intrinsic strength of π–hole interactions in this work. Normal vibrational modes do not give direct measurements of bond strength because of electronic and mass coupling. This results in delocalization of the normal modes in most cases. The electronic coupling is eliminated by solving the Wilson equation of spectroscopy [118] and transforming to normal coordinates. Konkoli and Cremer found that mass coupling can be removed by solving a mass–decoupled equivalent of the Wilson equation, which leads to LVMs. LVMs are associated with internal coordinates: bond length, bond angle, or dihedral angle [76], and lead to a direct relationship between the intrinsic strength of a bond and its k^a value [83]. For the first time, this theory is applied to π–hole interactions. LVM analysis was computed with the program COLOGNE2018 [119]. NBO populations were calculated using NBO6 [120–122]. Calculations of $\rho(\mathbf{r}_{CCP})$ and $\nabla^2\rho(\mathbf{r}_{CCP})$ were performed with the AIMAll program [123,124]. All DFT calculations were made with GAUSSIAN16 [125].

Table 1. Comparison of experimental *exp* normal mode vibrational frequencies ω_{exp}, with theoretical normal mode vibrational frequencies ω_μ for **1** computed at the ωB97X–D/aug-cc-pVTZ, ωB97X–D/aug-cc-pVQZ, ωB97X–D/def2-TZVPP, MP2/aug-cc-pVTZ , and MP2/def2-TZVPP levels of theory.

Mode	*exp* [117]	ωB97X–D/ aug-cc-pVTZ	ωB97X–D/ aug-cc-pVQZ	ωB97X–D/ def2-TZVPP	MP2/ aug-cc-pVTZ	MP2/ def2-TZVPP
H_2O ν_3 (asymmetric stretch)	3723.0	3811.0 (−2.3)	3821.4 (−2.6)	3822.2 (−2.6)	3745.8 (−0.6)	3769.9 (−1.2)
H_2O ν_1 (symmetric stretch)	3632.0	3710.3 (−2.1)	3722.4 (−2.4)	3722.2 (−2.4)	3629.7 (0.1)	3655.8 (−0.7)
H_2O ν_2 (bend)	1607.0	1570.2 (2.3)	1572.7 (2.2)	1568.7 (2.4)	1558.2 (3.1)	1570.1 (2.3)
C_6H_6 ν_{12} (C–C stretch)	1536.0	1511.4 (1.6)	1510.2 (1.7)	1509.7 (1.7)	1489.2 (3.1)	1495.5 (2.7)
C_6H_6 ν_{13} (C–F stretch)	999.0	991.8 (0.7)	991.5 (0.8)	990.0 (0.9)	971.4 (2.8)	976.1 (2.3)

ω_{exp} and ω_μ are reported in cm^{-1} and errors are given as % with respect to *exp* in parentheses next to each ω_μ. Scaling factors are as follows: 0.957 (ωB97X–D/aug-cc-pVTZ), 0.957 (ωB97X–D/aug-cc-pVQZ), 0.955 (ωB97X–D/def2-TZVPP), 0.953 (MP2/aug-cc-pVTZ), and 0.952 (MP2/def2-TZVPP) [126–132].

Table 2. Comparison of local vibrational mode LVM data for π–hole system **1**, where O···C$_6$ (acceptor···donor) represents the pure π–hole interaction between the acceptor O–atom and the geometric center of the C–atoms comprising the six–membered ring, O···C$_6$F$_6$ denotes similar as above but includes the six F–substituents of the π–hole donor, H···C$_6$ denotes one acceptor H–atom interacting with the geometric center of the six donor C–atoms, and H···C$_6$F$_6$ represents the aforementioned interaction with inclusion of the aryl F–substituents.

Parameter	r	k^a	ω^a
ωB97X–D/aug-cc-pVTZ			
O···C$_6$	3.121	0.090	108.1
O···C$_6$F$_6$	3.116	0.087	100.2
H···C$_6$	3.780	0.021	187.1
H···C$_6$F$_6$	3.775	0.020	185.7
ωB97X–D/aug-cc-pVQZ			
O···C$_6$	3.130	0.082	103.2
O···C$_6$F$_6$	3.125	0.080	95.7
H···C$_6$	3.787	0.020	185.6
H···C$_6$F$_6$	3.782	0.020	184.1
MP2/aug-cc-pVTZ			
O···C$_6$	2.981	0.087	106.3
O···C$_6$F$_6$	2.974	0.084	98.1
H···C$_6$	3.654	0.023	197.7
H···C$_6$F$_6$	3.646	0.023	195.8

bond lengths r are given in Å, LVM force constants k^a in mdyn/Å, and units for LVM frequencies ω^a are cm^{-1}.

Figure 2 illustrates how the special force constant k^a is defined for the special case of the π–hole interaction involving a six–membered ring as π–hole donor. k^a is defined via the direct interaction between the central O– or N–atom of the π–hole acceptor (position X$_1$ in Figure 2) and the geometric center of the six atoms composing the aryl ring of the π–hole donor (X$_2$ in Figure 2). A key feature of the LVM methodology is that the π–hole need not be at the X_2 geometric center of the ring. If this is the case, and the acceptor atom at X_1 is collinear with X_2 and the π–hole, the value of k^a will not change because the local modes of $X_1 \cdots X_2$ and $X_1 \cdots$ π–hole are normalized in the LVM theory formalism. In systems **R2, 1–4** and **11–12**, the ring atoms are all carbon; whereas in systems **6–7, 9–10,** and **13–14**, three N–atoms and three C–atoms are incorporated into the ring structure. In systems **5** and **8**, the six–membered rings are composed of four N–atoms and two C–atoms.

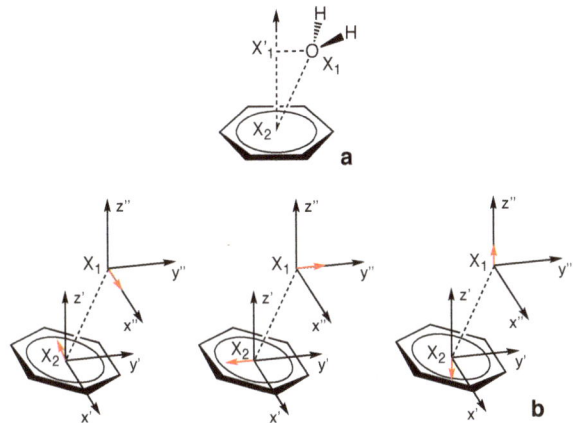

Figure 2. Schematic of how the special LVM force constant k^a is defined for the π–hole interaction involving a six–membered aromatic ring as π–hole donor, where X_1 is the location of the central atom of the acceptor molecule interacting directly with the π–hole located at X_2; shown is complex **2**.

3. Results/Discussion

3.1. Discussion of Model Chemistry

Table 1 shows experimental (*exp*) normal mode frequencies (ω_{exp}) and theoretical normal mode frequencies (ω_μ) for the water-hexafluorobenzene dimer (system **1**). Theoretical ω_μ were computed using Møller–Plesset perturbation theory of second order (MP2) and the ωB97X–D functional combined with aug–cc–pVTZ, aug–cc–pVQZ, and def2–TZVPP basis sets. In addition, scaling factors were applied to theoretical frequencies to correct for approximations to the full electronic configuration interaction and the harmonic approximation to the Morse potential [126–132]. In parentheses directly to the right of each theoretical frequency, are % error values calculated with respect to *exp*. It turns out that ωB97X–D/aug–cc–pVTZ calculations were in closest agreement with *exp*. MP2 calculations performed best for the highest frequencies, but were less accurate for low frequencies. The opposite is true of calculations carried out using ωB97X–D. The use of the def2-TZVPP basis set was computationally more efficient, but the aug–cc–pVTZ basis set significantly improved accuracy.

Table 2 compares LVM data calculated at the ωB97X–D/aug–cc–pVTZ, ωB97X–D/aug–cc–pVQZ, and MP2/aug–cc–pVTZ levels of theory for π–hole system **1**, where $O \cdots C_6$ denotes the pure π–hole interaction between the acceptor O–atom and the geometric center of the C–atoms composing the six–membered ring (acceptor$\cdots$donor). $O \cdots C_6 F_6$ is very similar to the interaction just described, except in this case the six F–substituents are included. $H \cdots C_6$ denotes the interaction between one acceptor H–atom and the geometric center of the six donor C–atoms; whereas $H \cdots C_6 F_6$ denotes a similar interaction, but with the six F–substituents included (analogous to the $O \cdots C_6 / O \cdots C_6 F_6$ comparison). The π–hole interactions in the remainder of this work are defined using the first notation

(O···C_6) in Table 2: the pure π–hole interaction between the central acceptor atom and the geometric center of the donor six-membered ring, not including aryl substituent atoms.

In comparison with the ωB97X–D/aug–cc–pVTZ calculations, adding a larger basis set (aug–cc–pVQZ quality) resulted in a modest r increase of 0.009 Å and a slight decrease in k^a of 0.008 mdyn/Å. On the other hand, MP2/aug–cc–pVTZ results gave significantly shorter r (by 0.140 Å and 0.149 Å), slightly weaker bond strength (by 0.003 mdyn/Å) compared to ωB97X–D/aug–cc–pVTZ, and slightly stronger bond strength (by 0.005 mdyn/Å) compared to ωB97X–D/aug–cc–pVQZ. This result is erratic in the case of MP2/aug–cc–pVTZ. The ωB97X–D/aug–cc–pVQZ level of theory has large computational cost with small increase of accuracy compared to ωB97X–D/aug–cc–pVTZ. Therefore, we have chosen in this study the ωB97X–D/aug–cc–pVTZ level of theory as a compromise between accuracy and computational efficiency. Note that for the remainder of this work, the terms 'k^a' and 'bond strength' are used interchangeably. In addition, the term secondary bonding interaction (SBI) refers to any interaction between a single atom of the acceptor molecule and a single atom of the donor molecule which contains a physically meaningful LVM.

3.2. Overall Findings and General Trends

Table 3 summarizes the LVM data of the π–hole interactions in **1–14** and two reference NCIs: **R1** (water dimer) and **R2** (Ar···C_6H_6). Figure 3 (top) shows molecular geometry of each system, r (shown in green), k^a (blue), ω^a (red), and symmetry point group (black) for **R1**, **R2**, and **1–14**. **R1** and **R2** have been incorporated to provide a frame of reference from well characterized compounds: The H_2O dimer represents complex containing a strong HB with non-negligible covalent character, and Ar···C_6H_6 represents a weak NCI [133]. Also in Figure 3 (bottom), selected NBO charges are given, where charges in green represent C–atoms, O–atomic charges are red, H–atomic charges are black, N–atomic charges are blue, and F–atomic charges are light blue. Bond length r and NBO charge on the acceptor O and N–atoms are plotted with respect to k^a in Figure 4a and Figure 4b, respectively. Shown as red plot points are interactions where H_2O is the acceptor (**1–4**), light blue points represent HCN acceptor systems (**5–7**), in green are NH_3 (**8–11**), blue points are the NO_3^- anion–π–hole interactions (**12–14**), and black points indicate **R1** and **R2**. This color convention is maintained in the subsequent plots.

Table 3. Summary of LVM data: π–hole interaction distances r, k^a, ω^a, charge transfer CT, and BSSE counterpoise corrected binding energies BE.

#	System	Point Group	r	k^a	ω^a	CT lp→ π–Hole	BE
R1	H_2O···HOH	C_s	1.936	0.171	553.3	−9.08	−4.98
R2	Ar···C_6H_6	C_{2v}	3.620	0.072	69.0	−0.10	−0.92
1	H_2O···C_6F_6	C_{2v}	3.121	0.090	108.1	−10.29	−2.57
2	H_2O···C_6F_5H	C_s	3.193	0.051	81.3	−7.72	−2.10
3	H_2O···$C_6F_4H_2$	C_{2v}	3.226	0.107	117.6	−5.66	−1.52
4	H_2O···$C_6F_3H_3$	C_s	3.359	0.086	105.5	−1.75	−2.03
5	HCN···$N_4C_2H_2$	C_{2v}	3.047	0.090	113.3	−30.99	−2.65
6	HCN···$N_3C_3H_3$	C_{3v}	3.154	0.051	85.2	−19.93	−1.75
7	HCN···$N_3C_3F_3$	C_{3v}	2.989	0.076	104.0	−45.02	−4.05
8	H_3N···$N_4C_2H_2$	C_s	3.062	0.125	133.7	−16.07	−3.87
9	H_3N···$N_3C_3H_3$	C_{3v}	3.170	0.144	143.7	−9.50	−2.54
10	H_3N···$N_3C_3F_3$	C_{3v}	3.026	0.185	162.8	−2.80	−5.37
11	H_3N···$C_6F_4H_2$	C_s	3.298	0.070	100.9	−8.24	−2.03
12	$[O_3N$···$C_6F_6]^-$	C_{3v}	3.078	0.228	181.7	−5.83	−12.00
13	$[O_3N$···$N_3C_3H_3]^-$	C_{3v}	3.128	0.169	155.7	−6.32	−6.03
14	$[O_3N$···$N_3C_3F_3]^-$	C_{3v}	2.955	0.276	198.6	−11.31	−13.03

Calculated at ωB97X–D/aug–cc–pVTZ level of theory. Units for reported data as follows: r in Å, k^a in mdyn/Å, ω^a in cm^{-1}, CT in milli-electron (m_e), and BE in kcal/mol.

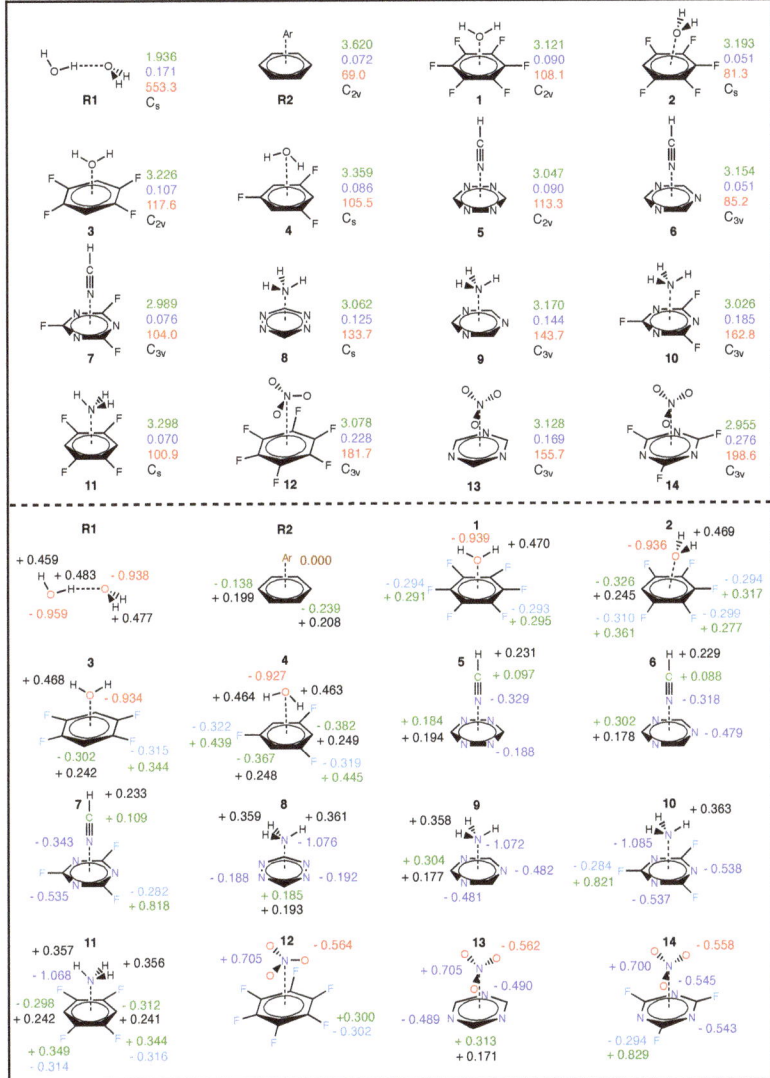

Figure 3. Schematics for **R1**, **R2**, and **1–14**, showing: (**top**) molecular geometries, distances r given in green font with units of Å, local vibrational mode LVM force constants k^a (blue font) given in mdyn/Å, corresponding LVM frequencies ω^a (red) given in cm^{-1}, point group (shown in black); and (**bottom**) selected NBO charges: C–atomic charges given in green, O–atomic charges in red, N–atomic charges in blue, F–atomic charges in light blue, and H–atomic charges are shown in black. NBO charges are given in A.U.

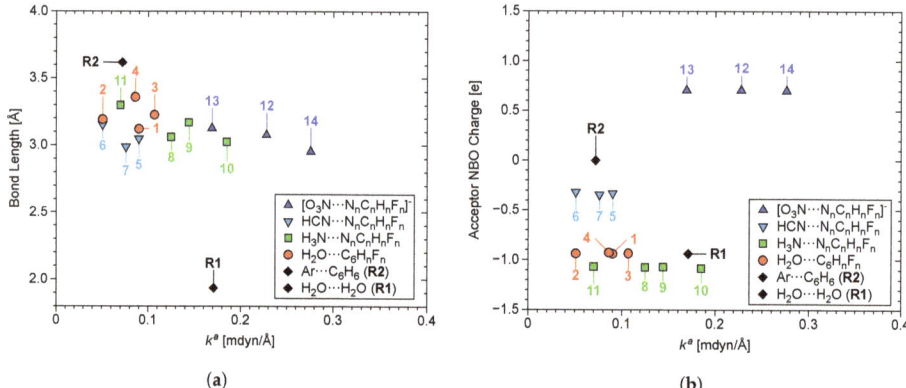

Figure 4. Calculated at the ωB97X–D/aug–cc–pVTZ level of theory, (a) r, and (b) NBO charges of the central acceptor atoms O and N; plotted with respect to k^a of π–hole interactions in **1–14**.

There is weak correlation at best between r and k^a, which becomes weaker by presence of **R1** and **R2**. The π–hole interaction length in **14** is 1.000 Å longer than the HB in **R1**, yet the former has a k^a value 0.100 mdyn/Å larger than the latter. The Ar···C_6H_6 interaction in **R2** is at least 0.200 Å longer than all 14 π–hole interactions, but is stronger than **2**, **6**, and **11**. Figure 5a,b show charge transfer (CT) and BE counterpoise corrected for basis set superposition error; both plotted with respect to k^a. CT was calculated as the transfer of charge between the acceptor lp–donor atom and the aryl ring. Both of these parameters correlate weakly with bond strength in terms of k^a, but BE and k^a show the best correlation of any properties considered in this work. Increase in magnitude of BE weakly correlates with increase in bond strength. The HB in **R1** has a k^a value three times larger than the weakest π–hole interactions (**2** and **6**). On the other hand, **14** contains the strongest π–hole interaction in this work with a k^a value 60% larger than k^a of the HB in **R1**.

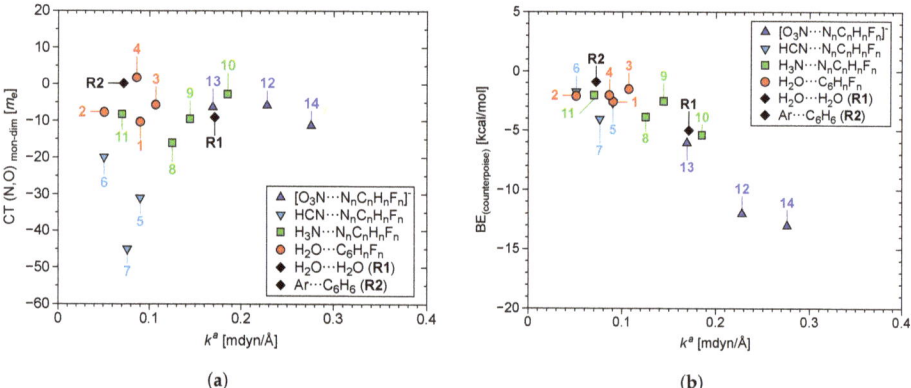

Figure 5. (a) CT (from central π–hole acceptor atom O in **1–4**, N in **5–14**, Ar in **R2**, and HB acceptor atom O in **R1** $\longrightarrow$ donor), and (b) BE, counterpoise corrected for basis set superposition error; both plotted with respect to k^a of π–hole interactions in **1–14**.

In Figure 6a, the Laplacian of the electron density ($\nabla^2\rho(r_{CCP})$), where CCP is a cage critical point encompassing N– or O–atoms from the acceptor and aryl C or N–atoms from the donor, is plotted with respect to k^a. $\nabla^2\rho(r_{CCP})$ tracks regions of local charge concentration/depletion [134]. $\nabla^2\rho(r_{CCP})$ increases with increasing strength of the π–hole interaction. In other words, increased local concentration of charge at the CCP corresponds to a stronger π–hole interaction. Figure 6b shows correlation between r of the HBs and their k^a values, where increased bond length corresponds to weakening of the HB. Figure 7 shows combined k^a values of all SBIs/HBs per π–hole system, plotted with respect to k^a of the π–hole interaction; where the larger quantity of stronger SBIs/HBs weakly correlate with stronger π–hole interactions. This correlation is weak because the HB can strengthen or weaken the π–hole, depending on the directionality of HB donation; a topic which is discussed further in Section 3.5.

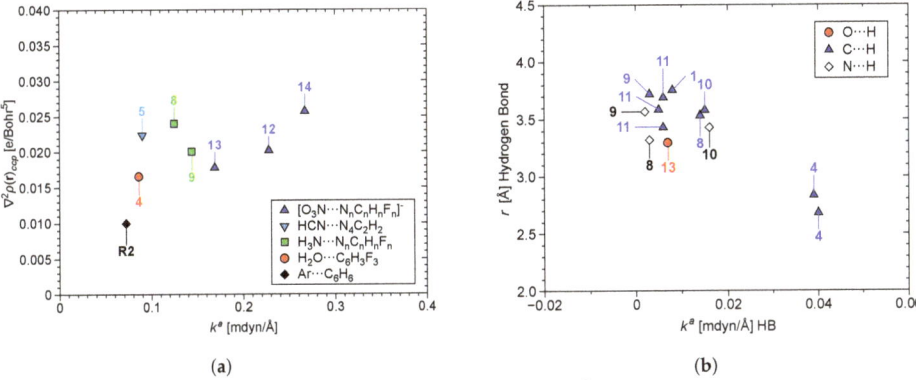

Figure 6. (a) The Laplacian of the electron density at the CCP ($\nabla^2\rho(\mathbf{r}_{CCP})$), and (b) r of the HBs; both plotted with respect to k^a of π–hole interactions in **1–14** and **R2**.

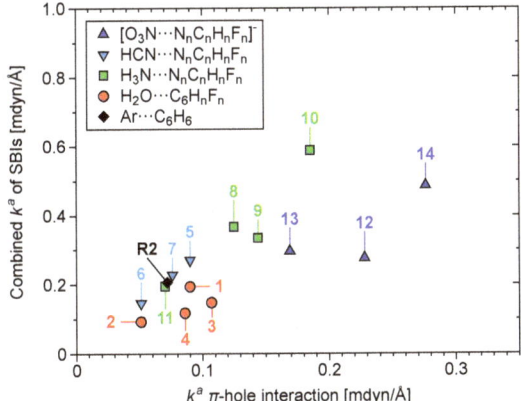

Figure 7. Combined k^a values of all SBIs (including HBs) plotted with respect to k^a of π–hole interactions in **1–14** and **R2**.

Table 4 summarizes LVM data for all HBs between donor/acceptor pairs. Figure 8 is a schematic of the atom labelling/numbering convention used in subsequent tables and figures. HBs were not found in systems **2–3, 5–7, 12,** and **14**. In **2**, the aryl C–atom bound to the lone H–substituent has a charge of -0.326 e, but the orientation of the water molecule eliminates the possibility of HB. The other five aryl C–atoms all carry positive charges with values between $+0.277$ e and $+0.361$ e. A bonding interaction between positive charges on aryl C–atoms and positive charges on acceptor H–atoms ($+0.469$ e) is not favored. Acceptor H–atoms in **3** do not form HBs because they are oriented such that they are not in plane with any of the aryl atoms or substituents and their distance from aryl C–atoms is maximized at the given conformation. It was expected that **5–7, 12,** and **14** would not form HBs for obvious reasons. Interestingly, **13** is the only example of HB between acceptor and aryl-substituents, where the three π–hole acceptor nitro O–atoms interact weakly with the three aryl H–atoms.

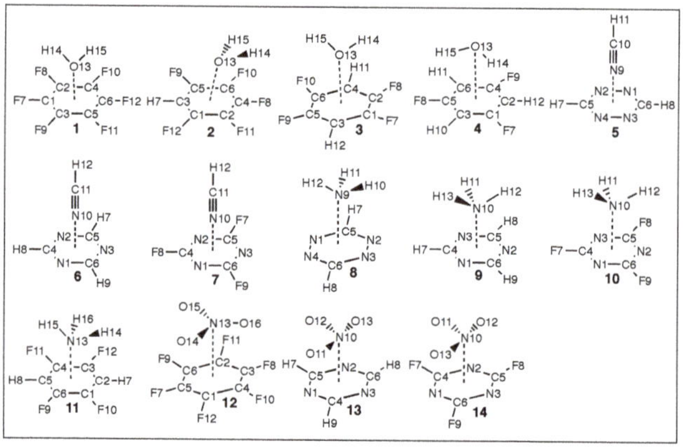

Figure 8. Schematic showing all atom numbers in **1–14**, for use as a reference to Tables 4 and 5.

Table 4. LVM analysis: r, k^a, and ω^a, for secondary bonding interactions SBI involving hydrogen atoms in **1, 4, 8–11**, and **13**.

#	Parameter	r	k^a	ω^a	Parameter	r	k^a	ω^a
1	H14⋯C1	3.755	0.008	124.2	H15⋯C6	3.755	0.008	124.2
4	H14⋯C1	2.834	0.039	267.5	H14⋯C4	2.834	0.039	267.5
	H14⋯C2	2.680	0.040	270.1	-	-	-	-
8	H10⋯C6	3.534	0.014	159.8	H11⋯N4	3.321	0.003	76.0
	H11⋯N1	3.321	0.003	76.0	H12⋯C5	3.534	0.014	159.8
9	H11⋯N3	3.567	0.002	60.6	H12⋯C6	3.721	0.003	72.8
	H11⋯C4	3.721	0.003	71.1	H13⋯N1	3.567	0.002	61.7
	H11⋯C5	3.721	0.003	70.4	H13⋯C4	3.721	0.003	72.3
	H12⋯N2	3.567	0.002	63.0	H13⋯C6	3.721	0.003	71.1
	H12⋯C5	3.721	0.003	73.3	-	-	-	-
10	H11⋯N3	3.431	0.016	169.2	H12⋯C6	3.580	0.015	163.3
	H11⋯C4	3.580	0.015	163.2	H13⋯N1	3.431	0.016	169.2
	H11⋯C5	3.580	0.015	163.3	H13⋯C4	3.580	0.015	163.1
	H12⋯N2	3.431	0.016	169.3	H13⋯C6	3.580	0.015	163.3
	H12⋯C5	3.580	0.015	163.1	-	-	-	-
11	H14⋯C1	3.432	0.006	108.4	H16⋯C2	3.585	0.005	91.9
	H14⋯C2	3.585	0.005	91.9	H16⋯C3	3.432	0.006	108.4
	H14⋯C6	3.691	0.005	91.0	H16⋯C4	3.691	0.005	91.0
13	O11⋯H9	3.297	0.007	108.8	O13⋯H8	3.297	0.007	108.6
	O12⋯H7	3.297	0.007	108.3	-	-	-	-

Units for LVM data are given as follows: r in Å, k^a in mdyn/Å, and ω^a in cm^{-1}.

Table 5 summarizes LVM data for all SBIs found in **1–14**, excluding HBs; where 13 of the 14 π–hole systems contains at as few as six non–HB SBIs (systems **1–3**, **5–11**, and **14**), 12 non-HB SBIs (**12**), and as many as 15 (system **13**) non–HB SBIs of the following type: C⋯O, C⋯N, or N⋯N; where the first atom listed (C/N) is from the π–hole donor and the second atom (O/N) is from the acceptor (donor⋯acceptor). In most cases, there is a LVM between the π–hole acceptor and all six atoms of the aryl ring; with **4** being the exception. The remainder of this section is divided into four subsections pertaining to significant factors for modulation of molecular geometry, bond strength, and the intrinsic nature of the π–hole interactions: (3.3) Aryl Substituent Effects, (3.4) Nature of the Aryl Rings, (3.5) Secondary Bonding Interactions and (3.6) Characterization of Normal Vibrational Modes.

Table 5. Summary of LVM data: r, k^a, and ω^a for secondary bonding interactions SBI not including hydrogen atoms, for **1–14**.

#	Parameter	r	k^a	ω^a	Parameter	r	k^a	ω^a
R2	Ar13···C1	3.877	0.031	76.0	Ar13···C4	3.877	0.031	76.0
	Ar13···C2	3.877	0.036	80.8	Ar13···C5	3.877	0.036	80.8
	Ar13···C3	3.877	0.036	80.8	Ar13···C6	3.877	0.036	80.8
1	O13···C1	3.414	0.027	81.5	O13···C4	3.415	0.031	87.7
	O13···C2	3.415	0.031	87.7	O13···C5	3.415	0.031	87.7
	O13···C3	3.415	0.031	87.7	O13···C6	3.414	0.027	81.5
2	O13···C1	3.551	0.012	53.6	O13···C4	3.320	0.027	81.8
	O13···C2	3.402	0.015	61.2	O13···C5	3.551	0.012	53.6
	O13···C3	3.637	0.012	54.7	O13···C6	3.402	0.015	61.2
3	O13···C1	3.515	0.022	74.4	O13···C4	3.515	0.022	74.4
	O13···C2	3.506	0.026	80.1	O13···C5	3.506	0.026	80.1
	O13···C3	3.506	0.026	80.1	O13···C6	3.506	0.026	80.1
5	N9···C1	3.342	0.044	102.9	N9···C4	3.342	0.044	102.9
	N9···C2	3.342	0.044	102.9	N9···C5	3.279	0.047	111.6
	N9···C3	3.342	0.044	102.9	N9···C6	3.279	0.047	111.6
6	N10···N1	3.437	0.024	75.9	N10···C4	3.405	0.025	80.6
	N10···N2	3.437	0.024	75.8	N10···C5	3.405	0.025	80.5
	N10···N3	3.437	0.024	76.6	N10···C6	3.405	0.024	80.2
7	N10···N1	3.288	0.038	95.8	N10···C4	3.240	0.038	100.1
	N10···N2	3.288	0.038	95.7	N10···C5	3.240	0.038	100.0
	N10···N3	3.288	0.038	95.7	N10···C6	3.240	0.038	100.1
8	N9···N1	3.316	0.059	119.5	N9···N4	3.316	0.059	119.5
	N9···N2	3.398	0.045	104.0	N9···C5	3.290	0.063	128.1
	N9···N3	3.398	0.045	104.0	N9···C6	3.290	0.063	128.1
9	N10···N1	3.452	0.049	108.5	N10···C4	3.419	0.056	121.2
	N10···N2	3.452	0.049	109.1	N10···C5	3.419	0.055	120.2
	N10···N3	3.452	0.048	108.0	N10···C6	3.419	0.054	119.4
10	N10···N1	3.222	0.074	133.8	N10···C4	3.273	0.076	140.9
	N10···N2	3.222	0.074	133.9	N10···C5	3.273	0.076	140.9
	N10···N3	3.222	0.074	133.8	N10···C6	3.273	0.076	140.9
11	N13···C1	3.557	0.021	74.6	N13···C4	3.587	0.022	76.3
	N13···C2	3.552	0.036	97.3	N13···C5	3.611	0.041	103.7
	N13···C3	3.557	0.021	74.6	N13···C6	3.587	0.022	76.3
12	N13···C1	3.375	0.031	90.6	O14···C1	3.149	0.015	61.2
	N13···C2	3.375	0.031	90.6	O14···C5	3.149	0.015	61.2
	N13···C3	3.375	0.031	90.6	O15···C2	3.149	0.015	61.2
	N13···C4	3.375	0.031	90.6	O15···C6	3.149	0.015	61.2
	N13···C5	3.375	0.031	90.6	O16···C3	3.149	0.015	61.2
	N13···C6	3.375	0.031	90.6	O16···C4	3.149	0.015	61.2
13	N10···N1	3.419	0.022	72.3	O11···C4	3.117	0.017	64.1
	N10···N2	3.419	0.022	72.4	O12···N1	3.394	0.016	59.7
	N10···N3	3.419	0.021	72.0	O12···N2	3.394	0.016	59.7
	N10···C4	3.375	0.021	73.8	O12···C5	3.117	0.017	64.0
	N10···C5	3.375	0.021	73.5	O13···N2	3.394	0.016	59.6
	N10···C6	3.375	0.021	73.7	O13···N3	3.394	0.016	59.5
	O11···N1	3.394	0.016	59.6	O13···C6	3.117	0.017	64.1
	O11···N3	3.394	0.016	59.4	-	-	-	-
14	N10···C4	3.205	0.134	187.7	O11···C4	2.936	0.028	83.6
	N10···C5	3.205	0.134	187.7	O11···C5	2.936	0.028	83.6
	N10···C6	3.205	0.134	187.7	O11···C6	2.936	0.028	83.6

Units for computational data are given as follows: r in Å, k^a in mdyn/Å, and ω^a in cm^{-1}.

3.3. Aryl Substituent Effects

Systems **1–4** are a good starting point to systematically analyze substituent effects. The donor in **1** is C_6F_6, the donor in **2** is C_6F_5H, $C_6F_4H_2$ in **3**, and $C_6F_3H_3$ in **4**. One effect is the physical response of acceptor to decreasing the number of aryl F-substituents. Each of the four water molecules in **1–4** is oriented quite differently from one another with respect to the aryl ring. **1** has C_{2v} symmetry, with the acceptor H–atoms pointing opposite the aryl ring. Each atom of the water molecule rests in plane with two aryl C–F groups positioned para to each other. Unexpectedly, the π–hole interaction in **1** is not particularly strong ($k^a = 0.090$ mdyn/Å) compared to the rest of H_2O acceptor group **2–4**, systems **5–14**, and even **R1** and **R2**. The six aryl F–substituents induce a sizable π–hole with large positive ESP which therefore should promote stronger π–hole interactions, but this effect is countered by a lack of cooperation between atoms of the aryl donor and atoms of the H_2O acceptor in forming SBIs [135]. Furthermore, there are two weak C$\cdots$H donor/acceptor SBIs ($k^a = 0.008$ mdyn/Å) in **1** (see Table 4).

Compared to **1**, the acceptor H–atoms in **2** are rotated nearly 90° to avoid repulsive forces from the donor H–atom. The aryl C–atom bound to H has a negative charge of -0.326 e; whereas the aryl–C atom para to the lone C–H bond has a charge of $+0.317$ e. In contrast, all C–atoms in **1** have positive charges (see Figures 3 and 8) of $+0.295$ e (C2 through C5) and $+0.291$ e (C1 and C6). The negative charge on the C–atom in **2** repels the electron rich acceptor O–atom toward the opposite end of the ring, resulting in the π–hole interaction distance increasing 0.072 Å compared to **1**. The O–atom is also no longer directly over the π–hole, which decreases orbital overlap. Instead, the O–atom is 0.407 Å closer to the C–atom para to C–H. Furthermore, the π–hole should migrate closer to the C–F group, and become weaker its ESP becomes more negative. The cumulative effect is that substitution of a single aryl F–atom for H disrupts the molecular symmetry, hinders the reactivity of the π–hole, and decreases k^a of the π–hole interaction in **2** by nearly 50% compared to **1**; the π–hole interaction in **2** is the weakest of the H_2O acceptor systems **1–4**. Although system **2** is an extreme case, where the other aryl rings of **1**, **3–4** are significantly more symmetric, there is clear indication that substituent effects involving the aryl ring can significantly weaken/strengthen the π–hole interaction and drastically alter the molecular geometry of the system. The strongest π–hole interaction among systems **1–4** occurs in **3**, which has C_{2v} symmetry with acceptor H–atoms still oriented away from the aryl ring. The water molecule forms a plane perpendicular to the two FC=CF bonds of the donor. The water O–H bonds in **3** ($k^a = 8.549$ mdyn/Å), are stronger than the O–H bonds in **1**, **2**, and **4** (between 8.532 and 8.547 mdyn/Å). This increase in O–H bond strength has a net stabilizing effect on the whole system, which extends to the π–hole interaction. The orientations of H_2O and the aryl F–substituents also benefit the π–hole interaction in system **3**; as any possible repulsive forces between the donor/accepter occur over maximum distances compared to **1**, **2**, and **4**, and the position of the π–hole is not affected due to the symmetry of the $C_6F_4H_2$ ring.

4 has 3 aryl F–substituents and 3 H–substituents, arranged symmetrically in an alternating pattern. Addition of the third H–substituent resulted in inversion of the acceptor H–atoms, which now point toward the aryl ring. H_2O is coplanar with para aryl C–H and C–F groups (C2–H12 and C5–F8). The acceptor atom H14 points downward toward C2, which is caused by the charge of -0.382 e on C2. The opposite occurs between acceptor–H15/donor–C5, where positive charges on each atom are repulsive. The H15$\cdots$C5 distance is 0.388 Å longer than H14$\cdots$C2 as a result. The π–hole interaction in **4** is only slightly weaker than **1** (by 0.004 mdyn/Å), an unexpected result based on substituent effects alone; as the O acceptor in **1** should interact much more strongly with its π–hole. However, other factors must be considered. The acceptor H14 in **4** can HB with the negatively charged C2 donor atom, yet is in close enough proximity to bind to C4. A third HB was found in **4**; all three HBs are of the C$\cdots$H–O type, and are among the strongest HBs in systems **1–14** (see Table 4). This factor is discussed with more detail in Section 3.5. Perhaps the most surprising substituent effect (or lack there of) is an absence of intermolecular interactions involving aryl–substituents and the π–hole acceptors. There is only one such SBI; it is in system **13** and is a O$\cdots$H–C type HB. This interaction is discussed further in Section 3.5.

3.4. Nature of the Aryl Rings

In **5–7**, the influence of SBIs is minimized, and each aryl ring contains four, three, and three N-atoms, respectively. **5** has C_{2v} symmetry while **6** and **7** have C_{3v} symmetry. N-substitution, atomic nature of the aryl ring, and three-fold symmetric F-substitution do not cause significant symmetry related changes in this case. However, it turns out that **5–7** have the weakest π–hole interaction strength on average compared to the H_2O acceptor (**1–4**), NH_3 acceptor (**8–11**), and NO_3^- acceptor (**12–14**) π–hole systems. One key difference between systems **5–7** and the systems just mentioned is the orientation and nature of the HCN acceptor in **5–7**, where the N-atom points downward toward the π–hole and the H-atom points in the opposite direction. This eliminates the possibility HB donation and decreases the overall possibility of SBIs. System **6** has one less aryl N-atom compared to **5**. The π–hole interaction in **5** is 0.107 Å shorter and has a k^a value nearly two times larger than the π–hole interaction in **6**. Incorporating N-atoms into the aryl ring appears to influence strength of the π–hole interaction more than F-substitution. The difference between donors of **6** and **7**, is a three fold F-substitution in the latter, which increases strength of the π–hole interaction by 0.024 mdyn/Å. Though significant, triple F-substitution is not able to modulate strength of the π–hole interaction as much as insertion/removal of aryl donor N-atoms, as is the case for **5** and **6**. Integration of a fourth N-atom to the aryl ring nearly doubles k^a of the interaction; whereas substituting three C–H groups for three C–F groups achieves an increase in interaction strength by approximately 50%. The NBO picture suggests the $N_4C_2H_2$ donor of **5** supports a more delocalized electronic density compared with the $N_3C_3H_3$ donor in **6** and $N_3C_3F_3$ in **7**. There is a CT of −30.99 milli-electrons (m_e) from the acceptor N-atom to the aryl ring in **5**, approximately −10 m_e more than in **6** but roughly −15 m_e less than CT in **7**. In addition to being the weakest interactions and not participating in HB, **5–7** also have the three largest CT values among **1–14**. Correlation between CT and k^a for **1–14** is very weak, but the general trend is that the π–hole interactions are stronger when CT gets closer to zero (see Figure 5a).

8–14 are not ideal for investigating how the nature of the aryl ring influences the π–hole interaction, given that the acceptors in these systems are ammonia and the nitrate anion. Each of the three acceptor H–/O–atoms are able to form SBIs, which make it difficult to assess both substituent effects and how addition of N-atoms into the aryl ring can influence the π–hole interaction. This is also evident in the case of **1–7**, where the acceptors each have one less atom than the acceptors in **8–14**. On the other hand, this makes **8–14** ideal for studying the effect of SBIs on π–hole interactions.

3.5. Secondary Bonding Interactions

Of all systems **1–14**, **4** is the only π–hole system completely void of SBIs between a non-hydrogen acceptor and non-hydrogen donor. However, the unusual orientation of H_2O in **4** puts H14 in close proximity to the C1−C2−C4 region of the aryl ring (see Figure 1 and Tables 4 and 5); where H14···C1, H14···C2, and H14···C4 lengths are 2.834, 2.680, and 2.834 Å, respectively. H14 interacts with all three aryl C-atoms, and the resultant HBs are among the strongest found in **1–14**. The effect of these HBs is stabilization and increased strength of the π–hole interaction by 0.035 mdyn/Å compared to system **2**, where the acceptor O-atom forms SBIs with the donor aryl C-atoms. **2** and **3** do not have any HBs, but all of their aryl C-atoms interact with the acceptor O-atom. **2** has the weakest π–hole interaction of **1–14**, which is largely due to the nature of the donor and the arrangement of the acceptor water molecule, but a contributing effect is that five of the six C···O interactions are among the weakest for **1–14** (see Table 5). The HBs in system **1**, are nearly the weakest (0.008 mdyn/Å) interactions found in **1–14**. Although these HBs do stabilize and increase strength of the π–hole interaction, stronger HBs will promote stronger π–hole interactions [136]. For example, system **4** contains the same type of C···H HB found in system **1**, but the k^a values of HBs in the former are 5 times larger than in the latter. Also, system **4** has three HBs while system **1** has two. The π–hole in **1** should be larger and have more positive ESP compared to **4**, due to the nature of the aryl substituents (six F-atoms in **1**, three F- and three H-atoms in **4**). In addition, the lp is more accessible in **1**, compared to **4** where the H-atoms point toward the aryl ring. Also, O13 is 0.238 Å closer to the π–hole in system **1**. Despite all of these factors, k^a of the π–hole interaction in **4** is within 0.004 mdyn/Å of the π–hole interaction in **1**. This a result of the comparatively strong HBs in system **4** providing stability and increasing strength of the π–hole interaction.

Although **5–7** do not provide information on the effect HBs have on the π–hole interaction, a clear picture emerges in terms of the role other SBIs play. In terms of strength of the π–hole interaction, the sequence is: **5 > 7 > 6**. This matches the trend in k^a of the N···N and C···N donor/acceptor SBIs. For C···N in **5**, $k^a = 0.044$ and 0.047 mdyn/Å, for N···N and C···N in **6**, $k^a = 0.024$ and 0.025 mdyn/Å, and for N···N/C···N in **7**, $k^a = 0.038$ mdyn/Å. This indicates that non-HB SBIs may play a cooperative role, where they help to strengthen π–hole interactions. However, HBs seem to have a more significant effect on the π–hole interaction comparatively.

8–11 are the only π–hole acceptors where each system participates in HB and SBIs between aryl C or N atoms and the central acceptor N-atom. In terms of π–hole interaction strength, the sequence is: **10 > 9 > 8 > 11**. The π–hole interaction in **8** should be stronger than **9** based on the nature of the aryl donor, but the ammonia H–atoms in **8** are staggered such that they are centered above the bonds encompassing the ring. H12 is oriented above an N–N bond, and the other two ammonia H–atoms orient above two of the aromatic C–N bonds. This results in **8** having fewer HBs compared to **9–11**. The N···N and C···N donor/acceptor interactions in **8** are slightly shorter and slightly stronger than comparable interactions in **9**. The same type of inter-monomer N···N and C···N interactions in system **10** are the strongest amongst the NH_3–acceptor group by at least 0.026 mdyn/Å for N···N and at least 0.055 mdyn/Å for C···N. Strength of the non–HB SBIs trends similarly to the π–hole interaction strength order: **10 > 9 ≈ 8 > 11**. This not the case with individual HB strength. However, when HB strength is considered as a sum of each individual HB per π–hole system, the collective HB strength matches the trend of π–hole interaction strength. HBs are effecting the system compared to the non-HB SBIs. **11** has the weakest π–hole interaction, the weakest collective HB strength, the weakest non–HB SBIs, the fewest N–aromatic atoms (zero), and the most F–substituents of **8–11**. Although SBIs are predominant in **8–11**, it turns out that N–aromatic atoms still play a major role in modulating bond strength; with N–aromatic systems having π–hole interaction k^a values increase 100% compared to the species with a C_6 ring. Though even less significant than the aforementioned, effects of F-substitution are again apparent when comparing systems **9** and **10**; where the F–substituents result in a 28% increase in k^a values.

12–14 are the only anion π–hole systems investigated in this work, and as expected, occupy the strong end of π–hole interaction spectrum. In **12**, NO_3^- has a staggered conformation with respect to the C_6F_6 ring which puts the three acceptor O–atoms are at maximal distances from all C and F donor–atoms. Of course, there is no possibility of HBs in **12**, but the negatively charged O–atoms interact with the positively charged aryl C–atoms. The acceptor N–atom also interacts with the aryl C–atoms. This is possible because N has a lone pair and NO_3^- has an excess of delocalized electrons. The π–hole interaction in **13** is substantially weaker than the interactions in **12** and **14**. Regardless, **12–14** have the three strongest π–hole interactions among **1–14**, while they have the weakest and fewest number of HBs among each acceptor group. All three acceptor O–atoms in **13** HB with the three H–substituents on the aryl ring. These are the only three HBs where the π–hole acceptor is also the HB acceptor. In every other case, the directionality of acceptor/donor in the HB is the reverse direction of the π–hole interaction. Because the HB donor in **13** is the aryl C−H, electronic density is transferred to the aryl ring and throughout the π–system [137]. This transfer will cause an increase in negative ESP at the π–hole, which in turn weakens the π–hole interaction. When the HB donor/acceptor roles are reversed, electronic density transfers from the aryl ring to the acceptor molecule. The depletion of electronic density from the aryl ring increases the positive ESP at the π–hole, and since the acceptor molecule now has more electronic density, it becomes a better electron donor. When the aryl ring is the HB acceptor, charge on the atoms involved will increase in the positive direction and negative charge is leaving the π–system. This explains why each HB except for the O···H−C interactions in

13 help increase the strength of the π–hole interaction; whereas **13** has a substantially weaker π–hole interaction compared to **12** and **14** which do not have any HBs.

3.6. Characterization of Normal Modes

In addition to providing k^a and ω^a and related local vibrational mode properties [80], the local mode analysis has led to a new way of analyzing vibrational spectra. The characterization of normal modes (CNM) procedure decomposes each normal vibrational mode into local mode contributions for a non-redundant set of LVMs by calculating the overlap between each local mode vector with this normal mode vector [77–79,82]. In this way, the character of each normal mode can be uniquely assessed [68,80,138]. In this work we performed a CNM decomposition for **R2** and **1** comparing in particular the contribution of the local vibrational π–hole-interaction mode to the lower frequency normal modes in both complexes. The corresponding decomposition plots are shown in Figures 9 and 10.

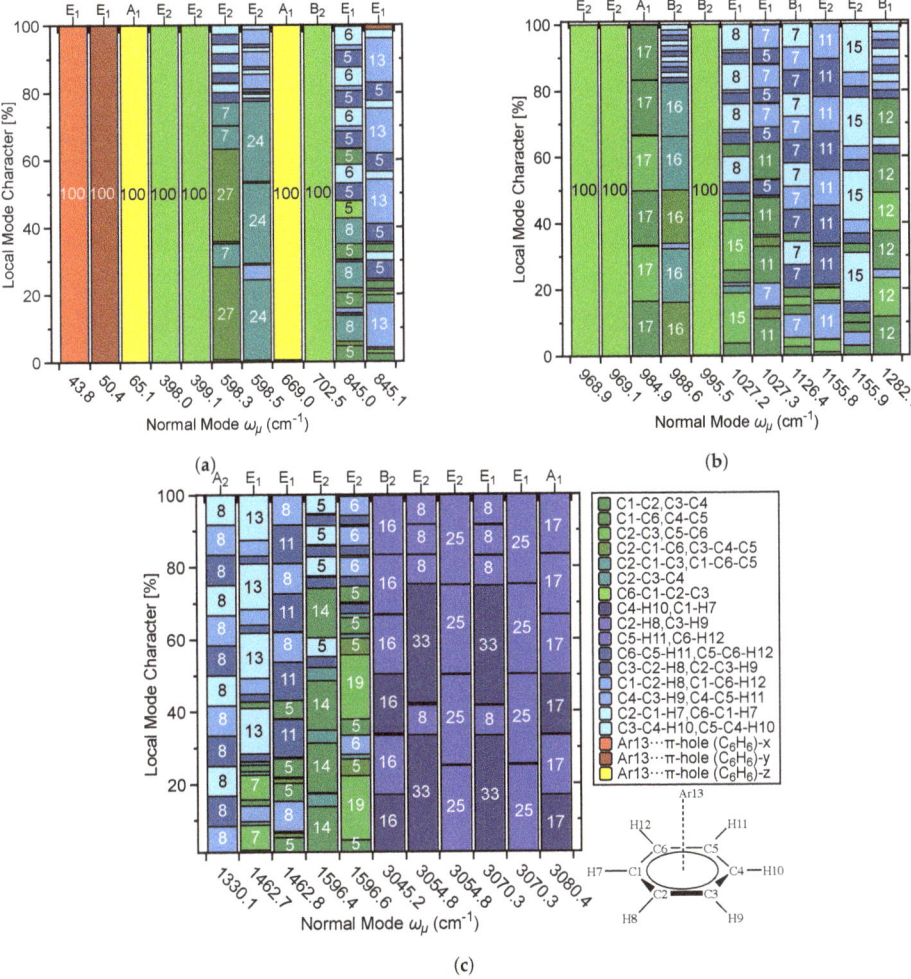

Figure 9. Decomposition of normal vibrational modes into % LVM contributions for the Ar···C$_6$H$_6$ dimer **R2**; (**a**) % LVM contributions to normal vibrational modes 1–11, (**b**) % LVM contributions to normal vibrational modes 12–22, and (**c**) % LVM contributions to normal vibrational modes 23–33.

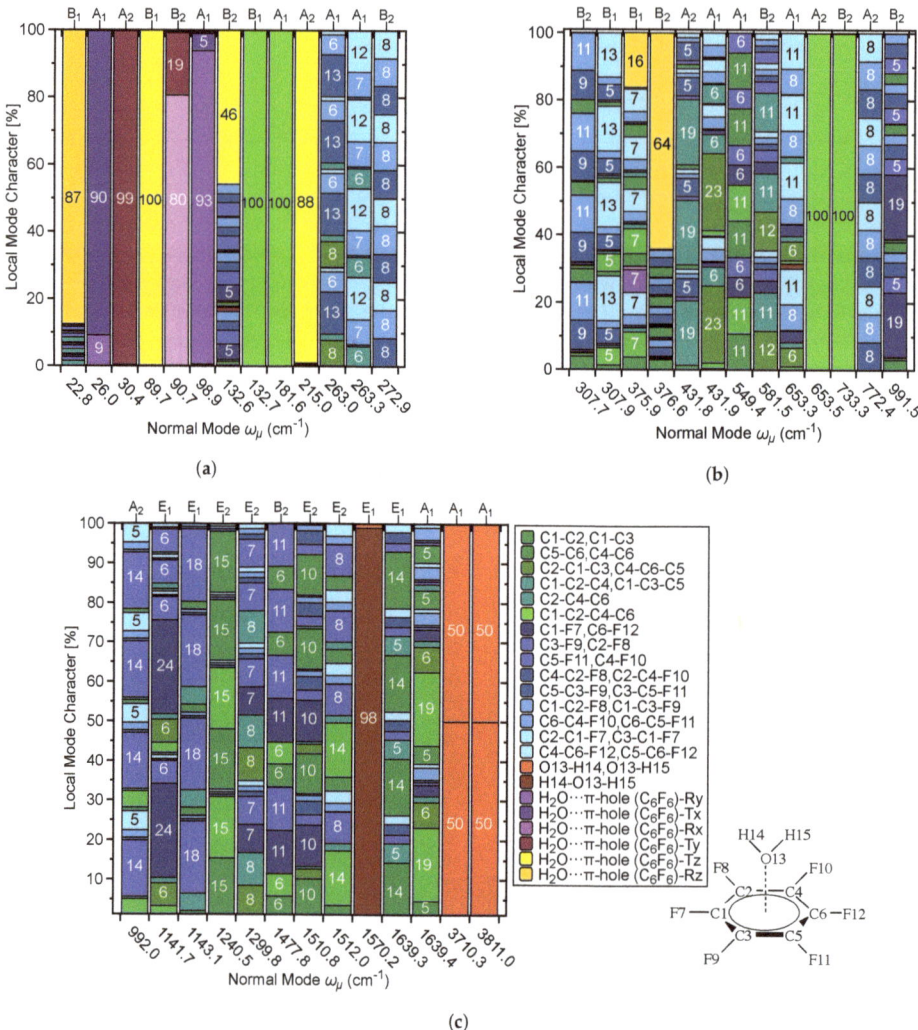

Figure 10. Decomposition of normal vibrational modes into % LVM contributions for the $H_2O \cdots C_6F_6$ π-hole system **1**; (**a**) % LVM contributions to normal vibrational modes 1–13, (**b**) % LVM contributions to normal vibrational modes 14–26, and (**c**) % LVM contributions to normal vibrational modes 27–39.

The set of local modes used for this purpose was chosen to include all inter-monomer local modes. As shown in Figure S1 of the Supplementary Materials there are 9 possible inter-monomer modes, 3 stretching motions (x, y, z direction), 3 rotations (x, y, z direction), and 3 anti-rotations (x, y, z direction). 6 of them are needed to define the set of inter-monomer modes. We generally use the 3 stretching motions labelled T_x, T_y, and T_z and 3 rotations R_x, R_y, R_z in the following. **R2** is a special case with one monomer being an atom reducing the number of inter-monomer modes to 3 translational modes (labelled as x, y, z).

3.6.1. Normal Modes Related to the π–Hole Interaction

In Figure 9a–c, normal modes ω_μ are decomposed into % LVM contributions for complex **R2**, and the corresponding CNM plots for complex **1** are given in Figure 10a–c. Figure 9a shows CNM for ω_μ 1–11 ($\omega_{\mu 1}$ through $\omega_{\mu 11}$) into % LVM contributions for **R2**, where 'x,' 'y,' and 'z' in the Figure legends denote translations of the Ar-atom in the x-, y-, and z-directions with respect to the benzene molecule, as described above. In standard orientation, the Ar$\cdots\pi$-hole interaction is in the z-direction. Correspondingly, the z-component of the three inter-monomer Ar$\cdots\pi$-hole LVM parameters represents the direct Ar$\cdots\pi$-hole interaction, i.e., this is the mode which corresponds to the special force constant k^a. It is shown in yellow color in the CMN plots in Figure 10a–c for quick reference.

For **R2**, $\omega_{\mu 1}$ through $\omega_{\mu 3}$ and $\omega_{\mu 8}$ are all 100% LVM character corresponding to inter-monomer vibrations, where $\omega_{\mu 3}$ at 65.1 cm^{-1} and $\omega_{\mu 8}$ at 669.0 cm^{-1} are dominated by the π-hole interaction in the z-direction, representing the stretching and contraction the argon atom with regard to the center of the benzene ring. $\omega_{\mu 3}$ is characterized by the translational motion of the Ar–atom perpendicular to the plane of the benzene ring, and $\omega_{\mu 8}$ represents wagging of the six benzene H–atoms towards and away from the Ar–atom. This mode also perturbs slightly the benzene C–atoms. Movies of the $\omega_{\mu 3}$ and $\omega_{\mu 8}$ vibrational modes are shown in the Supplementary Materials, see Table S1 for description. Collectively, it is all of the vibrations associated with $\omega_{\mu 3}$ and $\omega_{\mu 8}$ which are required to accurately describe the π-hole interaction. These findings clearly emphasize that the special force constant k^a as defined in this work is meaningful.

System **1** consists of 39 ω_μ and 39 LVMs, including six parameters describing the inter-monomer translations T_x, T_y, and T_z and the inter-monomer rotations R_x, R_y, R_z, introduced above, see also legend in Figure 10a–c. As with CNM for **R2**, the direct π-hole interaction occurs in the z-direction; therefore the z-components of the inter-monomer LVMs are of particular interest and and are represented with a light yellow bar for T_z in Figure 10a,b- and with the darker yellow color for R_z.

The z-components of the inter-monomer LVMs contribute to six of the normal modes: $\omega_{\mu 1}$ (22.8 cm^{-1}), $\omega_{\mu 4}$ (89.7 cm^{-1}), $\omega_{\mu 7}$ (132.6 cm^{-1}), $\omega_{\mu 10}$ (215.0 cm^{-1}), $\omega_{\mu 16}$ (375.9 cm^{-1}), and $\omega_{\mu 17}$ (376.6 cm^{-1}). The R_z LVM composes 87% of $\omega_{\mu 1}$, with C−F/C−C LVMs accounting for the remaining 13% (Figure 10a). The R_z component is much less significant for the π-hole interaction, but mixing of C-C, C-F, and R_z contributions to $\omega_{\mu 1}$ imply this mode likely relates more to the HBs found in **1**. The motion of $\omega_{\mu 1}$ involves rotation of acceptor H–atoms about the O–atom, parallel to the plane of the donor ring. Description of $\omega_{\mu 4}$ in **1** is comparable to $\omega_{\mu 3}$ of **R2**, which is translation of the acceptor molecule in the z-direction. This normal mode is of 100% T_z character. Therefore, it could be used in experimental spectra as quick identification of the π-hole interaction. The motion of the water molecule in $\omega_{\mu 4}$ perturbs the C_6F_6 slightly; whereas this does not occur in **R2**. The frequency of the former is also larger than the latter by 24.6 cm^{-1}.

The $\omega_{\mu 7}$ contains 46 % T_z character combined with small contributions from the C–F, C–C–F, and H–O–H LVMs. This mode is comparable to $\omega_{\mu 8}$ of **R2**: the z-direction wagging of the aryl substituents. As with the previous comparison, the wagging motion of the aryl–F atoms perturbs the acceptor molecule and the aryl C–atoms in **1**; this not the case for **R2**. On the other hand, similarly to **R2**, this mode is also important for a full description of the π-hole interaction. T_z accounts for 88% of $\omega_{\mu 10}$ and describes the translation of the aryl C–atoms in the z–direction. This mode strongly perturbs the acceptor H_2O molecule, and there is no comparable mode to this in **R2**. The $\omega_{\mu 10}$ is also a main component of the π-hole interaction, as the six aryl C–atoms move in phase and therefore translate the π-hole directly toward the acceptor. The $\omega_{\mu 16}$ and $\omega_{\mu 17}$ represent z-rotation of the four equivalent aryl C–atoms and z-rotation of the two equivalent aryl C–atoms, respectively. R_z compose 16% of $\omega_{\mu 16}$ and 64% of $\omega_{\mu 17}$. These modes do not strongly effect π-hole interactions. However, $\omega_{\mu 17}$ is related to the HBs between acceptor/donor, where the HB acceptor C–atoms rotate in the direction of the water H–atoms. This explains why the contribution from R_z is much larger for $\omega_{\mu 17}$. As was previously mentioned, the HBs with directionality opposite to that of the π-hole interaction effectively stabilize and increase strength of the π-hole interaction. This relationship is reflected in the CNM analysis. It is evident that the π-hole interaction in **1** is stronger than in **R2** based on the CNM because the inter-monomer LVMs T_z and R_z of **1** compose more of ω_μ, the comparable frequencies are larger tahnin **1**, and the vibrational modes are much more strongly coupled between monomers in **1**. Movies of the $\omega_{\mu 1}$, $\omega_{\mu 4}$, $\omega_{\mu 7}$, $\omega_{\mu 10}$, and $\omega_{\mu 17}$ vibrational modes are shown in the Supplementary Materials, see Table S1 for description.

3.6.2. Normal Modes Not Related to the π–Hole Interaction

The C6−C1−C2−C3 dihedral mode is the only LVM contributor to $\omega_{\mu 4}$ through $\omega_{\mu 5}$ and $\omega_{\mu 9}$. Modes 6–7 are composed of mainly angular C−C−C contributions with small components of the various C−C−H contributions. Mode 10 at 845.0 cm^{-1} consists of a nearly even mixture of C−C−C and C−C−H LVM contributions with small (5%) contribution from C6−C1−C2−C3; whereas mode 11 is C−C−H dominant with minor C−C−C, and C−H character. Figure 9b shows decomposition of ω_μ into % LVM contributions for $\omega_{\mu 12}$ through $\omega_{\mu 22}$ in **R2**. Again, the C6−C1−C2−C3 dihedral is the sole contribution to $\omega_{\mu 12}$ through $\omega_{\mu 13}$ and $\omega_{\mu 16}$. The six C−C LVMs compose $\omega_{\mu 14}$ and are the largest components of $\omega_{\mu 18}$ and $\omega_{\mu 22}$, with minor components being the C−C−H LVMs. C−C−C LVMs compose 80% of $\omega_{\mu 15}$ with C−C−H contributions accounting for the remaining 20%. At 1126.4 cm^{-1} through 1155.9 cm^{-1}, $\omega_{\mu 19}$ through $\omega_{\mu 21}$ are C−C−H LVM dominant with small contributions. In Figure 9c, the remaining ω_μ (23–33) are decomposed into % LVM contributions for **R2**. C−C−H LVMs are the major contributions to $\omega_{\mu 23}$ through $\omega_{\mu 25}$: 100% of $\omega_{\mu 23}$ at 1330.1 cm^{-1}, nearly 80% of $\omega_{\mu 24}$ at 1462.7 cm^{-1}, and 76% of $\omega_{\mu 25}$ at 1462.8 cm^{-1}. C−C LVM contributions steadily increase from $\omega_{\mu 24}$ through $\omega_{\mu 27}$ (1462.7 cm^{-1} through 1596.9 cm^{-1}), where % LVM contributions increase from 20 % of the former to nearly 70 % of the latter. The six highest ω_μ (28–33) span 3045.2 cm^{-1} through 3080.4 cm^{-1} and are composed entirely of C−H LVMs.

$\omega_{\mu 7}$ (132.6 cm^{-1}) and $\omega_{\mu 11}$ through $\omega_{\mu 15}$ (263.0 cm^{-1} through 307.9 cm^{-1}; see Figure 10a,b) consist mainly of C−F and C−C−F LVM contributions, with minor contributions from the $H_2O\cdots\pi$–hole interaction at 132.6 cm^{-1} and C−C/C−C−C LVMs at 263.0 cm^{-1} through 307.9 cm^{-1}. $\omega_{\mu 8}$, $\omega_{\mu 9}$, $\omega_{\mu 23}$, and $\omega_{\mu 24}$ are 100% C1−C2−C4−C6 character. $\omega_{\mu 14}$ and $\omega_{\mu 15}$ once again have LVM contributions from the $H_2O\cdots\pi$–hole interaction of 23% and 64%, respectively. From 431.8 cm^{-1} to 581.5 cm^{-1}, C−C and C−C−C LVMs are the major contributions, with C−F and C−C−F LVMs being minor components. After the C1−C2−C4−C6 modes at 653.5 cm^{-1} through 733.3 cm^{-1}, $\omega_{\mu 25}$ is completely C−C−F character and $\omega_{\mu 26}$ is largely C−F character with small C−C−F and C−C contributions. Figure 10c shows $\omega_{\mu 27}$ through $\omega_{\mu 39}$ for **1**, spanning 992.0 cm^{-1} through 3811.0 cm^{-1}. The 992.0 cm^{-1} through 1143.1 cm^{-1} region is C−F and C−C−F dominant, while $\omega_{\mu 31}$ at 1240 cm^{-1}, $\omega_{\mu 36}$ at 1639.3 cm^{-1}, and $\omega_{\mu 37}$ at 1639.4 cm^{-1} are mainly of C−C and C−C−C character. From 1299.8 cm^{-1} through 1512.0 cm^{-1} a mixture of C−F, C−C−F, C−C, and C−C−C LVMs compose the ω_μ. At 1570.2 cm^{-1} is the H_2O bending mode, and the H_2O stretching modes are at 3710.3 cm^{-1} (symmetric) and 3811.0 cm^{-1} (asymmetric). Overall, this discussion shows that the CMN feature offered by the local mode analysis provides a powerful tool for the detailed analysis of a vibrational spectrum.

4. Conclusions

In this work, the LVM analysis of Konkoli and Cremer was utilized to quantify strength of π–hole interactions in terms of a special local force constant k^a. This is the first work to quantify π–hole interactions in terms other than distance parameters r and binding/dissociation energies. Given the fact that the aforementioned parameters are not reliable descriptors of bond strength, our results provide a much needed perspective on the matter. In addition to quantification of π–hole interaction strength in terms of k^a, this work confirms an interplay between three key factors which can influence bond strength and can be insightful for the design of materials with specific properties. The three main factors influencing π–hole interaction strength in systems **1–14** are as follows: (1) aryl-substituent effects; where F–substituents polarization of aryl C–atoms which will encourage or discourage interactions between acceptor ligands and the aryl ring. Since these effects indirectly influence the π–hole interaction by affecting the nature of the aryl ring, aryl substituent effects are the least significant of the three effects; (2) the nature of the atoms which form the aryl ring, where presence of nitrogen can substantially increase strength of the π–hole interaction, where the more N the better; and (3) Presence of HBs and SBIs between π–hole acceptor/donor, where strength of the SBI correlates positively with strength of the π–hole interaction. HBs can have a substantial effect on strength of the π–hole interaction, depending on the directionality; where if the π–hole donor is the HB acceptor, strength of the π–hole interaction increases. Conversely, if HB donation is in the same direction as π–hole donation, the π–hole interaction will be weakened substantially. Future goals are to refine computational ω_μ harmonic scaling factors, and to expand this research on aryl π–hole interactions to a large number of systems, including halogen anions, CO, and OCH_3^- as acceptors.

Supplementary Materials: The following are available online at http://www.mdpi.com/2073-4352/10/7/556/s1, Figure S1: Description of the 9 possible local modes between the monomers of a dimeric system; Table S1: Description of the videos showing selected normal mode vibrations for **R2** and system **1**. The videos are uploaded as separate files; Table S2: Cartesian atomic coordinates of optimized equilibrium geometries for all model species.

Author Contributions: Conceptualization, S.Y., M.F., and E.K.; methodology, S.Y., M.F., Y.T., W.Z., and E.K.; validation, S.Y., M.F., Y.T., and E.K.; programming, W.Z.; formal analysis, S.Y.; investigation, S.Y.; resources, E.K.; data curation, S.Y.; writing—original draft preparation, S.Y.; writing—review and editing, M.F., and E.K.; visualization, S.Y.; supervision, E.K. and M.F.; funding acquisition, E.K. All authors have read and agreed to the published version of the manuscript.

Funding: This research was funded by National Science Foundation grant number CHE 1464906.

Acknowledgments: The authors thank SMU for providing computational resources. We thank Vytor Oliveira for helpful discussions.

Conflicts of Interest: The authors declare no conflicts of interest.

Abbreviations

The following abbreviations are used in this manuscript:

BE	Binding Energy
BSSE	Basis Set Superposition Error
CCP	Cage Critical Point
CNM	Characterization of Normal Modes
CT	Charge Transfer
DFT	Density Functional Theory
DE	Dissociation Energy
ESP	Electrostatic Potential
exp	Experimental
HB	Hydrogen Bond
lp	Lone–Pair
LVM	Local Vibrational Mode
MP2	Møller–Plesset Perturbation Theory of Second Order
NBO	Natural Bond Orbital
NCI	Noncovalent Interaction
SBI	Secondary Bonding Interaction

References

1. Murray, J.S.; Lane, P.; Clark, T.; Riley, K.E.; Politzer, P. σ–Holes, π–Holes and Electrostatically–Driven Interactions. *J. Mol. Model.* **2011**, *18*, 541–548. [CrossRef]
2. Politzer, P.; Murray, J.S.; Clark, T. Halogen Bonding: An Electrostatically–Driven Highly Directional Noncovalent Interaction. *Phys. Chem. Chem. Phys.* **2010**, *12*, 7748. [CrossRef] [PubMed]
3. Murray, J.S.; Politzer, P. The Electrostatic Potential: An Overview. *WIREs Comput. Mol. Sci.* **2011**, *1*, 153–163. [CrossRef]
4. Politzer, P.; Murray, J.S.; Clark, T. Halogen Bonding and other σ–Hole Interactions: A Perspective. *Phys. Chem. Chem. Phys.* **2013**, *15*, 11178. [CrossRef]
5. Wang, H.; Wang, W.; Jin, W.J. σ–Hole Bond vs π–Hole Bond: A Comparison Based on Halogen Bond. *Chem. Rev.* **2016**, *116*, 5072–5104. [CrossRef]

6. Frontera, A.; Bauzá, A. Concurrent Aerogen Bonding and Lone Pair/Anion–π Interactions in the Stability of Organoxenon Derivatives: A Combined CSD and Ab Initio Study. *Phys. Chem. Chem. Phys.* **2017**, *19*, 30063–30068. [CrossRef] [PubMed]
7. Mitra, M.; Manna, P.; Bauzá, A.; Ballester, P.; Seth, S.K.; Choudhury, S.R.; Frontera, A.; Mukhopadhyay, S. 3–Picoline Mediated Self-Assembly of M(II)−Malonate Complexes (M = Ni/Co/Mn/Mg/Zn/Cu) Assisted by Various Weak Forces Involving Lone Pair–π, π–π, and Anion⋯π–Hole Interactions. *J. Phys. Chem. B* **2014**, *118*, 14713–14726. [CrossRef]
8. Ran, J.; Hobza, P. On the Nature of Bonding in Lone Pair⋯π–Electron Complexes: CCSD(T)/Complete Basis Set Limit Calculations. *J. Chem. Theory Comput.* **2009**, *5*, 1180–1185. [CrossRef]
9. Foroutan-Nejad, C.; Badri, Z.; Marek, R. Multi–Center Covalency: Revisiting the Nature of Anion–π Interactions. *Phys. Chem. Chem. Phys.* **2015**, *17*, 30670–30679. [CrossRef]
10. Mooibroek, T.J. Coordinated Nitrate Anions can be Directional π–Hole Donors in the Solid State: A CSD Study. *CrystEngComm* **2017**, *19*, 4485–4488. [CrossRef]
11. Azofra, L.M.; Alkorta, I.; Scheiner, S. Noncovalent Interactions in Dimers and Trimers of SO_3 and CO. *Theor. Chem. Acc.* **2014**, *133*, 1586. [CrossRef]
12. Alkorta, I.; Elguero, J.; Frontera, A. Not Only Hydrogen Bonds: Other Noncovalent Interactions. *Crystals* **2020**, *10*, 180. [CrossRef]
13. Engdahl, A.; Nelander, B. A Matrix Isolation Study of the Interaction between Water and the Aromatic π–Electron System. *J. Phys. Chem.* **1987**, *91*, 2253–2258. [CrossRef]
14. Engdahl, A.; Nelander, B. A Matrix Isolation Study of the Benzene–Water Interaction. *J. Phys. Chem.* **1985**, *89*, 2860–2864. [CrossRef]
15. Gotch, A.J.; Zwier, T.S. Multiphoton Ionization Studies of Clusters of Immiscible Liquids. I. $C_6H_6-(H_2O)_n$, n = 1, 2. *J. Chem. Phys.* **1992**, *96*, 3388–3401. [CrossRef]
16. Suzuki, S.; Green, P.G.; Bumgarner, R.E.; Dasgupta, S.; Goddard, W.A.; Blake, G.A. Benzene Forms Hydrogen Bonds with Water. *Science* **1992**, *257*, 942–945. [CrossRef]
17. Gallivan, J.P.; Dougherty, D.A. Can Lone Pairs Bind to a π System? The Water⋯Hexafluorobenzene Interaction. *Org. Lett.* **1999**, *1*, 103–106. [CrossRef]
18. Danten, Y.; Tassaing, T.; Besnard, M. On the Nature of the Water–Hexafluorobenzene Interaction. *J. Phys. Chem. A* **1999**, *103*, 3530–3534. [CrossRef]
19. Raimondi, M.; Calderoni, G.; Famulari, A.; Raimondi, L.; Cozzi, F. The Benzene/Water/Hexafluorobenzene Complex: A Computational Study. *J. Phys. Chem. A* **2003**, *107*, 772–774. [CrossRef]
20. Egli, M.; Sarkhel, S. Lone Pair–Aromatic Interactions: To Stabilize or Not to Stabilize. *Acc. Chem. Res.* **2007**, *40*, 197–205. [CrossRef]
21. Baiocco, P.; Colotti, G.; Franceschini, S.; Ilari, A. Molecular Basis of Antimony Treatment in Leishmaniasis†. *J. Med. Chem.* **2009**, *52*, 2603–2612. [CrossRef] [PubMed]
22. Hoffmann, J.M.; Sadhoe, A.K.; Mooibroek, T.J. π–Hole Interactions with Various Nitro Compounds Relevant for Medicine: DFT Calculations and Surveys of the Cambridge Structural Database (CSD) and the Protein Data Bank (PDB). *Synthesis* **2019**, *52*, 521–528. [CrossRef]
23. Egli, M.; Gessner, R.V. Stereoelectronic effects of deoxyribose O4' on DNA conformation. *Proc. Natl. Acad. Sci. USA* **1995**, *92*, 180–184. [CrossRef]
24. Sarkhel, S.; Rich, A.; Egli, M. Water–Nucleobase "Stacking": H–π and Lone Pair–π Interactions in the Atomic Resolution Crystal Structure of an RNA Pseudoknot. *J. Am. Chem. Soc.* **2003**, *125*, 8998–8999. [CrossRef]
25. Belmont-Sánchez, J.C.; Ruiz-González, N.; Frontera, A.; Matilla-Hernández, A.; Castiñeiras, A.; Niclós-Gutiérrez, J. Anion–Cation Recognition Pattern, Thermal Stability and DFT–Calculations in the Crystal Structure of H_2dap[Cd(HEDTA)(H_2O)] Salt (H_2dap = H_2(N3,N7)–2,6–Diaminopurinium Cation). *Crystals* **2020**, *10*, 304. [CrossRef]
26. Varadwaj, A.; Marques, H.M.; Varadwaj, P.R. Nature of Halogen–Centered Intermolecular Interactions in Crystal Growth and Design: Fluorine–Centered Interactions in Dimers in Crystalline Hexafluoropropylene as a Prototype. *J. Comput. Chem.* **2019**, *40*, 1836–1860. [CrossRef] [PubMed]
27. Bauzá, A.; Sharko, A.V.; Senchyk, G.A.; Rusanov, E.B.; Frontera, A.; Domasevitch, K.V. π–Hole Interactions at Work: Crystal Engineering with Nitro–Derivatives. *CrystEngComm* **2017**, *19*, 1933–1937. [CrossRef]
28. Bauzá, A.; Frontera, A.; Mooibroek, T.J. NO_3^- Anions can Act as Lewis Acid in the Solid State. *Nat. Commun.* **2017**, *8*, 14522. [CrossRef]

29. Eliseeva, A.A.; Ivanov, D.M.; Novikov, A.S.; Kukushkin, V.Y. Recognition of the π–Hole Donor Ability of Iodopentafluorobenzene − a Conventional σ–Hole Donor for Crystal Engineering involving Halogen Bonding. *CrystEngComm* **2019**, *21*, 616–628. [CrossRef]
30. Franconetti, A.; Frontera, A.; Mooibroek, T.J. Intramolecular π–Hole Interactions with Nitro Aromatics. *CrystEngComm* **2019**, *21*, 5410–5417. [CrossRef]
31. Bauzá, A.; Frontera, A. σ/π–Hole Noble Gas Bonding Interactions: Insights from Theory and Experiment. *Coord. Chem. Rev.* **2020**, *404*, 213112. [CrossRef]
32. Bauzá, A.; Frontera, A. Theoretical Study on the Dual Behavior of XeO_3 and XeF_4 toward Aromatic Rings: Lone Pair–π versus Aerogen–π Interactions. *ChemPhysChem* **2015**, *16*, 3625–3630. [CrossRef]
33. Bauzá, A.; Frontera, A. π–Hole Aerogen Bonding Interactions. *Phys. Chem. Chem. Phys.* **2015**, *17*, 24748–24753. [CrossRef] [PubMed]
34. Bauzá, A.; Mooibroek, T.J.; Frontera, A. The Bright Future of Unconventional σ/π–Hole Interactions. *ChemPhysChem* **2015**, *16*, 2496–2517. [CrossRef] [PubMed]
35. Galmés, B.; Martínez, D.; Infante-Carrió, M.F.; Franconetti, A.; Frontera, A. Theoretical Ab Initio Study on Cooperativity Effects between Nitro π–Hole and Halogen Bonding Interactions. *ChemPhysChem* **2019**, *20*, 1135–1144. [CrossRef] [PubMed]
36. Galmés, B.; Franconetti, A.; Frontera, A. Nitropyridine–1–Oxides as Excellent π–Hole Donors: Interplay between σ–Hole (Halogen, Hydrogen, Triel, and Coordination Bonds) and π–Hole Interactions. *Int. J. Mol. Sci.* **2019**, *20*, 3440. [CrossRef] [PubMed]
37. Novikov, A.S.; Ivanov, D.M.; Bikbaeva, Z.M.; Bokach, N.A.; Kukushkin, V.Y. Noncovalent Interactions involving Iodofluorobenzenes: The Interplay of Halogen Bonding and Weak lp(O)···π–$Hole_{arene}$ Interactions. *Cryst. Growth Des.* **2018**, *18*, 7641–7654. [CrossRef]
38. Wheeler, S.E.; Houk, K.N. Are Anion/π Interactions Actually a Case of Simple Charge–Dipole Interactions? *J. Phys. Chem. A* **2010**, *114*, 8658–8664. [CrossRef]
39. Garau, C.; Frontera, A.; Quiñonero, D.; Russo, N.; Deyà, P.M. RI–MP2 and MPWB1K Study of π–Anion–' Complexes: MPWB1K Performance and Some Additivity Aspects. *J. Chem. Theory Comput.* **2011**, *7*, 3012–3018. [CrossRef] [PubMed]
40. Politzer, P.; Murray, J.S. Electrostatics and Polarization in σ– and π–Hole Noncovalent Interactions: An Overview. *ChemPhysChem* **2020**, *21*, 579–588. [CrossRef] [PubMed]
41. Politzer, P.; Murray, J.S.; Clark, T. Explicit Inclusion of Polarizing Electric Fields in σ– and π–Hole Interactions. *J. Phys. Chem. A* **2019**, *123*, 10123–10130. [CrossRef] [PubMed]
42. Lang, T.; Li, X.; Meng, L.; Zheng, S.; Zeng, Y. The Cooperativity between the σ–Hole and π–Hole Interactions in the ClO···$XONO_2$/XONO···NH_3(X=Cl, Br, I) Complexes. *Struct. Chem.* **2014**, *26*, 213–221. [CrossRef]
43. Zierkiewicz, W.; Michalczyk, M.; Wysokiński, R.; Scheiner, S. On the Ability of Pnicogen Atoms to Engage in both σ and π–Hole Complexes. HeteroDimers of $ZF_2C_6H_5$ (Z=P, As, Sb, Bi) and NH_3. *J. Mol. Model.* **2019**, *25*, 152. [CrossRef]
44. Gao, L.; Zeng, Y.; Zhang, X.; Meng, L. Comparative Studies on Group III σ–Hole and π–Hole Interactions. *J. Comput. Chem.* **2016**, *37*, 1321–1327. [CrossRef]
45. Guo, X.; Cao, L.; Li, Q.; Li, W.; Cheng, J. Competition between π–Hole Interaction and Hydrogen Bond in the Complexes of F_2XO (X = C and Si) and HCN. *J. Mol. Model.* **2014**, *20*, 2493. [CrossRef]
46. Katkova, S.A.; Mikherdov, A.S.; Kinzhalov, M.A.; Novikov, A.S.; Zolotarev, A.A.; Boyarskiy, V.P.; Kukushkin, V.Y. (Isocyano Group π–Hole)···[d_{Z^2}–M^{II}] Interactions of (Isocyanide) [M^{II}] Complexes, in which Positively Charged Metal Centers (d^8–M=Pt, Pd) Act as Nucleophiles. *Chem. Eur.* **2019**, *25*, 8590–8598. [CrossRef] [PubMed]
47. Liu, Z.F.; Chen, X.; Wu, W.X.; Zhang, G.Q.; Li, X.; Li, Z.Z.; Jin, W.J. 1,3,5–Trifluoro–2,4,6–triiodobenzene: A Neglected NIR Phosphor with Prolonged Lifetime by σ–Hole and π–Hole Capture. *Spectrochim. Acta A* **2020**, *224*, 117428. [CrossRef] [PubMed]
48. Varadwaj, P.R.; Varadwaj, A.; Marques, H.M. Does Chlorine in CH_3Cl Behave as a Genuine Halogen Bond Donor? *Crystals* **2020**, *10*, 146. [CrossRef]
49. Rozhkov, A.V.; Krykova, M.A.; Ivanov, D.M.; Novikov, A.S.; Sinelshchikova, A.A.; Volostnykh, M.V.; Konovalov, M.A.; Grigoriev, M.S.; Gorbunova, Y.G.; Kukushkin, V.Y. Reverse Arene Sandwich Structures Based upon π–Hole···[M^{II}](d^8M=Pt, Pd) Interactions, where Positively Charged Metal Centers Play the Role of a Nucleophile. *Angew. Chem. Int. Ed.* **2019**, *58*, 4164–4168. [CrossRef]

50. Prohens, R.; de Sande, D.; Font-Bardia, M.; Franconetti, A.; González, J.F.; Frontera, A. Gallic Acid Dimer As a Double π–Hole Donor: Evidence from X–ray, Theoretical Calculations, and Generalization from the Cambridge Structural Database. *Cryst. Growth Des.* **2019**, *19*, 3989–3997. [CrossRef]
51. Shukla, R.; Claiser, N.; Souhassou, M.; Lecomte, C.; Balkrishna, S.J.; Kumar, S.; Chopra, D. Exploring the Simultaneous σ–Hole/π–Hole Bonding Characteristics of a Br···π Interaction in an Ebselen Derivative via Experimental and Theoretical Electron–Density Analysis. *IUCrJ* **2018**, *5*, 647–653. [CrossRef] [PubMed]
52. Yang, F.L.; Yang, X.; Wu, R.Z.; Yan, C.X.; Yang, F.; Ye, W.; Zhang, L.W.; Zhou, P.P. Intermolecular Interactions between σ– and π–Holes of Bromopentafluorobenzene and Pyridine: Computational and Experimental Investigations. *Phys. Chem. Chem. Phys.* **2018**, *20*, 11386–11395. [CrossRef] [PubMed]
53. Yang, F.L.; Lu, K.; Yang, X.; Yan, C.X.; Wang, R.; Ye, W.; Zhou, P.P.; Yang, Z. Computational Investigations of Intermolecular Interactions between Electron–Accepting Bromo– and Iodo–Pentafluorobenzene and Electron–Donating Furan and Thiophene. *New J. Chem.* **2018**, *42*, 20101–20112. [CrossRef]
54. Zhang, J.; Hu, Q.; Li, Q.; Scheiner, S.; Liu, S. Comparison of σ–Hole and π–Hole Tetrel Bonds in Complexes of Borazine with TH_3F and F_2TO/H_2TO (T= C, Si, Ge). *Int. J. Quantum Chem.* **2019**, *119*, e25910. [CrossRef]
55. Zhang, Y.H.; Li, Y.L.; Yang, J.; Zhou, P.P.; Xie, K. Noncovalent Functionalization of Graphene via π–Hole···π and σ–Hole···π Interactions. *Struct. Chem.* **2019**, *31*, 97–101. [CrossRef]
56. Mikherdov, A.S.; Kinzhalov, M.A.; Novikov, A.S.; Boyarskiy, V.P.; Boyarskaya, I.A.; Avdontceva, M.S.; Kukushkin, V.Y. Ligation–Enhanced π–Hole···π Interactions Involving Isocyanides: Effect of π–Hole···π Noncovalent Bonding on Conformational Stabilization of Acyclic Diaminocarbene Ligands. *Inorg. Chem.* **2018**, *57*, 6722–6733. [CrossRef]
57. Kraka, E.; Cremer, D. Weaker Bonds with Shorter Bond Lengths. *Rev. Proc. Quim.* **2012**, *6*, 39–42. [CrossRef]
58. Setiawan, D.; Kraka, E.; Cremer, D. Hidden Bond Anomalies: The Peculiar Case of the Fluorinated Amine Chalcogenides. *J. Phys. Chem. A* **2015**, *119*, 9541–9556. [CrossRef]
59. Kraka, E.; Setiawan, D.; Cremer, D. Re–Evaluation of the Bond Length–Bond Strength Rule: The Stronger Bond Is not Always the Shorter Bond. *J. Comp. Chem.* **2015**, *37*, 130–142. [CrossRef]
60. Cremer, D.; Kraka, E. From Molecular Vibrations to Bonding, Chemical Reactions, and Reaction Mechanism. *Curr. Org. Chem.* **2010**, *14*, 1524–1560. [CrossRef]
61. Andrés, J.; Ayers, P.W.; Boto, R.A.; Carbó-Dorca, R.; Chermette, H.; Cioslowski, J.; Contreras-García, J.; Cooper, D.L.; Frenking, G.; Gatti, C.; et al. Nine questions on energy decomposition analysis. *J. Comput. Chem.* **2019**, *40*, 2248–2283. [CrossRef] [PubMed]
62. Zhao, L.; von Hopffgarten, M.; Andrada, D.M.; Frenking, G. Energy Decomposition Analysis. *WIREs Comput. Mol. Sci.* **2017**, *8*, 1–37. [CrossRef]
63. Stasyuk, O.A.; Sedlak, R.; Guerra, C.F.; Hobza, P. Comparison of the DFT-SAPT and canonical EDA Schemes for the energy decomposition of various types of noncovalent interactions. *J. Chem. Theory Comput.* **2018**, *14*, 3440–3450. [CrossRef]
64. Levine, D.S.; Head-Gordon, M. Energy decomposition analysis of single bonds within Kohn-Sham density functional theory. *Proc. Natl. Acad. Sci. USA* **2017**, *114*, 12649–12656. [CrossRef]
65. Lao, K.U.; Herbert, J.M. Energy Decomposition Analysis with a Stable Charge-Transfer Term for Interpreting Intermolecular Interactions. *J. Chem. Theory Comput.* **2016**, *12*, 2569–2582. [CrossRef] [PubMed]
66. Kraka, E.; Larsson, J.A.; Cremer, D. Generalization of the Badger Rule Based on the Use of Adiabatic Vibrational Modes. In *Computational Spectroscopy*; Grunenberg, J., Ed.; Wiley: New York, NY, USA, 2010; pp. 105–149.
67. Kalescky, R.; Zou, W.; Kraka, E.; Cremer, D. Local Vibrational Modes of the Water Dimer-Comparison of Theory and Experiment. *Chem. Phys. Lett.* **2012**, *554*, 243–247. [CrossRef]
68. Zou, W.; Kalescky, R.; Kraka, E.; Cremer, D. Relating Normal Vibrational Modes to Local Vibrational Modes: Benzene and Naphthalene. *J. Mol. Model.* **2012**, *19*, 2865–2877. [CrossRef]
69. Kalescky, R.; Kraka, E.; Cremer, D. Local Vibrational Modes of the Formic Acid Dimer–The Strength of the Double H–Bond. *Mol. Phys.* **2013**, *111*, 1497–1510. [CrossRef]
70. Wilson, E.B.; Decius, J.C.; Cross, P.C. *Molecular Vibrations*; McGraw-Hill: New York, NY, USA, 1955.
71. Woodward, L.A. *Introduction to the Theory of Molecular Vibrations and Vibrational Spectroscopy*; Oxford University Press: Oxford, UK, 1972.

72. Herzberg, G. *Molecular Spectra and Molecular Structure*, 2nd ed.; Reitell Press: New York, NY, USA, 2008; Volume I.
73. Herzberg, G. *Molecular Spectra and Molecular Structure. Volume II: Infrared and Raman Spectra of Polyatomic Molecules*; Krieger Publishing Co.: New York, NY, USA, 1991.
74. Herzberg, G.; Huber, K.P. *Molecular Spectra and Molecular Structure*; IV. Constants of Diatomic Molecules, Van Nostrand, Reinhold: New York, NY, USA, 1979.
75. Konkoli, Z.; Cremer, D. A New Way of Analyzing Vibrational Spectra. I. Derivation of Adiabatic Internal Modes. *Int. J. Quant. Chem.* **1998**, *67*, 1–9. [CrossRef]
76. Konkoli, Z.; Larsson, J.A.; Cremer, D. A New Way of Analyzing Vibrational Spectra. II. Comparison of Internal Mode Frequencies. *Int. J. Quant. Chem.* **1998**, *67*, 11–27. [CrossRef]
77. Konkoli, Z.; Cremer, D. A New Way of Analyzing Vibrational Spectra. III. Characterization of Normal Vibrational Modes in terms of Internal Vibrational Modes. *Int. J. Quant. Chem.* **1998**, *67*, 29–40. [CrossRef]
78. Konkoli, Z.; Larsson, J.A.; Cremer, D. A New Way of Analyzing Vibrational Spectra. IV. Application and Testing of Adiabatic Modes within the Concept of the Characterization of Normal Modes. *Int. J. Quant. Chem.* **1998**, *67*, 41–55. [CrossRef]
79. Cremer, D.; Larsson, J.A.; Kraka, E. New Developments in the Analysis of Vibrational Spectra on the Use of Adiabatic Internal Vibrational Modes. In *Theoretical and Computational Chemistry*; Parkanyi, C., Ed.; Elsevier: Amsterdam, The Netherlands, 1998; pp. 259–327.
80. Kraka, E.; Zou, W.; Tao, Y. Decoding chemical information from vibrational spectroscopy data: Local vibrational mode theory. *WIREs Comput. Mol. Sci.* **2020**, e1480. [CrossRef]
81. Kraka, E.; Cremer, D. Dieter Cremer's Contribution to the Field of Theoretical Chemistry. *Int. J. Quantum Chem.* **2019**, *119*, e25849. [CrossRef]
82. Zou, W.; Kalescky, R.; Kraka, E.; Cremer, D. Relating Normal Vibrational Modes to Local Vibrational Modes with the Help of an Adiabatic Connection Scheme. *J. Chem. Phys.* **2012**, *137*, 084114. [CrossRef] [PubMed]
83. Zou, W.; Cremer, D. C_2 in a Box: Determining its Intrinsic Bond Strength for the $X^1 \Sigma^+_g$ Ground State. *Chem. Eur. J.* **2016**, *22*, 4087–4097. [CrossRef]
84. McKean, D.C. Individual CH bond strengths in simple organic compounds: Effects of conformation and substitution. *Chem. Soc. Rev.* **1978**, *7*, 399. [CrossRef]
85. Kalescky, R.; Kraka, E.; Cremer, D. Identification of the Strongest Bonds in Chemistry. *J. Phys. Chem. A* **2013**, *117*, 8981–8995. [CrossRef]
86. Kraka, E.; Cremer, D. Characterization of CF Bonds with Multiple–Bond Character: Bond Lengths, Stretching Force Constants, and Bond Dissociation Energies. *ChemPhysChem* **2009**, *10*, 686–698. [CrossRef]
87. Setiawan, D.; Sethio, D.; Cremer, D.; Kraka, E. From Strong to Weak NF Bonds: On the Design of a New Class of Fluorinating Agents. *Phys. Chem. Chem. Phys.* **2018**, *20*, 23913–23927. [CrossRef]
88. Sethio, D.; Lawson Daku, L.M.; Hagemann, H.; Kraka, E. Quantitative Assessment of B−B−B, B−H_b−B, and B−H_t Bonds: From BH_3 to $B_{12}H_{12}^{2-}$. *ChemPhysChem* **2019**, *20*, 1967–1977. [CrossRef] [PubMed]
89. Oliveira, V.; Kraka, E.; Cremer, D. The Intrinsic Strength of the Halogen Bond: Electrostatic and Covalent Contributions Described by Coupled Cluster Theory. *Phys. Chem. Chem. Phys.* **2016**, *18*, 33031–33046. [CrossRef] [PubMed]
90. Oliveira, V.; Kraka, E.; Cremer, D. Quantitative Assessment of Halogen Bonding Utilizing Vibrational Spectroscopy. *Inorg. Chem.* **2016**, *56*, 488–502. [CrossRef] [PubMed]
91. Oliveira, V.; Cremer, D. Transition from Metal–Ligand Bonding to Halogen Bonding Involving a Metal as Halogen Acceptor: A Study of Cu, Ag, Au, Pt, and Hg Complexes. *Chem. Phys. Lett.* **2017**, *681*, 56–63. [CrossRef]
92. Yannacone, S.; Oliveira, V.; Verma, N.; Kraka, E. A Continuum from Halogen Bonds to Covalent Bonds: Where Do λ^3 Iodanes Fit? *Inorganics* **2019**, *7*, 47. [CrossRef]
93. Oliveira, V.; Cremer, D.; Kraka, E. The Many Facets of Chalcogen Bonding: Described by Vibrational Spectroscopy. *J. Phys. Chem. A* **2017**, *121*, 6845–6862. [CrossRef]
94. Oliveira, V.; Kraka, E. Systematic Coupled Cluster Study of Noncovalent Interactions Involving Halogens, Chalcogens, and Pnicogens. *J. Phys. Chem. A* **2017**, *121*, 9544–9556. [CrossRef]
95. Setiawan, D.; Kraka, E.; Cremer, D. Strength of the Pnicogen Bond in Complexes Involving Group VA Elements N, P, and As. *J. Phys. Chem. A* **2014**, *119*, 1642–1656. [CrossRef]

96. Setiawan, D.; Kraka, E.; Cremer, D. Description of Pnicogen Bonding with the help of Vibrational Spectroscopy - The Missing Link Between Theory and Experiment. *Chem. Phys. Lett.* **2014**, *614*, 136–142. [CrossRef]
97. Setiawan, D.; Cremer, D. Super–Pnicogen Bonding in the Radical Anion of the Fluorophosphine Dimer. *Chem. Phys. Lett.* **2016**, *662*, 182–187. [CrossRef]
98. Sethio, D.; Oliveira, V.; Kraka, E. Quantitative Assessment of Tetrel Bonding Utilizing Vibrational Spectroscopy. *Molecules* **2018**, *23*, 2763. [CrossRef] [PubMed]
99. Freindorf, M.; Kraka, E.; Cremer, D. A Comprehensive Analysis of Hydrogen Bond Interactions Based on Local Vibrational Modes. *Int. J. Quant. Chem.* **2012**, *112*, 3174–3187. [CrossRef]
100. Tao, Y.; Zou, W.; Jia, J.; Li, W.; Cremer, D. Different Ways of Hydrogen Bonding in Water - Why Does Warm Water Freeze Faster than Cold Water? *J. Chem. Theory Comput.* **2016**, *13*, 55–76. [CrossRef] [PubMed]
101. Tao, Y.; Zou, W.; Kraka, E. Strengthening of Hydrogen Bonding With the Push–Pull Effect. *Chem. Phys. Lett.* **2017**, *685*, 251–258. [CrossRef]
102. Makoś, M.Z.; Freindorf, M.; Sethio, D.; Kraka, E. New Insights into Fe–H_2 and Fe–H^- Bonding of a [NiFe] Hydrogenase Mimic – A Local Vibrational Mode Study. *Theor. Chem. Acc.* **2019**, *138*, 76.
103. Makoś, M.Z.; Zou, W.; Freindorf, M.; Kraka, E. Metal-Ring Interactions in Actinide Sandwich Compounds: A Combined Normalized Elimination of the Small Component and Local Vibrational Mode Study. *Mol. Phys.* **2020**, in press.
104. Zhang, X.; Dai, H.; Yan, H.; Zou, W.; Cremer, D. B–H π Interaction: A New Type of Nonclassical Hydrogen Bonding. *J. Am. Chem. Soc.* **2016**, *138*, 4334–4337. [CrossRef]
105. Zou, W.; Zhang, X.; Dai, H.; Yan, H.; Cremer, D.; Kraka, E. Description of an Unusual Hydrogen Bond Between Carborane and a Phenyl Group. *J. Organometal. Chem.* **2018**, *865*, 114–127. [CrossRef]
106. Burianova, V.K.; Bolotin, D.S.; Mikherdov, A.S.; Novikov, A.S.; Mokolokolo, P.P.; Roodt, A.; Boyarskiy, V.P.; Dar'in, D.; Krasavin, M.; Suslonov, V.V.; et al. Mechanism of Generation of *Closo-* Amidrazones. Intramol. Non- B−H$\cdots\pi$(Ph) Interact. Determ. Stab. Config. Around Amidrazone C=N Bond. *New J. Chem.* **2018**, *42*, 8693–8703. [CrossRef]
107. Alkorta, I.; Rozas, I.; Elguero, J. An Attractive Interaction between the π–Cloud of C_6F_6 and Electron-Donor Atoms. *J. Org. Chem.* **1997**, *62*, 4687–4691. [CrossRef]
108. Alkorta, I.; Rozas, I.; Elguero, J. Interaction of Anions with Perfluoro Aromatic Compounds. *J. Am. Chem. Soc.* **2002**, *124*, 8593–8598. [CrossRef]
109. Mascal, M.; Armstrong, A.; Bartberger, M.D. Anion-Aromatic Bonding: A Case for Anion Recognition by π–Acidic Rings. *J. Am. Chem. Soc.* **2002**, *124*, 6274–6276. [CrossRef] [PubMed]
110. Quiñonero, D.; Garau, C.; Rotger, C.; Frontera, A.; Ballester, P.; Costa, A.; Deyà, P.M. Anion–π Interactions: Do They Exist? *Angew. Chem. Int. Ed.* **2002**, *41*, 3389–3392. [CrossRef]
111. Chai, J.D.; Head-Gordon, M. Systematic Optimization of Long–Range Corrected Hybrid Density Functionals. *J. Chem. Phys.* **2008**, *128*, 084106. [CrossRef] [PubMed]
112. Chai, J.D.; Head-Gordon, M. Long–Range Corrected Hybrid Density Functionals with Damped Atom–Atom Dispersion Corrections. *Phys. Chem. Chem. Phys.* **2008**, *10*, 6615–6620. [CrossRef] [PubMed]
113. Dunning, T.H. Gaussian Basis Sets for use in Correlated Molecular Calculations. I. The Atoms Boron through Neon and Hydrogen. *J. Chem. Phys.* **1989**, *90*, 1007–1023. [CrossRef]
114. Kendall, R.A.; Dunning, T.H.; Harrison, R.J. Electron Affinities of the First–Row Atoms Revisited. Systematic Basis Sets and Wave Functions. *J. Chem. Phys.* **1992**, *96*, 6796–6806. [CrossRef]
115. Woon, D.E.; Dunning, T.H. Gaussian Basis Sets for use in Correlated Molecular Calculations. III. The Atoms Aluminum through Argon. *J. Chem. Phys.* **1993**, *98*, 1358–1371. [CrossRef]
116. Woon, D.E.; Dunning, T.H. Gaussian Basis Sets for use in Correlated Molecular Calculations. IV. Calculation of Static Electrical Response Properties. *J. Chem. Phys.* **1994**, *100*, 2975–2988. [CrossRef]
117. Amicangelo, J.C.; Irwin, D.G.; Lee, C.J.; Romano, N.C.; Saxton, N.L. Experimental and Theoretical Characterization of a Lone Pair–π Complex: Water–Hexafluorobenzene. *J. Phys. Chem. A* **2012**, *117*, 1336–1350. [CrossRef] [PubMed]
118. Wilson, E.B.; Decius, J.C.; Cross, P.C. *Molecular Vibrations: The Theory of Infrared and Raman Vibrational Spectra*; McGraw-Hill: New York, NY, USA, 1955.
119. Kraka, E.; Zou, W.; Filatov, M.; Tao, Y.; Grafenstein, J.; Izotov, D.; Gauss, J.; He, Y.; Wu, A.; Konkoli, Z.; et al. COLOGNE2018. 2018. Available online: http://www.smu.edu/catco (accessed on 25 March 2020).

120. Reed, A.E.; Curtiss, L.A.; Weinhold, F. Intermolecular Interactions from a Natural Bond Orbital, Donor–Acceptor Viewpoint. *Chem. Rev.* **1988**, *88*, 899–926. [CrossRef]
121. Reed, A.E.; Weinstock, R.B.; Weinhold, F. Natural Population Analysis. *J. Chem. Phys.* **1985**, *83*, 735–746. [CrossRef]
122. Reed, A.E.; Weinhold, F. Natural Localized Molecular Orbitals. *J. Chem. Phys.* **1985**, *83*, 1736–1740. [CrossRef]
123. Bader, R.F.W. A Quantum Theory of Molecular Structure and Its Applications. *Chem. Rev.* **1991**, *91*, 893–928. [CrossRef]
124. Keith, T.A. AIMAll (Version 17.11.14). 2017. Available online: aim.tkgristmill.com (accessed on 25 March 2020).
125. Frisch, M.J.; Trucks, G.W.; Schlegel, H.B.; Scuseria, G.E.; Robb, M.A.; Cheeseman, J.R.; Scalmani, G.; Barone, V.; Petersson, G.A.; Nakatsuji, H.; Li, X.; et al. *Gaussian16 Revision B.01*; Gaussian Inc.: Wallingford, CT, USA, 2016.
126. Byrd, E.F.C.; Sherrill, C.D.; Head-Gordon, M. The Theoretical Prediction of Molecular Radical Species: A Systematic Study of Equilibrium Geometries and Harmonic Vibrational Frequencies. *J. Phys. Chem. A* **2001**, *105*, 9736–9747. [CrossRef]
127. Coolidge, M.B.; Marlin, J.E.; Stewart, J.J.P. Calculations of Molecular Vibrational Frequencies using Semiempirical Methods. *J. Comput. Chem.* **1991**, *12*, 948–952. [CrossRef]
128. Galabov, B.; Yamaguchi, Y.; Remington, R.B.; Schaefer, H.F. High Level Ab Initio Quantum Mechanical Predictions of Infrared Intensities. *J. Phys. Chem. A* **2002**, *106*, 819–832. [CrossRef]
129. Halls, M.D.; Velkovski, J.; Schlegel, H.B. Harmonic Frequency Scaling Factors for Hartree–Fock, S–VWN, B–LYP, B3–LYP, B3–PW91 and MP2 with the Sadlej pVTZ Electric Property Basis Set. *Theor. Chem. Acc.* **2001**, *105*, 413–421. [CrossRef]
130. Morse, M.D. Clusters of Transition–Metal Atoms. *Chem. Rev.* **1986**, *86*, 1049–1109. [CrossRef]
131. Irikura, K.K.; Johnson, R.D.; Kacker, R.N. Uncertainties in Scaling Factors for Ab Initio Vibrational Frequencies. *J. Phys. Chem. A* **2005**, *109*, 8430–8437. [CrossRef]
132. Scott, A.P.; Radom, L. Harmonic Vibrational Frequencies: An Evaluation of Hartree–Fock, Møller–Plesset, Quadratic Configuration Interaction, Density Functional Theory, and Semiempirical Scale Factors. *J. Phys. Chem.* **1996**, *100*, 16502–16513. [CrossRef]
133. Faeder, J. A Distributed Gaussian Approach to the Vibrational Dynamics of Ar–Benzene. *J. Chem. Phys.* **1993**, *99*, 7664–7676. [CrossRef]
134. Nanayakkara, S.; Kraka, E. A New Way of Studying Chemical Reactions: A Hand-in-hand URVA and QTAIM Approach. *Phys. Chem. Chem. Phys.* **2019**, *21*, 15007–15018. [CrossRef]
135. Wei, Y.; Li, Q.; Li, W.; Cheng, J.; McDowell, S.A.C. Influence of the Protonation of Pyridine Nitrogen on Pnicogen Bonding: Competition and Cooperativity. *Phys. Chem. Chem. Phys.* **2016**, *18*, 11348–11356. [CrossRef] [PubMed]
136. Bauzá, A.; Frontera, A. Competition between Lone Pair–π, Halogen–π and Triel Bonding Interactions Involving BX_3 (X=F, Cl, Br and I) Compounds: An Ab Initio Study. *Theor. Chem. Acc.* **2017**, *136*, 37. [CrossRef]
137. Vácha, R.; Marsalek, O.; Willard, A.P.; Bonthuis, D.J.; Netz, R.R.; Jungwirth, P. Charge Transfer between Water Molecules As the Possible Origin of the Observed Charging at the Surface of Pure Water. *J. Phys. Chem. Lett.* **2011**, *3*, 107–111. [CrossRef]
138. Verma, N.; Tao, Y.; Zou, W.; Chen, X.; Chen, X.; Freindorf, M.; Kraka, E. A Critical Evaluation of Vibrational Stark Effect (VSE) Probes with the Local Vibrational Mode Theory. *Sensors* **2020**, *20*, 2358. [CrossRef]

© 2020 by the authors. Licensee MDPI, Basel, Switzerland. This article is an open access article distributed under the terms and conditions of the Creative Commons Attribution (CC BY) license (http://creativecommons.org/licenses/by/4.0/).

MDPI
St. Alban-Anlage 66
4052 Basel
Switzerland
Tel. +41 61 683 77 34
Fax +41 61 302 89 18
www.mdpi.com

Crystals Editorial Office
E-mail: crystals@mdpi.com
www.mdpi.com/journal/crystals

www.ingramcontent.com/pod-product-compliance
Lightning Source LLC
LaVergne TN
LVHW070744100526
838202LV00013B/1296